Nanoscience: The Science of the Small

Nanoscience: The Science of the Small

Edited by **Rich Falcon**

NY RESEARCH
P R E S S

New York

Published by NY Research Press,
23 West, 55th Street, Suite 816,
New York, NY 10019, USA
www.nyresearchpress.com

Nanoscience: The Science of the Small
Edited by Rich Falcon

International Standard Book Number: 978-1-63238-496-6 (Hardback)

Printed in the United States of America.

Contents

Preface

This book has been a concerted effort by a group of academicians, researchers and scientists, who have contributed their research works for the realization of the book. This book has materialized in the wake of emerging advancements and innovations in this field. Therefore, the need of the hour was to compile all the required researches and disseminate the knowledge to a broad spectrum of people comprising of students, researchers and specialists of the field.

Nanoscience is the study of extremely small matter particularly on molecular, super molecular and atomic scale. These matters are used across different disciplines namely physics, biology, chemistry, engineering and materials science. The current areas of research in nanoscience ranges from carbon nanotubes, nanorods, nanomedicine, nanopillars employed in solar cells, nanoparticles in biological imaging, to nanomaterials used in drug delivery. This book is a complete source of knowledge on the present status of this important field. While understanding the long-term perspectives of the topics the book makes an effort in highlighting their impact as a modern tool for the growth of the discipline. The rapid progress happening in this field across the globe has been extensively covered in this book. For someone with an interest and eye for detail, this book covers the most significant topics of nanoscience.

At the end of the preface, I would like to thank the authors for their brilliant chapters and the publisher for guiding us all-through the making of the book till its final stage. Also, I would like to thank my family for providing the support and encouragement throughout my academic career and research projects.

Editor

Role of Chromium Intermediate Thin-Film on the Growth of Silicon Oxide (SiO$_x$) Nanowires

Anima Johari[1], Anoopshi Johari[2], Vikas Rana[1], M. C. Bhatnagar[1]

[1]CARE, Physics Department, IIT Delhi, New Delhi, India
[2]THDC Institute of Hydropower Engineering and Technology, Tehri, India
Email: animajohari@gmail.com, anoopshi.akg@gmail.com, vikas.rana@care.iitd.ac.in, mukesh@physics.iitd.ac.in

Abstract

In the present work, one-dimensional nanostructures of silicon oxide (SiO$_x$) have been synthesized by thermal annealing method with and without chromium thin film on silicon substrate. The synthesis was carried out at different process temperatures ranging from 1000°C to 1100°C by using gold/chromium (Au/Cr) catalysts stack layer on the Si substrate in nitrogen (N$_2$) ambience. The as-synthesized SiO$_x$ nanostructures have tetragonal rutile structure and show polycrystalline nature. The SEM images reveal wire-like nanostructures on the substrate with and without chromium thin film. Under the catalytic reaction of the gold/chromium metal, the density of SiO$_x$ nanowires is enhanced, since the Cr layer serves as a diffusion barrier for the diffusion of the gold downwards into the Si substrate. The vapor-liquid solid (VLS) growth mechanism is found to be dominant in the growth of SiO$_x$ nanowires. Furthermore, X-Ray diffraction microscopy (XRD) and Photoluminescence spectroscopy (PL) analysis conclude the defect free growth of the SiO$_x$ nanowires on gold/chrome/silicon substrate.

Keywords

SiO$_x$ Nanowires; Catalyst Assisted Growth; Gold; Chromium; Thermal Annealing

1. Introduction

Amorphous silicon oxide (SiO$_x$) nanowires have many potential applications in blue light emitters, optical sensors [1] and reinforcing composites [2]. These nanowires are generally grown with transition metal catalysts (gold, iron, palladium etc.) at an elevated temperature. Liu *et al.* used Fe as a catalyst for the growth of the SiO$_x$ nanowires [3]. Jiang *et al.* produced the SiO$_x$ nanowires by using iron-cobalt-nickel (Fe-Co-Ni) alloy nanoparticles as the catalyst and showed that they had a strong blue-green emission [4]. Zhang *et al.* displayed that the SiO$_x$ nanowires cloud can be formed on tin balls by chemical vapor deposition via vapor-liquid-solid (VLS) process [5]. Wang *et al.* reported that the amorphous SiO$_x$ nanowires could be grown on the Si substrate by using platinum (Pt) as a catalyst [6]. Lin *et al.* synthesized the amorphous SiO$_x$ nanowires from silicon monoxide powder under super critically hydrothermal conditions [7]. Park *et al.* used gold (Au) and palladium-gold (Pd-Au) thin film as the catalyst for the growth of amorphous SiO$_x$ nanowires. These nanowires were grown via Solid-Liquid-Solid (SLS) mechanism [8]. However, thermal annealing is the simplest method for the growth of SiO$_x$ nanowires. During the thermal annealing, a thin layer of the gold on the Si substrate is heated at growth of the high temperature (~1100°C) in the presence of inert gas environment. At this temperature, some amount of the gold diffuses into the Si substrate. This reduced the density of the catalyst nanoparticles on the Si surface and resulted into a lower density of the nanowires. To enhance the density of nanowires, the diffusion of the gold into the Si substrate must be retarded. The gold diffusion can be retarded either by reducing the growth temperature or by inserting a barrier layer in the Au/Si catalyst system.

In present work, we have synthesized SiO$_x$ nanostructures by thermal evaporation method with and without chromium thin film as a catalyst on Si substrate. During the growth of SiO$_x$ nanowires, the diffusion of the gold into the Si substrate is retarded by inserting a thin layer of chromium (Cr) metal in the Au/Si substrate. The growth of nanowires was carried out with the gold/chromium/silicon (Au/Cr/Si) substrate at different temperatures ranging from 1000°C to 1100°C. To investigate the effect of the Cr layer, the nanowire growth was also carried out in the Cr/Si sample at 1100°C and 1150°C.

The as-synthesized products were analyzed with Scanning Electron Microscopy (SEM), X-Ray diffraction microscopy (XRD), Energy Dispersive X-ray Spectroscopy (EDX), Transmission Electron Microscopy (TEM) and Photoluminescence spectroscopy (PL) for observing the effect of chromium thin film on the structural morphology, crystal structure, composition and optical properties of silicon oxide (SiO$_x$) nanostructures.

2. Experiments Details

N-type (100) Si wafer was used as a substrate for the growth of SiO$_x$ nanowires. Before depositing the catalyst films of Au and Cr, the Si substrate was atomically cleaned by using Isopropyl alcohol (IPA) with ultrasonic vibrations, a mixture solution of H$_2$SO$_4$ and H$_2$O$_2$, and 10% HF solution for removing the native SiO$_2$. The gold (Au) and chromium (Cr) thin films were deposited on the Si surface in radio-frequency (RF) sputtering chamber at the pressure of 1×10^{-6} mbar and 100 watt power. The as-deposited thin layer of the Au and the Cr acts as catalyst for the growth of SiO$_x$ nanowires. Two types of samples were fabricated which consists of Au/Si and Au/Cr/Si substrates. Successively, these samples were loaded into the maximum temperature zone of horizontal tubular furnace. The furnace temperature was maintained at various temperatures ranging from 1000°C to 1100°C. The nanowires were grown by thermal annealing of Au (20 nm)/Si and the Au (10 nm)/Cr (10 nm)/Si substrates in nitrogen (N$_2$) ambience. The synthesis was carried out at atmospheric pressure. The as-synthesized products were characterized by Scanning Electron Microscopy (SEM), X-Ray diffraction microscopy (XRD), Energy Dispersive X-ray Spectroscopy (EDX), Transmission Electron Microscope (TEM) and photoluminescence spectroscopy (PL) for observing the effect of chromium thin film on the structural morphology, crystal structure and composition and optical properties.

3. Results and Discussion

Figure 1 shows the typical SEM images of the Au (20 nm)/Si and Au (10 nm)/Cr (10 nm)/Si substrates annealed at 1000°C for 40 min in N$_2$ ambience. Under these conditions, there was no evidence of the nanowire growth. However, the Au catalyst layer on both substrates is agglomerated into nanoparticles after the thermal annealing. The density of nanoparticles is higher in the Au (10 nm)/Cr (10 nm)/Si as compared to the Au (20 nm)/Si substrate. These nanoparticles serve as nucleation sites for the growth of nanowires. To initiate the growth of nanowire, process temperature is further increased to 1100°C.

Figure 2 shows SEM images of the Au (20 nm)/Si and the Au (10 nm)/Cr (10 nm)/Si substrates annealed at 1100°C for 40 min in N$_2$ ambience. At this temperature, both substrates show the growth of nanowires. However, the nanowires are homogeneously deposited over large area on the Au/Cr/Si substrate. This is due to the

(a) (b)

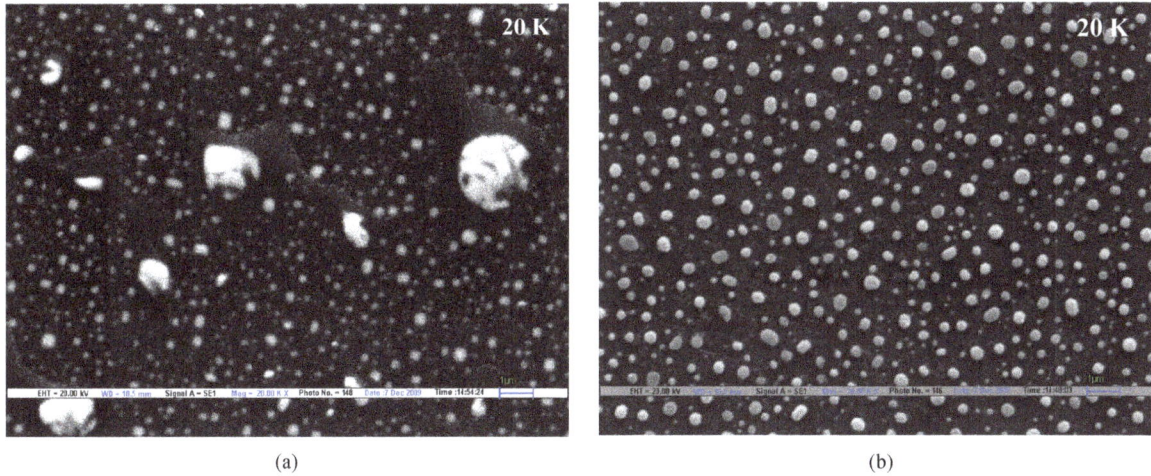

Figure 1. SEM images of the SiO$_x$ nanowires synthesized at 1000°C on: (a) Au/Si, (b) Au/Cr/Si sample.

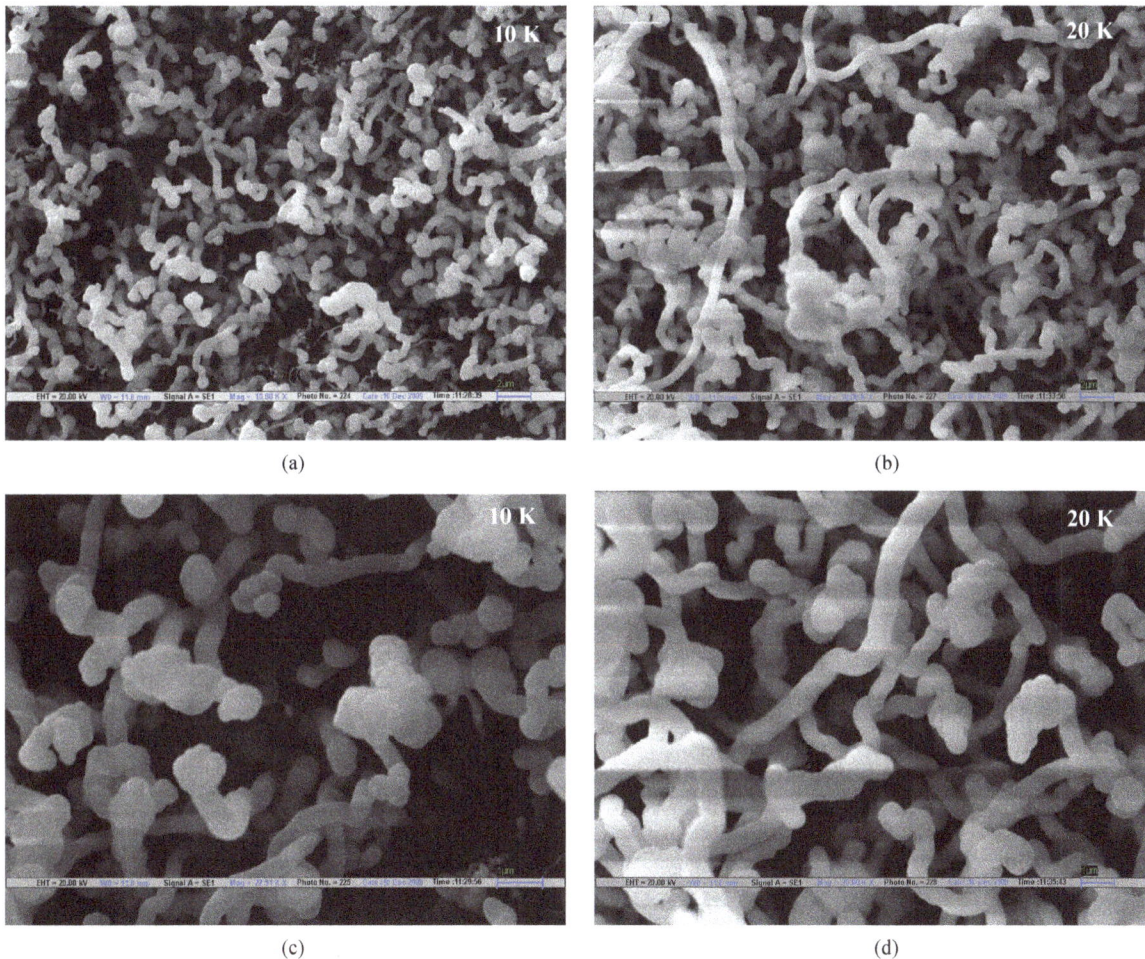

(a) (b)

(c) (d)

Figure 2. SEM images of the SiO$_x$ nanowires synthesized at 1100°C on: (a) and (b) Au/Si, (c) and(d) Au/Cr/Si sample.

formation of higher density of the Au nanoparticles on the Au/Cr/Si substrate. The typical length of the SiO$_x$ nanowires is several tens of micrometers while the width is in the nanometer range.

Further, to investigate the role of the Cr layer, the Si substrate with 10 nm thick Cr layer were annealed in N$_2$ ambience at 1100°C and 1150°C for 40 min. At these temperatures, the Cr thin film agglomerates into nanoparticles on Si substrate but the samples do not show any sign of the nanowire growth. Thus, it can be concluded that the Cr layer does not act as a catalyst and only the Au metal layer serves as catalyst in the growth of nanowires.

The XRD pattern (**Figure 3**) reveals the overall crystal structure and phase purity of the as-synthesized products on Au/Si and Au/Cr/Si substrates annealed at 1100˚C for 40 min in the N_2 ambience. Most of the diffraction peaks can be indexed to the orthorhombic structure of SiO_2. No characteristic peaks of impurities, such as other oxides, were observed. The strong and sharp reflection peaks suggest that the well-crystallized SiO_x products were successfully obtained through the present synthesis method. Using Scherrer's formula, the average crystallite size was found to be about 300 nm.

Figure 4 shows HRTEM (**Figure 4 (a)**) and EDX spectra (**Figure 4 (b)**) of the as-synthesized products on Au (10 nm)/Cr (10 nm)/Si substrate annealed at 1100˚C for 40 min in N_2 ambience. The nanowire diameter was estimated in the range of 300 nm. The associated EDX analysis confirms that the synthesized products are composed of only Si and oxygen and no metal (Au and Cr) traces were observed in the nanowire. However, these metal nanopaticles may be present on the tip of the nanowire. This confirms that these metals only catalyze the growth of nanowires. The C and Cu related signals are due to the contamination of C while preparing HRTEM specimens and due to the presence of Cu grids respectively.

Figure 5 shows room temperature PL spectra of the SiO_x nanowires grown on the Au (20 nm)/Si and the Au (10 nm)/Cr (10 nm)/Si substrate at 1100˚C. Both samples show a sharp strong ultraviolet (UV) near band edge emission at 380 nm and Au (20 nm)/Si substrate exhibit emissions at 400 nm and 440 nm also. The emission at 400 nm is due to intrinsic diamagnetic defect centers [8] whereas Photoluminescence emission peaks at 380 nm and 440 nm is due to the oxygen deficiency [9] in the growth of nanowires. These results were also confirmed by the XRD which shows very high intensity peaks on the Au/Cr/Si substrate in comparison with the Au/Si substrate.

From above experiments, the nanowires growth mechanism is concluded in the following way: For the Au/Si as well as Au/Cr/Si substrate, no nanowires growth was observed at process temperatures of 1000˚C, shown in **Figure 1**. The nanowire growth was only observed at the elevated temperature of 1100˚C as shown in **Figure 2**. A higher density of nanowire was observed for the Au/Cr/Si catalyst than the Au/Si catalyst system. This phenomenon can be understood in a way that the Cr layer acts as a diffusion barrier for the Au and stops the inward diffusion of the Au into the Si and causes the higher density of the Au nanoparticles, which catalyze the nanowire growth. This results the higher density of the SiO_x nanowires on Au/Cr/Si substrate in comparison with the Au/Si substrate.

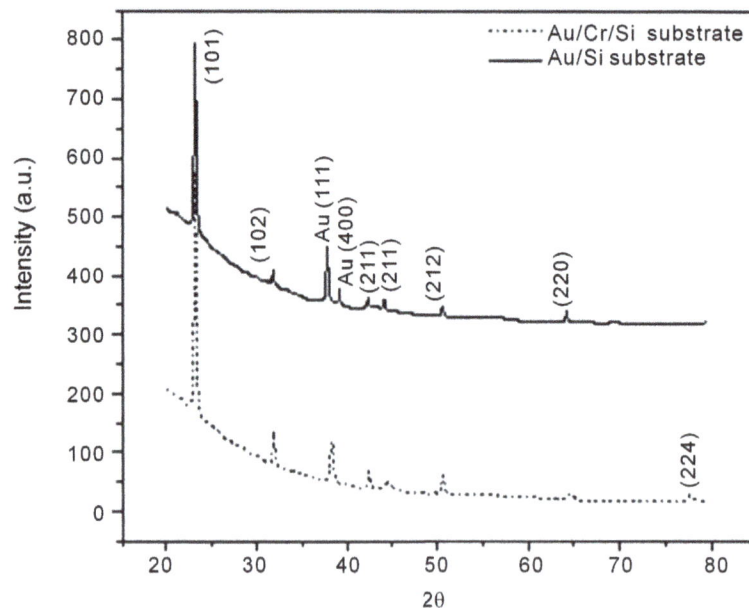

Figure 3. XRD image of the SiO_x nanowires synthesized at 1100˚C in N_2 ambience.

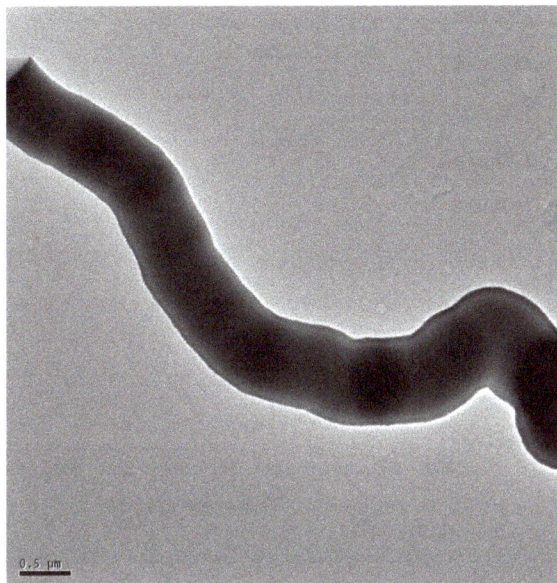

(a)

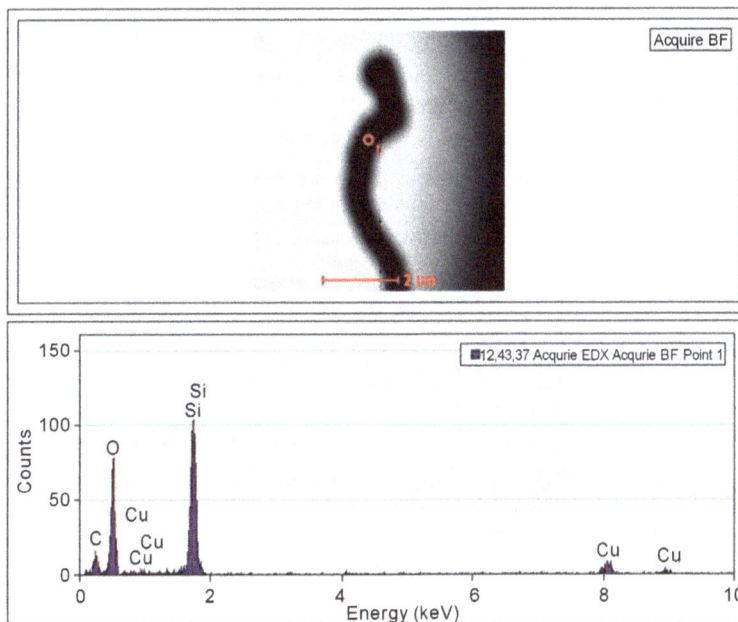

(b)

Figure 4. (a) HRTEM and (b) EDX images of the SiO_x nanowires synthesized at 1100°C on the Au/Cr/Si sample.

In this growth process, the Au film agglomerates into the nanoparticles at relatively lower temperature than the Cr layer. This is due to the lower melting point of the Au film (melting point ≈ 1064°C) in comparison to the Cr film (melting point ≈ 1857°C) [4] [10]. Afterwards, the Si atom diffuses outward through the boundary between Cr particles and colloids with Au particles, the SiO_x nanowires are then formed due to super-saturation of the Si in the Au nanoparticles and their reaction with the ambience oxygen. Therefore, the Au layer is only responsible for the nanowire growth and the Cr layer serves only as a diffusion barrier.

4. Conclusion

We have synthesized the silicon oxide (SiO_x) nanowires on gold coated and gold/chrome coated Si substrate by

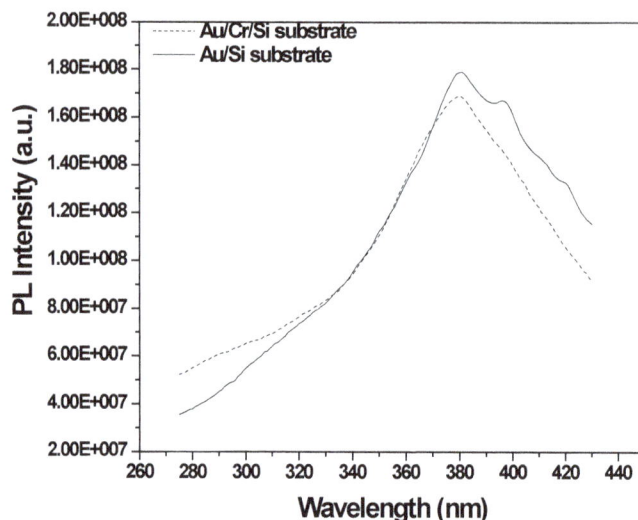

Figure 5. Room Temperature Photoluminescence spectra of the SiO_x nanowires.

thermal annealing process. The higher density of the SiO_x nanowires is attained at process temperature of 1100°C by using the thin layer of the Cr metal on the Au/Si substrate. This is due to the fact that the Cr layers stop the inward diffusion of the Au into the Si and cause the higher density of the Au nanoparticles which catalyze the nanowire growth. The surface morphology study concludes the growth of nanowires. The synthesized nanostructures have orthorhombic structure and polycrystalline in nature. In the PL spectra of SiO_x nanowires, we have not observed any peak corresponding to oxygen deficiency on the Au/Cr/Si substrate whereas oxygen deficiency peaks exist on the Au/Si substrate. This confirms that we have synthesized defect free silicon oxide (SiO_x) nanowires by using chromium as an intermediate layer.

References

[1]   Tong, L., Lou, J. and Gattas, R.R. (2005) Assembly of Silicon Nanowires on Silica Aerogels for Microphotonic Devices. *Nano Letters*, **5**, 259-262. http://dx.doi.org/10.1021/nl0481977

[2]   Yan, X.Q., Zhou, W.Y. and Sun, L.F. (2005) The Influence of Hydrogen on the Growth of Gallium Catalyzed Silicon Oxide Nanowires. *Journal of Physics and Chemistry of Solids*, **66**, 701-705. http://dx.doi.org/10.1016/j.jpcs.2004.06.021

[3]   Liu, W.-L., Hseih, S.-H., Chen, C.H. and Chen, W.-J. (2009) Effect of Fe Metal on the Growth of Silicon Oxide Nanowires. *International Journal of Minerals, Metallurgy and Materials*, **16**, 317-321. http://dx.doi.org/10.1016/S1674-4799(09)60057-1

[4]   Jiang, Z., Xie, T. and Yuan, B.Y. (2005) Synthesis of Core-Shell Nanowires of FeCoNi Alloy Core with Silicon Oxide Layers. *Applied Physics A*, **81**, 477-479. http://dx.doi.org/10.1007/s00339-005-3279-0

[5]   Zhang, J., Xu, B. and Yang, Y. (2006) Catalyzed Assisted Growth of Well-Aligned Silicon Oxide Nanowires. *Journal of Non-Crystalline Solids*, **352**, 2859-2862. http://dx.doi.org/10.1016/j.jnoncrysol.2006.02.088

[6]   Wang, C.Y., Chan, L.H. and Xiao, D.Q. (2006) Mechanism of Solid-Liquid-Solid on the Silicon Oxide Nanowire Growth. *Journal of Vacuum Science & Technology B*, **24**, 613. http://dx.doi.org/10.1116/1.2172953

[7]   Lin, L.W., Tang, Y.H. and Li, X.X. (2007) Water-Assisted Synthesis of Silicon Oxide Nanowires Under Supercritically Hydrothermal Conditions. *Journal of Applied Physics*, **101**, Article ID: 014314. http://dx.doi.org/10.1063/1.2404092

[8]   Wang, X.J., Dong, B. and Zhou, Z. (2009) Preparation and Photoluminescence of High Density SiO_x Nanowires with Fe_3O_4 Nanoparticles Catalyst. *Materials Letters*, **63**, 1149-1152. http://dx.doi.org/10.1016/j.matlet.2009.01.084

[9]   Chen, X.Y., Lu, Y.F., Tang, L.J., Wu, Y.H., Cho, B.J., Xu, X.J., Dong, J.R. and Song, W.D. (2005) Annealing and Oxidation of Silicon Oxide Films Prepared by Plasma-Enhanced Chemical Vapor Deposition. *Journal of Applied Physics*, **97**, Article ID: 014913. http://dx.doi.org/10.1063/1.1829789

[10]   Park, H.K., Yang, B.L., Kim, S.W., *et al.* (2007) Formation of Silicon Oxide Nanowires Directly From Au/Si and Pd-Au/Si Substrates. *Physica E*, **37**, 158-162. http://dx.doi.org/10.1016/j.physe.2006.08.003

X-Ray Analysis by Williamson-Hall and Size-Strain Plot Methods of ZnO Nanoparticles with Fuel Variation

Yendrapati Taraka Prabhu[1*], Kalagadda Venkateswara Rao[1],
Vemula Sesha Sai Kumar[1], Bandla Siva Kumari[2]

[1]Centre for Nano Science and Technology, IST, Jawaharlal Nehru Technological University, Hyderabad, India
[2]Botany Department, Andhra Loyola College, Vijayawada, India
Email: [*]prabusj@gmail.com

Abstract

In this paper, a simple and facile surfactant assisted combustion synthesis is reported for the ZnO nanoparticles. The synthesis of ZnO-NPs has been done with the assistance of non-ionic surfactant TWEEN 80. The effect of fuel variations and comparative study of fuel urea and glycine have been studied by using characterization techniques like X-ray diffraction (XRD), transmission electron microscope (TEM) and particle size analyzer. From XRD, it indicates the presence of hexagonal wurtzite structure for ZnO-NPs. Using X-ray broadening, crystallite sizes and lattice strain on the peak broadening of ZnO-NPs were studied by using Williamson-Hall (W-H) analysis and size-strain plot. Strain, stress and energy density parameters were calculated for the XRD peaks of all the samples using (UDM), uniform stress deformation model (USDM), uniform deformation energy density model (UDEDM) and by the size-strain plot method (SSP). The results of mean particle size showed an inter correlation with W-H analysis, SSP, particle analyzer and TEM results.

Keywords

Surfactant Assisted Combustion; X-Ray Diffraction (XRD); Transmission Electron Microscope (TEM); Particle Analyzer

1. Introduction

In many areas of chemistry, physics and material science transition metal oxides with nano structure have at-

---

[*]Corresponding author.

tracted substantial interest during the last few years because of their novel optical and electrical properties as well as semiconductor crystals with a large binding energy (60 meV) [1]. In a variety of applications, Zinc oxide nanoparticles are used as photocatalyst [2], catalyst [3], antibacterial treatment [4] and UV absorption. Various physical methods such as vapor phase transparent process [5], pulse laser deposition [6] [7], vapor transparent deposition and chemical vapor deposition [8] have been developed for preparation of nano ZnO. These days, Sol-gel method is one of the known procedures for the preparation of metal oxide nanoparticles [9] which is based on the hydrolysis of reactive metal precursor.

Deviations from perfect crystallinity extend infinitely in all directions which lead to broadening of the diffraction peaks. The crystallite size and lattice strain are the two main properties which could be extracted from the peak width analysis. Due to the formation of polycrystalline aggregates [10], the crystallite size of the particle is not the same as the particle size. The crystal imperfections could be measured from the distributions of lattice constants. The basis of strain also includes contact or sinter stress, grain boundary, triple junction, stacking faults and coherency stress [11]. In different ways, Bragg peak is affected by crystallite size and lattice strain which increase the peak width and intensity shifting the 2θ peak position accordingly. The crystallite size varies as $1/\cos\theta$ and stain varies as $\tan\theta$ from the peak width. The size and strain effects on peak broadening are known from the above difference of 2θ. W-H analysis is an integral breadth method. Size-induced and strain-induced broadenings are known by considering the peak width as a function of 2θ [12].

In this paper firstly, a simple method for the synthesis of ZnO-NPs by surfactant assisted combustion synthesis is disused. It was found that this method is a quick, mild, energy-efficient and eco-friendly route to producing ZnO nanoparticles. Secondly, comparative studies of the mean particle size of ZnO-NPs from TEM measurements and from the powdered XRD are dealt with. Using William Hall modified form strain, uniform deformation model (UDM), uniform stress deformation model (USDM), uniform deformation energy-density model (UDEDM) and the size-strain plot method (SSP) provided information on the stress-strain relation, and the strain ε as a function of energy density (u) was estimated.

2. Experimental Details

2.1. Instrumentation

The crystal phases of the synthesized powders were determined by X-ray diffraction (XRD, Bruker D & Advance, Germany) using CuKα as radiation source (40 kV, step size 0.02, scan rate 0.5 min^{-1}, $20° \leq 2\theta \leq 80°$). The particle size is measured by Nano Particle Size Analyzer (SZ-100 Nanoparticle, Horiba, Germany). The surface morphology of ZnO nanoparticles were studied with transmission electron microscope (Tecnai 20 T G2 (FEI)).

2.2. Preparation of ZnO Nanoparticles

The starting materials such as zinc nitrate and non-ionic surfactant are taken in same amounts for both the samples and thus changing the fuels glycine and urea. Freshly prepared aqueous solutions of the chemicals were used for the synthesis of nanoparticles. At room temperature the chemicals are added one by one with the 0.1 M solution of zinc nitrate, 0.15 M solution of glycine for first sample and urea for the second sample with 0.025 M solution of non-ionic surfactant. The mixture of chemicals was then heated on a hot plate in separate beakers which led the chemical mixture to self-combustion. After combustion the final precipitate is subjected to calcinations for 1 hr at 400°C. Thus we successfully obtained a pure ZnO nano powders for different fuels in this synthesis.

3. Results and Discussion

3.1. XRD

Figure 1 shows the XRD patterns of as-syntheised ZnO nanoparticles by surfactant assisted combustion with fuels glycine and urea. All the diffraction peaks can be assigned to hexagonal phase with Wurtzite structure with space group (P63mc), JCPDS card No.36-1415 and unit cell parameters a = b = 0.3249 nm and c = 0.5206 nm. The crystallite size is calculated from full width at half maximum (FWHM) of the peaks (1 0 0) (0 0 2) (1 0 1) (1 0 2) (1 1 0) (1 0 3) (1 1 2) and (2 0 1). The Bragg peak breadth is a combination of both instrument- and sample

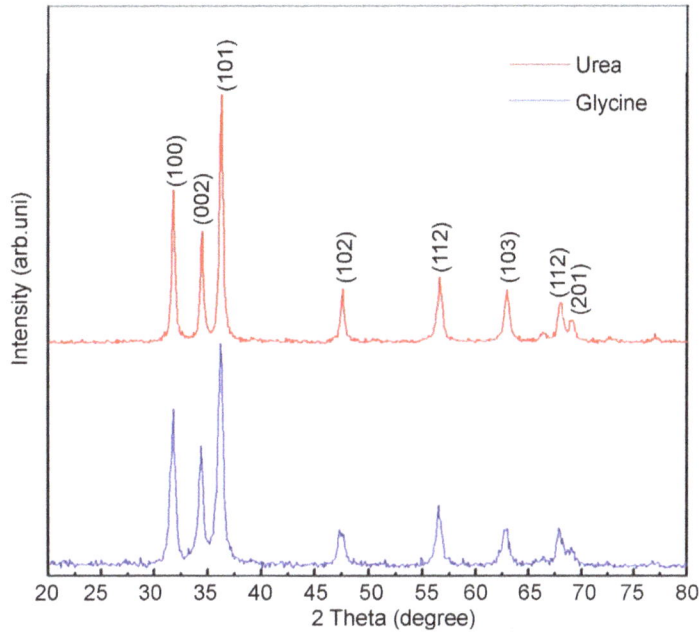

Figure 1. The XRD pattern for the nano ZnO with fuels Gylcine and urea.

dependent effects. To remove these aberrations, it is needed to assemble a diffraction pattern from the line broadening of a standard material such as silicon to determine the instrumental broadening. The instrument-corrected broadening [13] β_D corresponding to the diffraction peak of ZnO was estimated using the relation. The lattice constants of ZnO with varied fuels are given in the **Table 1**.

$$\beta_D^2 = \left[\beta_{measures}^2 - \beta_{instrumental}^2 \right] \tag{1}$$

$$D = \frac{k\lambda}{\beta_D \cos\theta} \Rightarrow \cos\theta \frac{k\lambda}{D}\left(\frac{1}{\beta_D}\right) \tag{2}$$

3.2. Williamson-Hall Methods

Crystal imperfections and distortion of strain-induced peak broadening are related by $\varepsilon \approx \beta_s/\tan\theta$. There is an extraordinary property of Equation (2) which has the dependency on the diffraction angle θ. Scherrer-equation follows a $1/\cos\theta$ dependency but not $\tan\theta$ as W-H method. This basic difference was that both microstructural causes small crystallite size and microstrain occur together from the reflection broadening. Depending on different θ positions the separation of size and strain broadening analysis is done using Williamson and Hall. The following results are the addition of the Scherrer equation and $\varepsilon \approx \beta_s/\tan\theta$.

$$\beta_{hkl} = \beta_S + \beta_D \tag{3}$$

$$\beta_{hkl} = \left(\frac{k\lambda}{D\cos\theta}\right) + 4\varepsilon \tan\theta \tag{4}$$

Rearranging Equation (4) gives:

$$\beta_{hkl} = \left(\frac{k\lambda}{D}\right) + 4\varepsilon \sin\theta \tag{5}$$

Here Equation (5) stands for UDM where it is assumed that stain is uniform in all crystallographic directions. $\beta\cos\theta$ was plotted with respect to $4\sin\theta$ for the peaks of ZnO with varied fuels. Strain and particle size are calculated from the slope and y-intercept of the fitted line respectively. From the lattice parameters calculations it was observed that this strain might be due to the lattice shrinkage. The UDM analysis results are shown in **Figure 2**.

Table 1. The structure parameters of ZnO nanoparticles with fuels glycine and urea.

Fuel	Unit Cell Parameters/nm		Cell Volume (nm³)	Size (nm)	c/a ratio
	a	c			
Glycine	3.252	5.214	47.755	12.89	1.6032
Urea	3.252	5.208	47.701	36.77	1.6015

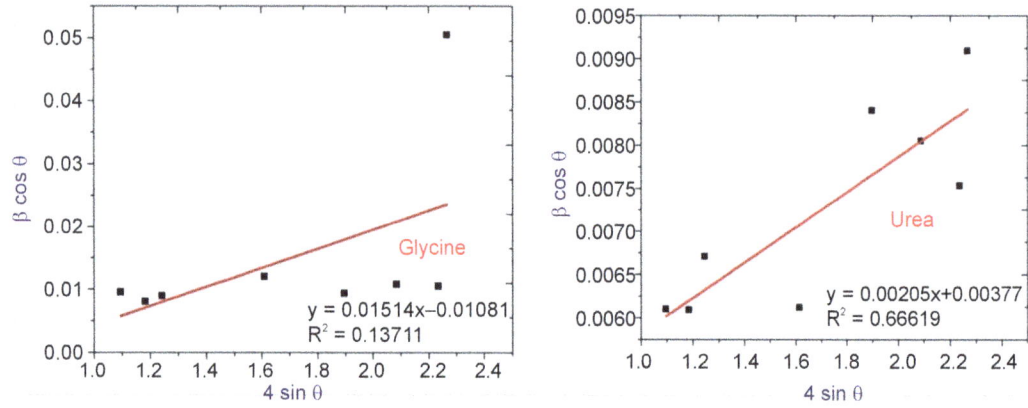

Figure 2. The W-H analysis of ZnO nanoparticles with fuels glycine and urea assuming UDM. Fit to the data, the strain is extracted from the slope and the crystalline size is extracted from the y-intercept of the fit.

From Uniform Stress Deformation Model (USDM) strain is calculated from the Hook's Law maintaining linea proportionality between stress and strain by $\sigma = Y\varepsilon$, where σ is the stress and Y is the Young's modulus. This Hook's law is valid for a significantly small strain. Supposing a small strain to be present in the ZnO with varied fuels, Hooke's law can be used here. Applying the Hooke's law approximation to Equation (5) yields:

$$\beta_{hkl}\cos\theta = \left(\frac{k\lambda}{D}\right) + \left(\frac{4\sigma\sin\theta}{Y_{hkl}}\right) \tag{6}$$

For a hexagonal crystal, Young's modulus is given by the following relation [11] (Equation (7)). where S_{11}, S_{13}, S_{33}, S_{44} are the elastic compliances of ZnO with values of 7.858×10^{-12}, -2.206×10^{-12}, 6.940×10^{-12}, 23.57×10^{-12} m²·N⁻¹, respectively [11]. Young's modulus, Y, for hexagonal ZnO was calculated as ≈130 GPa. $4\sin\theta/Y_{hkl}$ and $\beta\cos\theta$ were taken on x-axis and y-axis respectively. The USDM plots for ZnO with varied fuels are shown in the **Figure 3**. The stress is calculated from the slope.

$$Y_{hkl} = \left(\frac{\left[h^2 + \frac{(h+2k)^2}{3} + \frac{(al)^2}{c}\right]^2}{S_{11}\left(h^2 + \frac{(h+2k)^2}{3}\right)^2 + S_{33}\left(\frac{al}{c}\right)^4 + (2S_{13} + S_{44})\left(h^2 + \frac{(h+2k)^2}{3}\right)\left(\frac{al}{c}\right)^2}\right) \tag{7}$$

The energy density of a crystal was calculated from a model called Uniform Deformation Energy Density Model (UDEDM). From Equation (8) we need to implicit that crystals are to be homogeneous and isotropic nature. The energy density u can be calculated from $u = \left(\varepsilon^2 Y_{hkl}\right)/2$ using Hooke's law. The Equation (8) can be modified according the energy and strain relation.

$$\beta_{hkl} = \left(\frac{k\lambda}{D}\right) + \left(4\sin\theta\left(\frac{2u}{Y_{hkl}}\right)^{1/2}\right) \tag{8}$$

$4\sin\theta\left(2u/Y_{hkl}\right)^{1/2}$ and $\beta_{hkl}\cos\theta$ were taken on x-axis and y-axis respectively. From the slope the anisotropic energy density u was calculated and the crystallite size D from the Y-intercept from **Figure 4**. We know that $\sigma = \varepsilon Y$ and $u = \left(\varepsilon^2 Y_{hkl}\right)/2$ the stress σ was calculated as $u = \left(\varepsilon^2 Y_{hkl}\right)/2$.

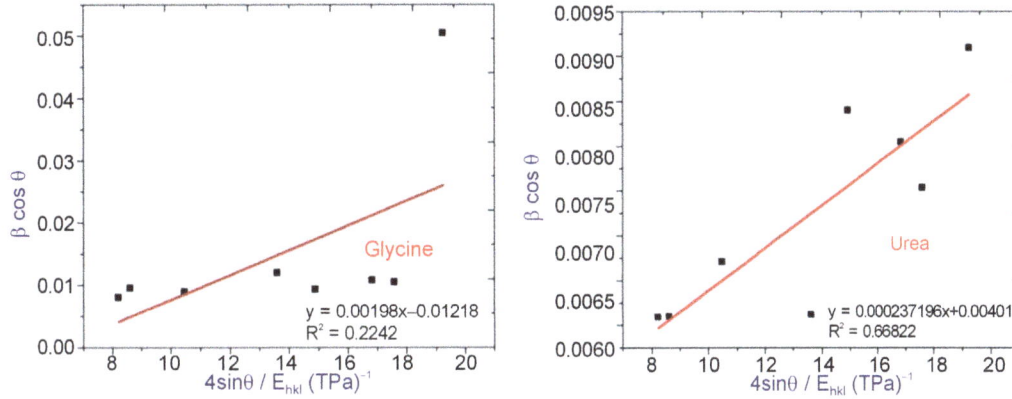

Figure 3. The modified form of W-H analysis assuming USDM for ZnO with fuels glycine and urea. Fit to the data, the stress is extracted from the slope and the crystalline size is extracted from the y-intercept of the fit.

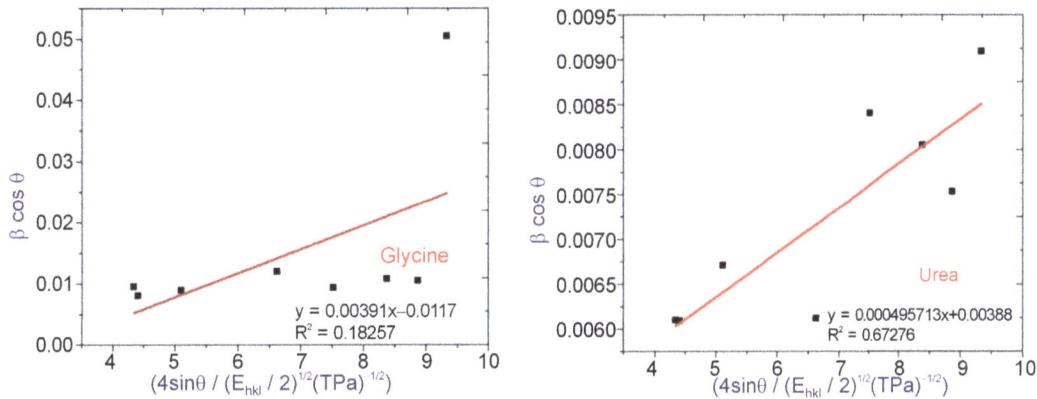

Figure 4. The modified form of W-H analysis assuming UDEDM for ZnO with fuels glycine and urea. Fit to the data, the density of energy is extracted from the slope and the crystalline size is extracted from the y-intercept of the fit.

3.3. Size-Strain Plot Method

Williamson-Hall plot has illustrated that line broadening was basically isotropic. Due to microstrain contribution the diffracting domains were isotropic. Size-strain parameters can be obtained from the "size-strain plot" (SSP). This has a benefit that less importance is given to data from reflections at high angles. In this estimation, it is assumed that profile is illustrated by "strain profile" by a Gaussian function and the "crystallite size" by Lorentzian function [14]. Hence we have

$$\left(d_{hkl}\beta_{hkl}\cos\theta\right)^2 = \frac{k}{D}\left(d_{hkl}^2\beta_{hkl}\cos\theta\right)+\left(\frac{\varepsilon}{2}\right)^2 \qquad (9)$$

where k is a constant, shape of the particles for spherical particles it is given as 3/4. In **Figure 5**, $d_{hkl}^2\beta_{hkl}\cos\theta$ and $\left(d_{hkl}\beta_{hkl}\cos\theta\right)^2$ were taken on x-axis and y-axis respectively for all peaks of ZnO-NPs with the wurtzite hexagonal phase from $2\theta = 20°$ to $2\theta = 80°$. The particle size is calculated from the slope linearly fitted data and the root of the y-intercept gives the strain.

3.4. TEM Method

From TEM results size and morphology of ZnO particles are analyzed and represented in **Figures 6** and **7**. Image reveals that the samples are with the average size of 20 - 30 nm which is in good agreement with that estimated by Scherer formula. In the **Table 2** the results attained from Scherrer method, UDM, USDM, UDEDM, SSP models and TEM are summarized. SAED pattern is shown in **Figure 8**.

Table 2. Geometric parameters of ZnO nanoparticles with fuels glycine and urea.

Fuel	Williamsom-Hall Method									Size-Strain Plot Method			
	UDM		USDM			UDEDM							
	D	ε no	D	ε no	σ	D	ε no	σ	u	D	ε no	σ	u
	(nm)	Unit $\times 10^{-3}$	(nm)	Unit $\times 10^{-3}$	(Mpa)	(nm)	Unit $\times 10^{-3}$	(Mpa)	(kJm^{-3})	(nm)	Unit $\times 10^{-3}$	(Mpa)	(kJm^{-3})
Glycine	12.82	1.51	11.37	1.77	168.24	11.84	1.84	198.21	81.3	12.14	1.808	199.35	92.54
UREA	36.75	0.205	34.56	0.212	123.54	35.72	0.9422	136.71	53.1	35.86	0.439	139.24	52.69

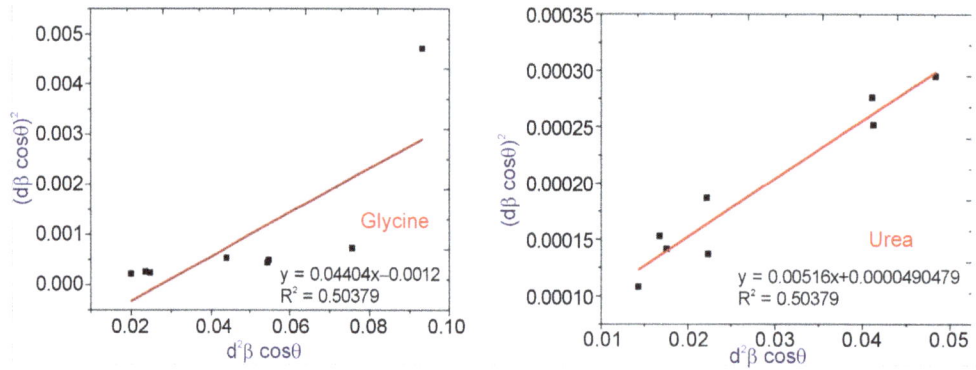

Figure 5. The SSP plot of ZnO with fuels glycine and urea. The particle size is achieved from the slop of the liner fitted data and the root of y-intercept gives the strain.

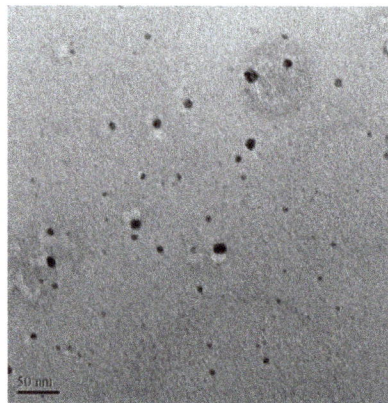

Figure 6. TEM micrographs of ZnO nanoparticles with fuel glycine.

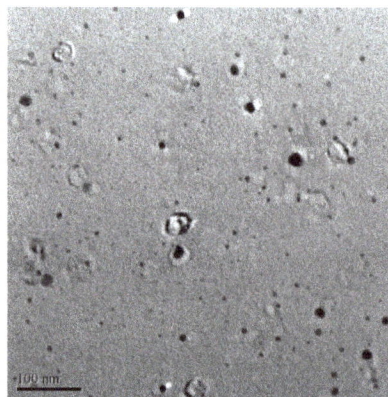

Figure 7. TEM micrographs of ZnO nanoparticles with fuel urea.

3.5. Particle Analyzer

Using Nano Particle Analyzer (SZ100) the size of the nanopowders is measured. The average particle sizes for samples with fuels glycine and urea are shown with histogram in **Figures 9** and **10**. All the results from particle analyzer are in good agreement with the XRD results of crystallite sizes.

Figure 8. SAED pattern of ZnO.

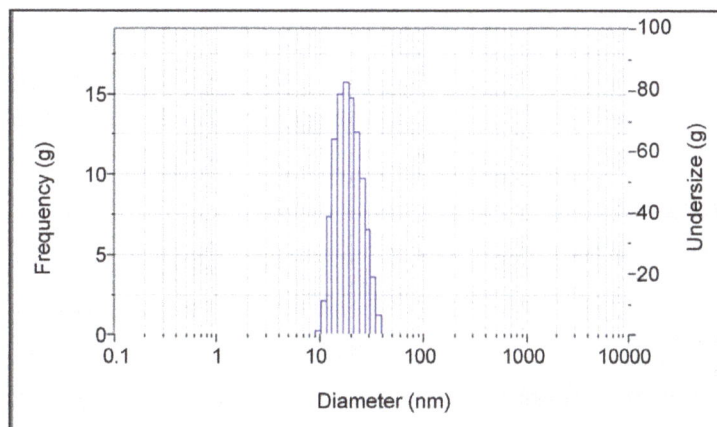

Figure 9. Particle analyzer histograms of fuel glycine of ZnO powders.

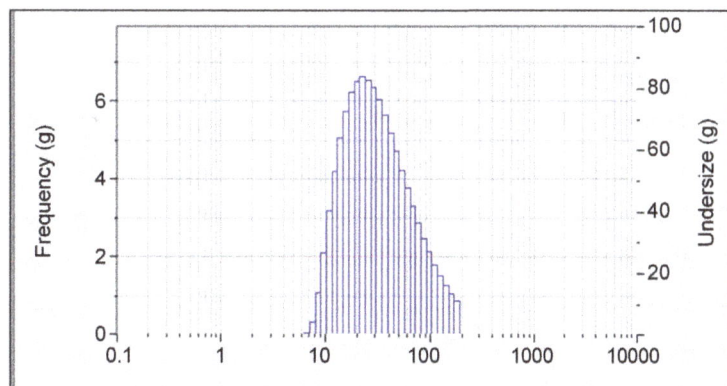

Figure 10. Particle analyzer histograms of fuel urea of ZnO powders.

4. Conclusion

In summary, we have successfully synthesized porous nanocrystalline ZnO powder by novel surfactant assisted combustion method. Systematic variation of fuels resulted in nanoparticles of the smallest size of crystallite size 12.82 nm and 36.42 nm for glycine and urea fuels respectively which were   in good agreement with TEM results. When the crystallite size decreased for sample with fuel glycine cell volume, c/a ratio and micro strain increased. Similarly as the crystallite size increased for sample with fuel urea cell volume, c/a ratio and micro strain decreased. This broadening was analyzed by the Scherrer formula, modified forms of W-H analysis and the size-strain plot method. From the results, it was observed that the strain value decreased but the particle size increased as doping concentration increased. The TEM results were in good agreement with the results of the W-H and the SSP methods.

References

[1]   Hu, J.T., Odom, T.W. and Liebe, C.M. (1999) Chemistry and Physics in One Dimension: Synthesis and Properties of Nanowires and Nanotubes. *Accounts of Chemical Research*, **32**, 435-445. http://dx.doi.org/10.1021/ar9700365

[2]   Annapoorani, R., Dhananjeyan, M.R. and Renganathan, R. (1997) An Investigation on ZnO Photocatalysed Oxidation of Uracil. *Journal of Photochemistry and Photobiology A*: *Chemistry*, **111**, 215-221. http://dx.doi.org/10.1016/S1010-6030(97)00170-6

[3]   Huang, W.-J., Fang, G.-C. and Wang, C.-C. (2005) A Nanometer-ZnO Catalyst to Enhance the Ozonation of 2,4,6-Trichlorophenol in Water. *Colloids and Surfaces A*: *Physicochemical and Engineering Aspects*, **260**, 45-51. http://dx.doi.org/10.1016/j.colsurfa.2005.01.031

[4]   Sánchez, L., Peral, J. and Domènech, X. (1996) Degradation of 2,4-Dichlorophenoxyacetic Acid by *in Situ* Photogenerated Fenton Reagent. *Electrochimica Acta*, **41**, 1981-1985. http://dx.doi.org/10.1016/0013-4686(95)00486-6

[5]   Chen, B.J., Sun, X.W., Xu, C.X. and Tay, B.K. (2004) Growth and Characterization of Zinc Oxide Nano/Micro-Fibers by Thermal Chemical Reactions and Vapor Transport Deposition in Air. *Physica E*, **21**, 103-107. http://dx.doi.org/10.1016/j.physe.2003.08.077

[6]   Nakata, Y., Okada, T.and Maeda, M. (2002) Deposition of ZnO Film by Pulsed Laser Deposition at Room Temperature. *Applied Surface Science*, **197**, 368-370. http://dx.doi.org/10.1016/S0169-4332(02)00426-9

[7]   Yoo, Y.-Z., Jin, Z.-W., Chikyow, T., Fukumura, T., Kawasaki, M. and Koinuma, H. (2012) S Doping in ZnO Film by Supplying ZnS Species with Pulsed-Laser-Deposition Method. *Applied Physics Letters*, **81**, 3798. http://dx.doi.org/10.1063/1.1521577

[8]   Li, Y.J., Duan, R., Shi, P.B. and Qin, G.G. (2004) Synthesis of ZnO Nanoparticles on Si Substrates Using a ZnS Source. *Journal of Crystal Growth*, **260**, 309-315. http://dx.doi.org/10.1016/j.jcrysgro.2003.08.041

[9]   Yang, X.-L., Dai, W.-L., Chen, H., Cao, Y., Li, H.X., He, H.Y. and Fan, K.N. (2004) Novel Efficient and Green Approach to the Synthesis of Glutaraldehyde over Highly Active W-Doped SBA-15 Catalyst. *Journal of Catalysis*, **229**, 259-263. http://dx.doi.org/10.1016/j.jcat.2004.10.009

[10]   Ramakanth, K. (2007) Basic of Diffraction and Its Application. I.K. International Publishing House Pvt. Ltd., New Dehli.

[11]   Zhang, J.-M., Zhang, Y., Xu, K.-W. and Ji, V. (2006) General Compliance Transformation Relation and Applications for Anisotropic Hexagonal Metals. *Solid State Communications*, **139**, 87-91. http://dx.doi.org/10.1016/j.ssc.2006.05.026

[12]   Suranarayana, C. and Norton, M.G. (1998) X-Ray Diffraction: A Practical Approach. Springer, New York. http://dx.doi.org/10.1007/978-1-4899-0148-4

[13]   Rogers, K.D. and Daniels, P. (2002) An X-Ray Diffraction Study of the Effects of Heat Treatment on Bone Mineral Microstructure. *Biomaterials*, **23**, 2577-2585. http://dx.doi.org/10.1016/S0142-9612(01)00395-7

[14]   Nye, J.F. (1985) Physical Properties of Crystals: Their Representation by Tensors and Matrices. Oxford, New York.

Nano-Scale Modelling and Simulation of Metal Wiredrawing by Using Molecular Dynamics Method

Ken-ichi Saitoh[1*], Youhei Sameshima[2], Syuhei Daira[2,3]

[1]Department of Mechanical Engineering, Kansai University, Suita, Japan
[2]Graduate School of Science and Engineering, Kansai University, Suita, Japan
[3]Denso Techno, Co. Ltd., Obu, Japan
Email: *saitou@kansai-u.ac.jp

Abstract

In this paper, molecular dynamics (MD) simulations of nano-sized wiredrawing are performed. The wiredrawing is a traditional plastic working method, but there has not been any insight to develop it in a nano-sized scale. Therefore, to materialize the concept of the nano-sized wiredrawing, a numerical modelling is pursued at first in this paper, and the interatomic potential, a crystalline orientation, the drawing condition realized by a die geometry are thoroughly investigated. In particular, to reduce the friction between a wire and a die, a simple friction model for the MD analysis is newly proposed, where the interatomic interaction is adequately modified by a single factor ω. Then, the fruitful results are obtained by using $\omega = 0.1$. We checked the availability of such nano-sized MD simulation by constructing a two-dimensional wiredrawing model, at first. The analysis of atomic stress during drawing is also assessed. It is useful to use invariant of the atomic stress tensor, such as hydrostatic stress (average stress, σ_m) or von Mises equivalent stress (σ_{eq}). The former is related to the phase transformation from the body-centered-cubic (bcc) structure to the face-centered-cubic (fcc) one, which is found in the present MD simulation. It is observed that an initial α-iron crystal with bcc structure changes partially into the fcc phase. It is recognized that the phase transformation is caused by the positive hydrostatic stress values, which is occurring especially inside the die region. We observed that a lot of dislocation core structures occur in wiredrawing process and their existence and evolution are well related to the equivalent stress values.

Keywords

Molecular Dynamics, Iron and Steel, Wiredrawing, Plastic Working, Nanotechnology, Materials

---

*Corresponding author.

Processing, Crystalline Structure

---

1. Introduction

There are many plastic working processes that are important for industry. Above all, drawing (wiredrawing) process is extensively used to fabricate axisymmetric metallic bars and wires into a prescribed diameter. It is interesting that the wiredrawing is basically using only a die. So far, for example, drawn wires made up of iron and steel have been applied to many industrial products, such as tire codes, piano (music) wire, and suspension wire of huge bridge. Today, materials are trending toward very narrow and small size. For the purpose of utilizing wire structures in the field of MEMS (Micro Electro Mechanical System) or small-sized electronic/computer devices, the dimension of wire material should be as small as possible. From now on, a diameter required in such purpose will become less than one micrometer, *i.e.* less than several hundreds of nano-meters. Certainly, there is another cutting-edge method to synthesize very tiny wire using nano-structured template [1]. But, it has to be based on chemical reaction process and spontaneous crystal growth technique, so it is not so easy to control the fabricating process. Therefore, as an effective method for the production of very narrow and small wires, the traditional plastic working technique using the die may be applied, by downsizing the wiredrawing process into a nano-meter scale. Indeed, it is, of course, just an idea with possibility now, but in near future it will be categorized as a prospective extension of the traditional wiredrawing method. So far, the smallest steel wire made by commercial drawing has just reached to several micrometers in diameter. Accordingly, "nanoscale wiredrawing" is one of prospective technologies. However, now, we do not have any evidence for feasibility of nano-meter wire drawing technique. For this reason, we need first to study the possibility of this new technique, and we should predict the material behaviors by numerical simulation.

In this study, the molecular dynamics (MD) method, which is one of theoretical and computational techniques for materials modelling and simulation, is applied to the nano-sized wiredrawing process. In numerical modelling, except for MD method, usually, finite element (FE) analysis, which is definitely based on some mathematical elastic-plastic continuum theory, has been widely applied to investigate the wiredrawing process. In recent years, since information about the crystalline change of metallic materials during wiredrawing is very crucial, the FE analysis including crystal plasticity theory (CPFE) [2] [3] is well being applied to the study. Though the FE analysis can produce many practical results, its material model must be constructed in mathematical fashion and the interaction between the die and the wire is still inevitably ambiguous. Namely, microscopic physics of a wire material is not sufficiently modeled by the conventional FE or even precise CPFE framework. The conventional FE is also unsuitable to analyze the texture and defects which are being nucleated and developing in the wire. To the contrary, the MD method is capable of modelling a wire material from atomic-scale physics and so the method has an advantage in predicting microscopic phenomena. It is well known that there are some drawbacks of MD, however. At the present, the largest three-dimensional computational size of atomic system treatable in the MD method is just limited to the range around several hundreds nano-meters. However, as stated above, there is a trend to make smaller metallic wire and this fact requires us to simulate the phenomena from the atomic scale to obtain a clear insight about wiredrawing. For those reasons, the present authors suppose that it is very worthwhile using the MD method to analyze the wiredrawing process.

So far, plasticity or plastic working process has been investigated by the MD method as follows. As for analyses of plastic working application, for example, the rolling [4] and the cutting [5] of metallic materials have been studied. As for unknown phenomena concerning plasticity, to name a few, dynamic recrystallization in Cu-Al system [6], pencil glide dislocation motion in bcc (body-centered cubic) crystal [7], and martensitic phase transformation of alloy [8], have been studied. Unfortunately yet, there has not been any MD study concerning wiredrawing. In the wiredrawing process, the force and the contact conditions subjected to the wires are tremendously strong. Therefore, the wiredrawing is one of strong plastic working techniques, where the material is subjected to very strong multi-axial stress and strain. A similar viewpoint can be found in the MD analysis concerning severe plastic deformation (SPD) processes [9], many of which have been experimentally invented as novel plastic processing method. Also, the nano-sized imprint (nano-imprint) technique of copper has been attempted in the MD simulation [10], while the material implemented there is generally a soft polymer or a resin. Once again, to our knowledge, there has not yet been any MD study of nano-sized wiredrawing.

In this study, iron and steel wires are focused on, for they are extensively used with the wiredrawing in industry. In iron and steel, an impurity addition such as carbon, etc., is usually carried out. Consequently, both strength and ductility are enhanced at the same time, thanks to an eutectic phase called a pearlite phase. The pearlite phase is composed of ferrite (α-Fe) and cementite (Fe$_3$C) crystal as sub-structures inside. In the present MD simulation, however, some drastic simplification in the computational model is inevitable, especially for the modelling of interatomic potential function. So, at this point, it is reasonable to use a single crystal model comprising just the ferrite crystal and then a reliable potential function just between iron atoms will be successfully adopted.

In this study, an atomic-scale wiredrawing model is constructed for the MD simulation. The new knowledge of atomic mechanism in the wiredrawing which is presented in this study is useful for the future micro-scale or nano-scale wiredrawing process. Ideally, the wiredrawing includes an axisymmetric condition and so a full three-dimensional model would be optimum. But, for the simplification of the MD model, a two-dimensional drawing model using periodic boundary condition is adequate now and here. The realistic crystal rotation and the motion of lattice defects will be precisely captured by upcoming study of a full three-dimensional model [11].

As one of topics in our study of MD modelling, a friction model is proposed and is evaluated. As frequently observed in MD analyses, a friction force between nano-sized solids becomes tremendously large, because of larger relative contacting area between surfaces. In our model, the interatomic potential between the die and the wire atoms is modified to conduct the drawing smoothly with a moderate lubricating condition. Since the present friction model requires somewhat parametric study, the first evaluation is provided here.

Understanding the stress or strain distribution inside a drawn wire, including the residual stress distribution, is important in discussing the result of wiredrawing process. In the history of studies of the wiredrawing [12]-[16], hydrostatic and (von Mises) equivalent stress components are key factors which should be focused on. This paper shows that the atomic motion and behavior of lattice defects are closely related to their stress distribution. We also affirm the result of crystalline phase transition between bcc and fcc (face-centered cubic) in microscopic scale. This behavior has not yet observed experimentally in the drawing of steel wires. The present study shows that the mechanism of phase transition is also closely related to the stress distribution inside the wire.

2. Theory and Methods

2.1. Molecular Dynamics

Molecular dynamics (MD) is a microscopic numerical simulation technique, where the time evolution of atomic or molecular system is obtained. The interatomic interaction between atoms or molecules is prescribed. The equations of motions are numerically integrated, with a certain time increment (usually in the order of femtosecond), to update positions and velocities of individual atoms. Using the atomic trajectories in the system, the development of crystal structure and material phase is captured. Accordingly, dynamics of phase transition can be simulated directly, and transport coefficients such as diffusion constant can be estimated effectively.

The equations of motion are given by Equation (1),

$$m_i \left(\mathrm{d}^2 \boldsymbol{r}_i \big/ \mathrm{d}t^2 \right) = \boldsymbol{F}_i = -\partial E_{\mathrm{tot}} \big/ \partial \boldsymbol{r}_i , \tag{1}$$

where m_i, $\boldsymbol{r}_i$, $\boldsymbol{F}_i$, and t are atomic mass, position, interatomic force, and time, respectively. The MD simulation is sometimes called a deterministic method, which is different from any stochastic method such as Monte Carlo method. The interatomic force is straightforwardly determined as the derivative of the potential energy of the total system E_{tot}. Though the MD method provides a high-resolution view of the material, the calculated size is only limited to very tiny portion of the material. In addition, the interatomic potential function is just an approximation of real material system. However, those drawbacks are being somewhat resolved by introducing *ab-initio* (first principle) method involving quantum mechanical formulation and also by utilizing a parallel algorithm to enable very larger-scale computations.

2.2. Interatomic Potential Function of Iron and Steel

The crystal structure of iron is bcc in the standard temperature and pressure, while it changes to fcc in high temperature or stress application. Accordingly, most of potential functions are formulated in the bcc structure. The

potential function of the embedded atom method (EAM) is adopted in the present study. The EAM can repro-
duce the many-body effect around a metallic atom. We use the Finnis-Sinclair type potential (FS potential) [17],
which has been extensively used in other MD studies, because of its high accuracy in system energy and atomic
behavior of iron system. In the FS potential formulation, the total energy of the system is given by,

$$E_{\text{tot}} = E_{\text{m}} + E_{\text{p}} = \sum_{i=1}^{N} AF(\rho_i) + (1/2)\sum_{i=1}^{N}\sum_{j\neq i}^{N} \phi(r_{ij}),\tag{2}$$

where E_{m} means the many-body term, ρ_i is an electron density made up of interactions from surrounding atoms,
and $F(\rho)$ means a functional of the electron density. The second term E_{p} is a summation of usual pairwise interactions $\phi(r)$.

2.3. The Concept of Nano-Scale MD Wiredrawing Model

Generally speaking, the wiredrawing is one of traditional plastic working processes, in which bar or rod is
forced to go through the die, and is eventually given a narrower diameter (*i.e.* it is shrunk in diameter and is
stretched in the longitudinal direction). Consequently, the wire material is given higher strength and ductility as
well as a fine surface finish. The purpose of the present study is to build a nano-meter-sized simulation model
for the wiredrawing process. The idea that a hollow die is used in the wiredrawing is straightforwardly adopted.
Therefore, whether it will be actually better choice or not is, in fact, left to the future experimental study.

2.4. Nano-Scale MD Wiredrawing Model

It is true that a three-dimensional model is more precise for the wiredrawing as shown in **Figure 1(b)** which is
being used in another study [11] [18]. But, in three-dimension, understanding the behavior of crystal transition
and defects is quite complicated yet. Therefore, in the present study, an approximated two-dimensional model of
the wiredrawing, as shown in **Figure 1(a)**, is constructed and is applied to the MD simulation. The MD calculation is carried out in three-dimensional space, but the periodic boundary condition is imposed on in one direction

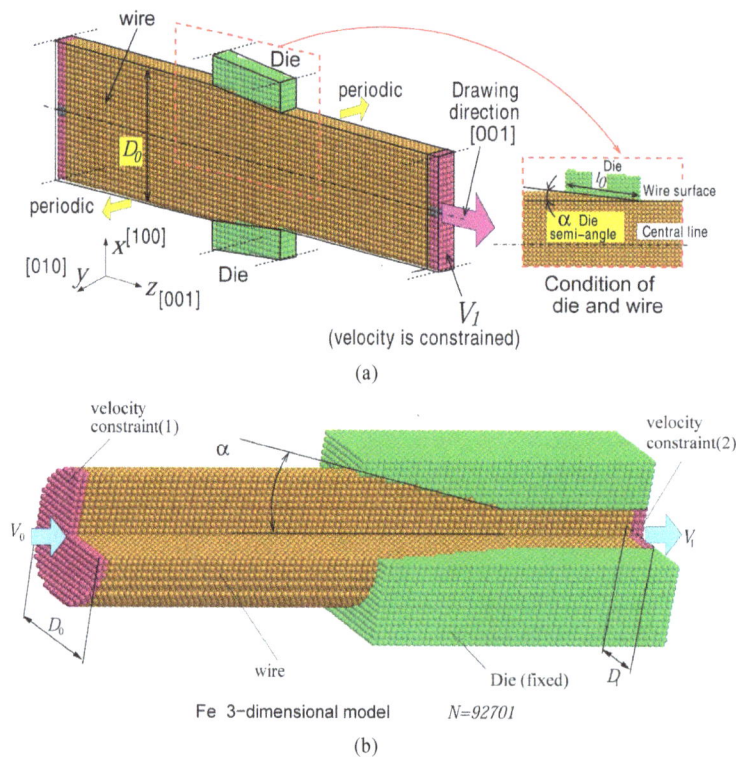

Figure 1. Molecular dynamics models for wiredrawing. (a) An example of the
present 2-d model(a model of the drawing in [100] direction); (b) Another
example of 3-d model [11] [18] (this is not calculated here).

(in this case, y direction) which is perpendicular to the drawing direction (in this case, z direction). This model seems like the rolling process, but, since the deformation in periodic direction (y direction) is constrained, the stress state tends to be tri-axial similar to the wiredrawing.

The die semi-angle is 5.0 degrees and the length of the die slope is 5.0 nm, which are fixed through the present study. In fact, the die semi-angle in between 5 and 7 degrees is proved to be optimum in axisymmetric drawing by analytic consideration of theoretical plasticity. As recognized well, the choice of crystalline orientation relative to the drawing direction affects the mechanical property of the drawn wire. Therefore, [100] [111] orientations are compared here. All the calculation conditions are summarized in **Table 1**.

2.5. Construction of Friction Model

The plastic working technique usually copes with the problem of the friction between the die and the wire material [19]. Since the relative contact area is larger in nano-sized scale, the effect of friction becomes larger. It is supposed that the future nanotechnology should overcome the friction problem as a primal subject. The developing study of nano-imprint technique is one of the challenges [10].

In this study, it is proposed that the friction is reduced by modifying the potential function between the die and the wire atoms as shown in **Figure 2**. The interatomic interaction is somewhat needed to make the friction,

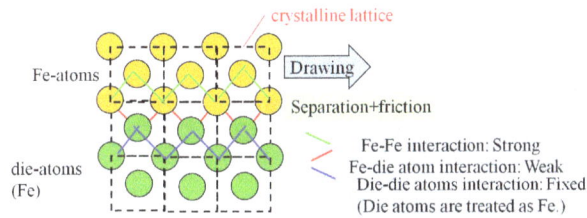

Figure 2. The concept of friction model of MD in wiredrawing process (Interaction as to die atoms are that of Fe atoms).

Table 1. Calculation conditions and materials properties.

Property	Unit	Value
The model of drawing in [100] direction		
The number of total atoms		91,885
The number of fixed atoms (wire)		1362
The number of fixed atoms (dies)		15,968
The model of drawing in [110] direction		
The number of total atoms		91,919
The number of fixed atoms (wire)		973
The number of fixed atoms (dies)		16,132
The model of drawing in [111] direction		
The number of total atoms		91,911
The number of fixed atoms (wire)		973
The number of fixed atoms (dies)		16,089
Size $x \times y \times z$	nm	$9.8796 \times 9.8796 \times 27.5184$
Lattice constant a_0 (α-Fe)	nm	0.28665
Length of die land l_0	nm	5.00
Diameter D_0	nm	6.8796 (=24 a_0)
Die semi-angle α	deg.	5.00
Outlet velocity V_1	m/s	50.0 (fast)/10.0 (slow)
Time increment Δt	fs	2.0
Temperature T	K	10.0

but it is reduced not to adhere each other. The general formulation is shown as follows. For example for pairwise potentials, the cohesive energy (interatomic binding energy) is determined by the intensity of the potential function itself. Therefore, the potential function is reduced by an arbitrary factor ω, which is less than unity ($\omega < 1$). At this point, the factor ω is completely ad-hoc parameter and it should be fitted by ab-initio calculation or experimental result in the future. So far, we have evaluated the adequate value of ω in our MD simulation. As an consequence, it is found that ω should be less than 0.1, and then it shows good results for most of wiredrawing conditions.

Generally, the original potential function $V(r)$ is modified by Equation (3).

$$V_a(r) = \omega \cdot V(r) \tag{3}$$

The many-body potential is composed of the many-body term and the pairwise term, as shown in Equation (2). Interatomic bonding is largely contributed by the many-body term and it is usually a monotonically increasing function of the electron density $\rho(r)$. Thus, for the FS potential in this study, $\rho(r)$ is directly multiplied by the factor ω introduced above. The expression of the FS potential including the friction condition is given by Equation (4) together with Equation (2),

$$\rho_{i,a} = \begin{cases} \omega \cdot \rho(r) & (\text{wire-die}) \\ \rho(r) & (\text{wire-wire}) \end{cases} = \sum_{j=1}^{N(\text{wire-die})} \omega \cdot \psi(r_{ij}) + \sum_{j=1}^{N(\text{wire-wire})} \psi(r_{ij}), \tag{4}$$

where $\psi(r)$ is a pairwise function for the electron density. The $\omega < 0.1$ is found to show a relevant friction in the many-body case as well.

2.6. Analyzing Method of the MD Output

CNA Method: In order to detect a crystal type and its orientation, the common neighbor analysis (CNA) can be used effectively [20] [21]. By using the CNA, one can identify each atom as fcc, bcc or the other crystalline structures. Besides, by the CNA, we can determines atoms near a lattice defect such as dislocation, grain boundary or surface.

Atomic Stress: It is possible to calculate the atomic stress tensor (atomic stress) in MD calculation. The atomic stress tensor is derived along with the continuum concept of stress tensor. When an individual atomic volume is assumed, then strain energy stored inside the volume is evaluated from interatomic potential. The energy is mathematically differentiated by the strain tensor which assumes homogeneous deformation and is expressed by the distance between pair of atoms. This means that atomic stress corresponds to increase of the elastic strain energy per unit volume. Thus, the general formulation of atomic stress of atom i is defined by Equation (5).

$$\sigma_i^{\alpha\beta} = \{1/(2\Omega)\} \sum_{j=1}^{N} \left(\partial E_i(r)/\partial r \right)\big|_{r=r_{ij}} \cdot r_{ij}^\alpha r_{ij}^\beta / r_{ij}, \tag{5}$$

where $r_{ij}^\alpha = r_i^\alpha - r_j^\alpha$ is the difference of position vectors between i and j atoms, r_{ij} is their distance. $E_i(r)$ is an effective energy contribution from i atom. Ω is the atomic volume for the reference structure. Generally speaking, the summation of many-body energy is unable to be divided into the energy of each atoms. However, for FS potential, we can do it only in considering an effective potential,

$$E_i(r_{ij}) = F(\rho_i) + \sum_{j=1, j \neq i}^{N} \phi(r_{ij}) \tag{6}$$

In finite temperature, atomic stress inevitably contains large fluctuation due to the small number of atoms. So, it needs to be averaged in volume or in time interval to give continuous distribution. The averaged stresses are displayed without the atomic subscript i, like $\sigma^{\alpha\beta}$, and we use these values throughout this paper and call them only "stress".

Once the components of the stress with regard to a set of coordinate axes (x, y, z) are obtained, effective stress invariants are calculated from them. They are, for instance, hydrostatic stress (mean stress) σ_m and von Mises equivalent stress σ_{eq}. σ_m is also the first invariant of stress tensor, defined by,

$$\sigma_m = \left(\sigma^{xx} + \sigma^{yy} + \sigma^{zz} \right)/3 = -p, \tag{7}$$

where p means hydrostatic pressure. The σ_m represents a tri-axial condition of the stress and it is related to the volume change. For example in the wiredrawing process, with increase of the σ_m in positive regime (*i.e.* tensile stress, $\sigma_m > 0$), an internal fracture (also called central burst or "cuppy" failure in the actual process) tends to take place. On the other hand, the σ_{eq} is related to the second invariant of stress. It combines the deviatric stress components and is defined by,

$$\sigma_{eq} = \sqrt{\left\{\left(\sigma^{xx} - \sigma^{yy}\right)^2 + \left(\sigma^{yy} - \sigma^{zz}\right)^2 + \left(\sigma^{zz} - \sigma^{xx}\right)^2\right\}\bigg/2 + 3\left\{\left(\sigma^{xy}\right)^2 + \left(\sigma^{yz}\right)^2 + \left(\sigma^{zx}\right)^2\right\}}, \qquad (8)$$

Physically, the σ_{eq} value reflects a nucleation event of dislocation and its motion in a crystal. In the atomic system in MD simulation, the σ_{eq}, which is evaluated locally in an atomic system, exhibits the intensity of driving force for the nucleation or the mobility of dislocations.

3. Results and Discussion

3.1. Effect of the Factor ω on the Friction Behavior

Wiredrawing simulations are performed with the variety of friction factors, $\omega = 1.0, 0.1, 0.01$ and 0.001, and the dependency on friction factor ω is investigated. The maximum stress value for components σ^{zz}, σ_m, and σ_{eq}, which are averaged over all atoms in the wire, are shown in **Figure 3(a)**. Moreover, **Figure 3(b)** shows the time-averaged stress components. It is clear that the larger ω is, the larger the stresses are. It means that resistance of the wire material during the drawing process, is totally raised by larger ω near unity. The resistance is possibly caused by the friction at the wire surface between the die and the wire atoms as well as the plastic deformation occurring inside the wire. If we assume that the resistance due to plasticity provides a constant resistance, the friction at the wire surface is responsible for the change in the total resistance in the drawing process. Besides, the results for $\omega = 0.01$ and $\omega = 0.001$ are not so different and their difference in stresses is almost zero. Thus, we can apply the friction model with the factor $\omega = 0.001$ in all the MD simulations presented here.

3.2. Stress Distribution and Its Change in Wiredrawing Process

3.2.1. Drawing Stress
Figure 4 shows the value of σ^{zz} which is the normal stress component in the direction of wiredrawing (*i.e.* σ^{zz} is a drawing stress) and σ^{xx} which is in the perpendicular direction and not in the direction of periodicity of the model. After a start-up period when the length 1.5 nm of the wire has been drawn, the stress component σ^{xx} tends to be negative (compressive) and then becomes stationary. On the other hand, the drawing stress, σ^{zz}, tends to be positive (tensile) and becomes stationary. The value of σ^{zz} reaches 6.0 GPa at maximum, the magnitude of

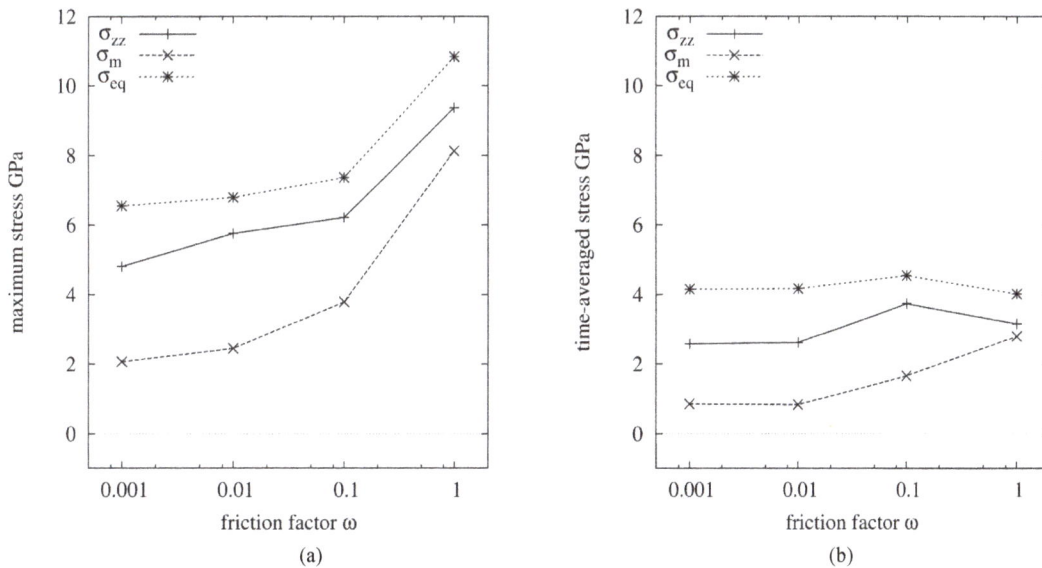

Figure 3. Relation between friction factor ω and (a) maximum stress; (b) time-averaged stress in wiredrawing.

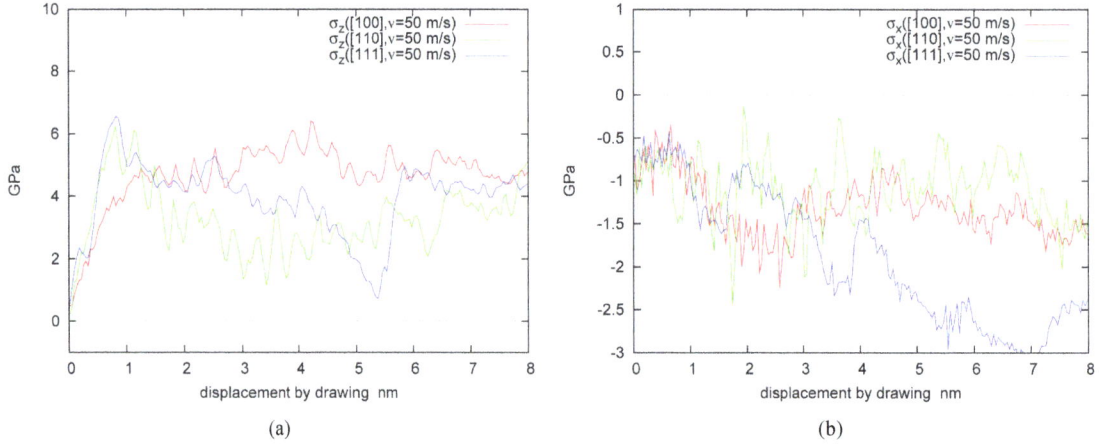

Figure 4. Time transition of stresses occurring in wire (drawing direction σ^{zz}/perpendicular direction σ^{xx} (one of not periodic directions), case: $v = 50$ m/s). (a) σ^{zz}; (b) σ^{xx}.

which is about twice the maximum value of the magnitude of σ^{xx}. As for comparison among crystalline orientations, the wiredrawing in [100] direction shows the largest stress values among three. If we assume the same effect of the friction at the surface, this fact means that the crystal arranged in [100] orientation shows the largest resistance to the plastic deformation among three orientations.

3.2.2. Hydrostatic Stress
In order to comprehend characteristics of the stress state in the wire, it is highly reasonable to look into both hydrostatic stress, σ_m, and equivalent stress, σ_{eq}. The normal stress components, σ^{xx}, σ^{yy} or σ^{zz} may be either tensile or compressive, but an averaged tri-axial stress state, i.e. σ_m tends to be in tension or in compression. The hydrostatic stress, σ_m, corresponds only to an elastic deformation, whereas equivalent stress, σ_{eq}, is derived from the deviatric stress tensors which are a residue of stress tensor in excluding isotropic contribution. Accordingly, the atomistic value of σ_{eq} shows a measure of the plastic working which those atoms have been experienced during the wiredrawing.

Time transition of σ_m and σ_{eq} during drawing is visualized in **Figure 5**. It is understood that σ_m is always positive (tensile stress). However, when the drawing speed is slower (condition (B)), the magnitude of σ_m is converging into zero. This means that, in the case with a higher drawing speed (such as $v = 50$ m/s, **Figure 5(a)**), the deformation velocity in plasticity is defeated by the traveling speed of the wire. As a result, the plastic deformation has not fully been carried out even after the wire comes out of the die region. To the contrary, in a slower drawing case, there is a plenty of time for the atoms to undertake a plastic deformation, including a nucleation of lattice defect such as dislocation. The whole crystalline structure is relaxed and the hydrostatic stress σ_m drops soon after that, in this case. As shown in **Figure 5(b)**, the plastic deformation can proceed, not increasing the stress σ_{eq} and keeping the magnitude of σ_m at a low level.

Figure 6 shows the stress distribution when it is viewed in the cross-section of the wire (projected onto xz plane). This is the case of wiredrawing with [100]-orientation and $v = 50$ m/s. Those pictures are arranged as follows: (A) normal stress in drawing direction σ^{zz}, (B) hydrostatic (mean) stress σ_m, and (C) equivalent stress σ_{eq}, from the top to the bottom. Also, the pictures at 40, 80 and 120 ps are arranged from the left to the right.

The drawing stress σ^{zz} possesses a positive value after passing through the die region, whereas the trailing part shows a negative value (compressive). These plus and minus in the stress value seems balanced in the total region of the wire. The mean stress σ_m shows a larger value, especially near the exit of the die region, and it is decreasing with increase of the distance from the exit of the die. Besides, a v-shaped region of the compressive stress exists at the entrance of the die region. Since the wire is always compressed by the inclined surface of the die, it is reasonable that the atoms in the wire have always compressive stress.

Equivalent stress σ_{eq} shows a large value at the center of the die region. The stress value of the wire is increasing with increase of the distance from the exit of the die. It seems natural that the σ_{eq} is large around the region in which the wire material is experiencing a plastic deformation. As an example, an isolated dislocation structure is observed at the lower part in the cross-sectional picture which is taken outside the die region, as

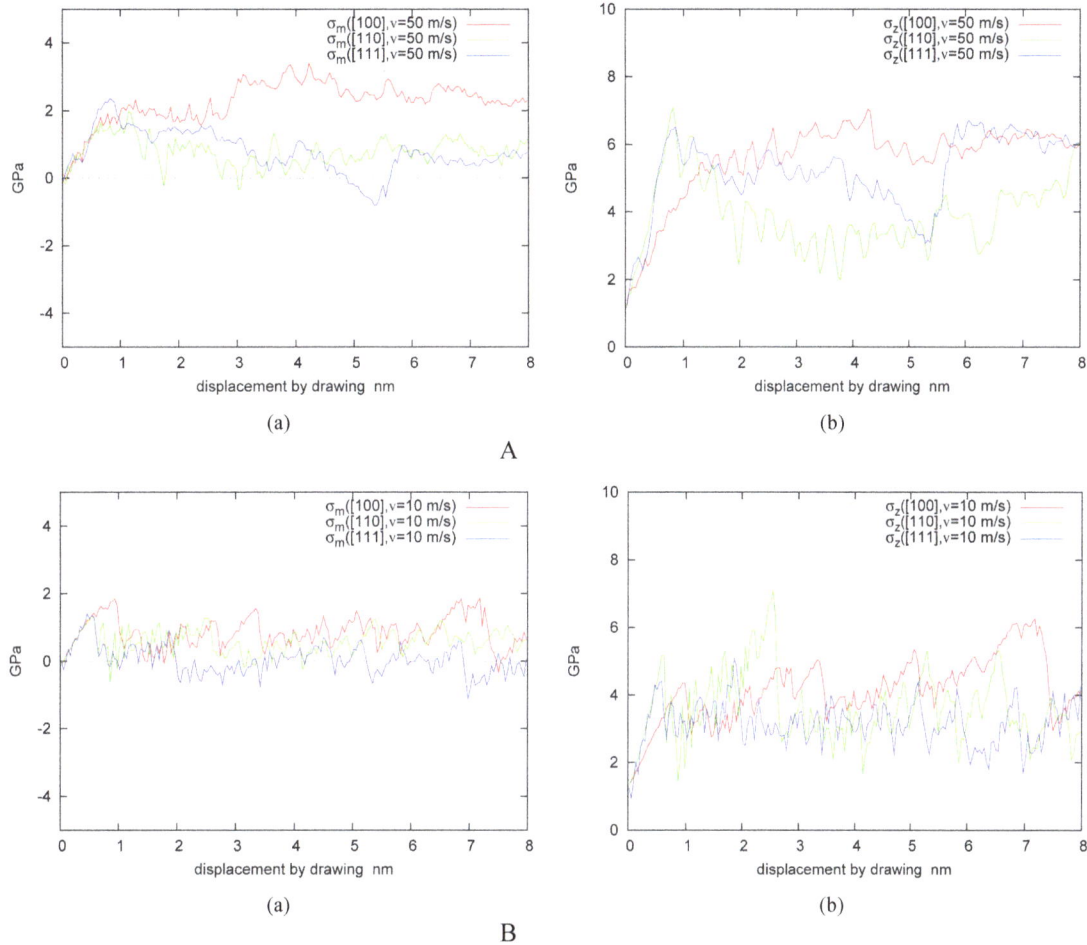

Figure 5. Time transition of stresses occurring in wire (hydrostatic component σ_m and equivalent stress component σ_{eq}). A) Fast drawing velocity $v = 50$ m/s, (a) hydrostatic stress σ_m; (b) equivalent stress σ_{eq}. B) Slow drawing velocity $v = 10$ m/s, (a) hydrostatic stress σ_m; (b) equivalent stress σ_{eq}.

shown in the figure at 120 ps. Discussion concerning the structure and the dynamics of dislocations will be shown in the subsequent section in relation to the CNA analysis.

3.3. Results of Crystal Change by Common Neighbor Analysis (CNA)

Figure 7 shows the distribution and its transition of the crystalline state which is analyzed by common neighbor analysis (CNA) during wiredrawing. It is found that, when the drawing velocity is high, there occurs a phase transformation from bcc into fcc in some part of the wire. This fact will be discussed in more detail later in this paper. For a slow drawing velocity, this type of phase transition certainly takes place, but the amount of it is much smaller compared with the fast drawing. Anyway, for both velocities, however, the transformed fcc phase does not stay alive so long time. Most of the transformed fcc phase disappears and changes back into a bcc phase spontaneously. This return event of the crystalline phase is observed, after the transformed fcc region of the wire just exits from the die region. It is guessed that the driving force needed to the phase transformation is no longer applied outside the die region.

Figure 8 shows the time transition of the ratio of fcc atoms to the whole wire atoms. As can be seen in the figures, the phase transformation takes place instantly after the drawing starts. There is a tendency that a higher drawing velocity promotes more atoms which are relating to the phase transformation.

Figure 9 shows a selected region of the atomic configurations including a dislocation core, which are analyzed by the CNA method. The arbitrarily selected time is 120 ps after the wiredrawing started.

This case is for the drawing of [100]-orientation. The series of pictures in **Figure 9** are displayed to compare

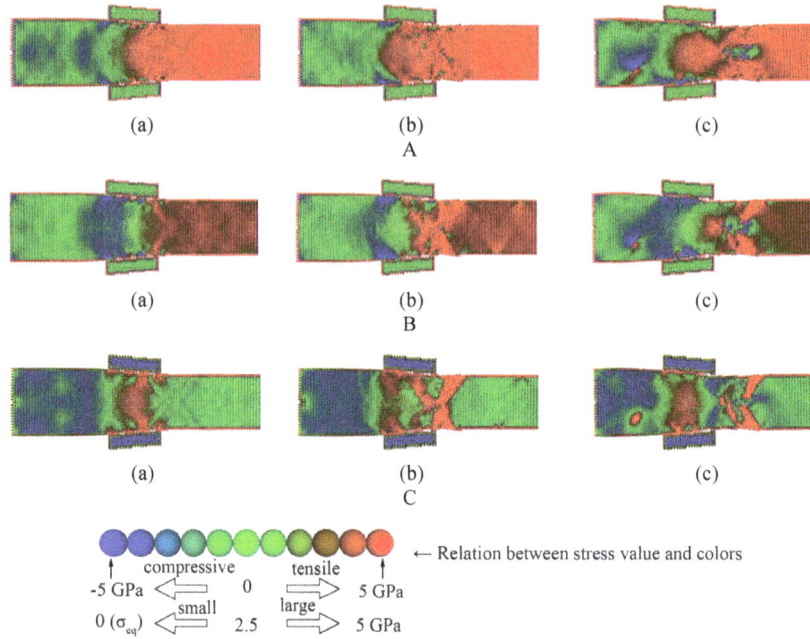

Figure 6. Stress distribution (σ^{zz}, σ_m, σ_{eq} in drawing process ($v = 50$ m/s, the drawing in [100] direction)). A) σ^{zz} drawing stress, (a) 40 ps, (b) 80 ps, (c) 120 ps; B) σ_m hydrostatic (mean) stress, (a) 40 ps, (b) 80 ps, (c) 120 ps; C) σ_{eq} von Mises equivalent stress, (a) 40 ps, (b) 80 ps, (c) 120 ps.

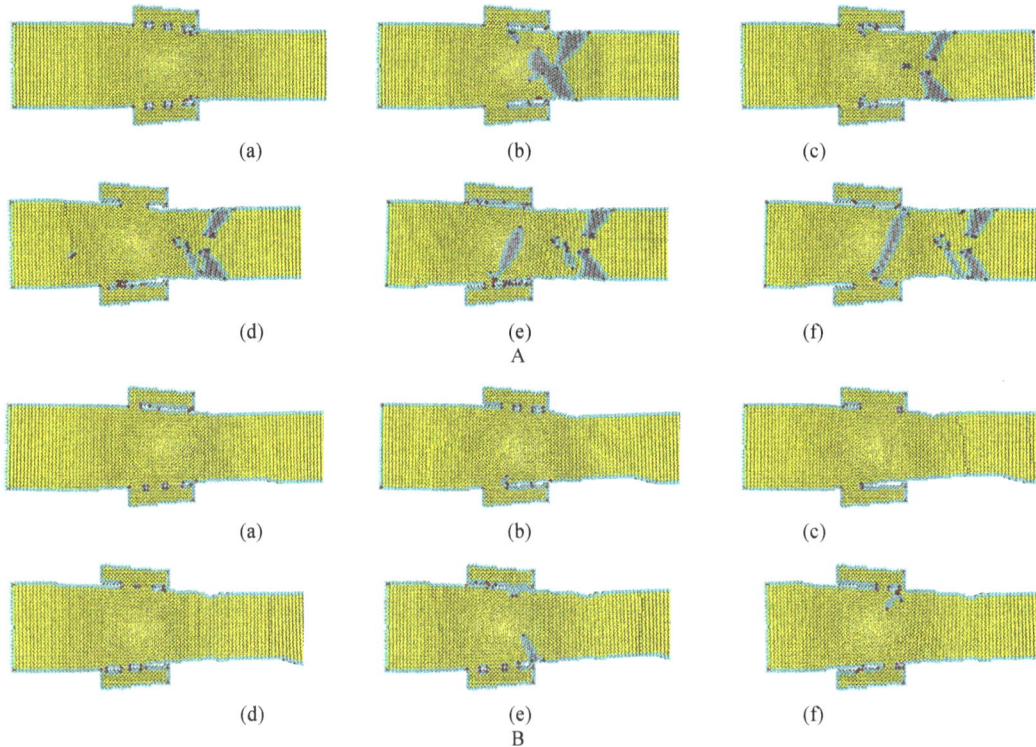

Figure 7. Distribution and its transition of atomic crystalline state: results of CNA analysis during the wire-drawing ($v = 50$ m/s and 10 m/s, the drawing in [100] direction). The colors are being used as follows: yellow = bcc atom, gray = fcc atom, blue = low-coordination atom (surface or amorphous), and brown = high-coordination (defected) atom. A) $v = 50$ m/s (fast), (a) 40 ps, (b) 80 ps, (c) 100 ps, (d) 120 ps, (e) 140 ps, (f) 160 ps; B) $v = 10$ m/s (fast), (a) 40 ps, (b) 80 ps, (c) 100 ps, (d) 120 ps, (e) 140 ps, (f) 160 ps.

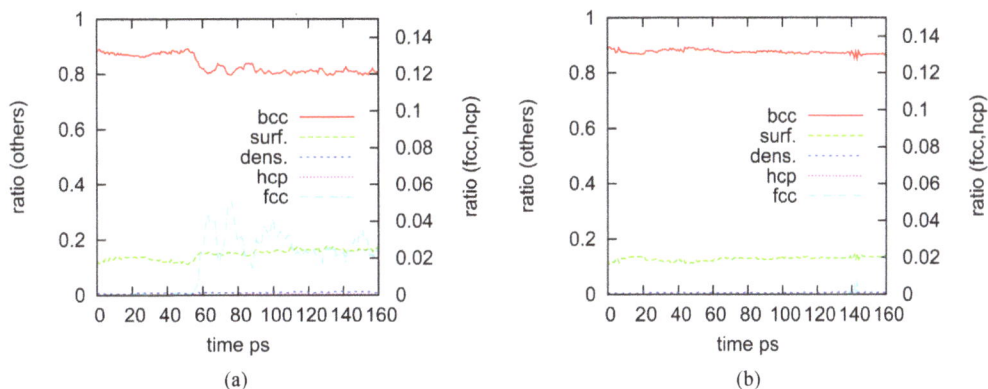

Figure 8. Time transition of fcc-atom ratio (the case of the drawing in [100] direction). (a) $v = 50$ m/s (fast); (b) $v = 10$ m/s (slow).

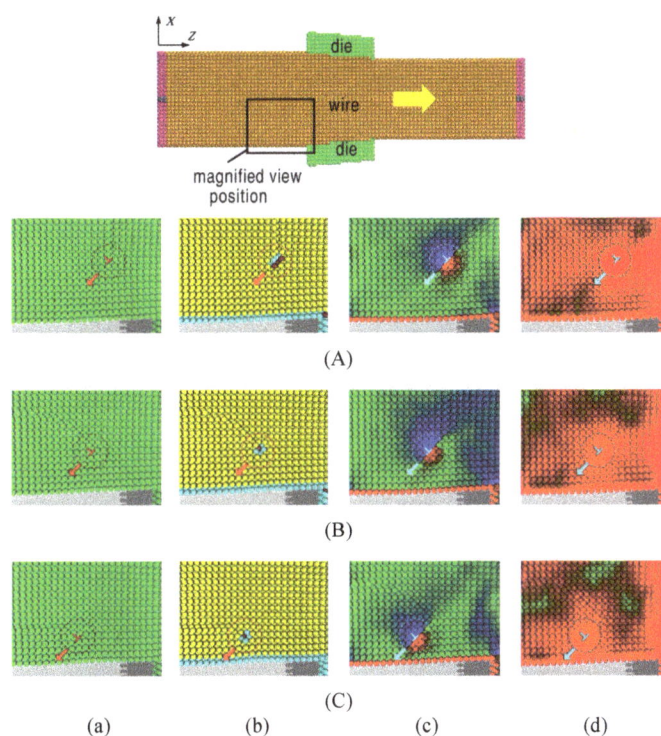

Figure 9. Comparison of visualization pictures around an isolated dislocation structure (at $v = 50$ m/s, in the case of the drawing in [100] direction). (A) at $t = 119.20$ ps; (B) at $t = 120.00$ ps; (C) at $t = 120.80$ ps. (a) atomic arrangement; (b) CNA analysis; (c) hydrostatic (σ_m); (d) equivalent stress (σ_{eq}).

between the atomic configuration, CNA analysis, the distribution of hydrostatic stress, and the distribution of equivalent stress, respectively. It is understood, from the atomic arrangement and the stress distribution, that this dislocation contains edge and screw components at the same time. The magnitude of the stress value is especially larger in the dislocation core. The dislocation is moving from upper-right to lower-left in angle of 45 degrees with regard to the horizontal axis in these pictures. By viewing pictures of the CNA analysis, it is confirmed that this dislocation is an isolated one, and it is atomistically composed of both less-coordinated and excessive-coordinated atoms which are located above and beneath the slip plane, respectively.

The hydrostatic stress of the atoms located above and beneath the slip plane have the opposite sign each other. The equivalent stress exhibits a large value widely around the dislocation core. The surface of the wire is seen at the bottom of these figures. When the dislocation reaches at the surface, the lattice imperfection due to the dislocation structure resolves naturally into a perfect crystal, where the wire surface is contacting strongly to the die surface. These MD results of the wiredrawing show that the events concerning generation, movement and annihilation of dislocation are all performed in the nano-sized scale and in a very short time.

Figure 10 shows stress components (hydrostatic stress σ_m, equivalent stress σ_{eq} and drawing stress σ^{zz}), all of which are averaged over atoms in each CNA categories. In the condition shown here (the [100]-oriented drawing and the drawing velocity of $v = 50$ m/s), it is observed that the phase transition into the fcc crystal occurs on a large area in this cross-section. All the stress values indicated in **Figure 10** are averaged and normalized by the total average. At that moment, the ratio of categories is as follows: bcc = 85%, fcc = 3%, hcp (hexagonal close-packed) = 0%, surface (less coordinated) = 12%, and grain boundaries and defects (excessively coordinated) $\leq$ 1%. The fcc phase occupies relatively a large volume, but the fcc atoms there do not transform into a hcp phase (the hcp phase means regional stacking fault of the fcc crystal) at all. The hydrostatic stress is generally larger in the fcc-transformed region. The equivalent stress is also generally larger in the disordered (defected) region, such as the phase boundary around the fcc-transformed region, dislocation, or grain boundary.

From these facts, it is concluded that the hydrostatic stress component is responsible for the bcc → fcc phase transition. It can be said that the iron atomic system in our MD model undergoes a process of the severe plastic deformation (SPD). It also understood that such a strong plasticity condition can be realized by a nano-sized wiredrawing condition.

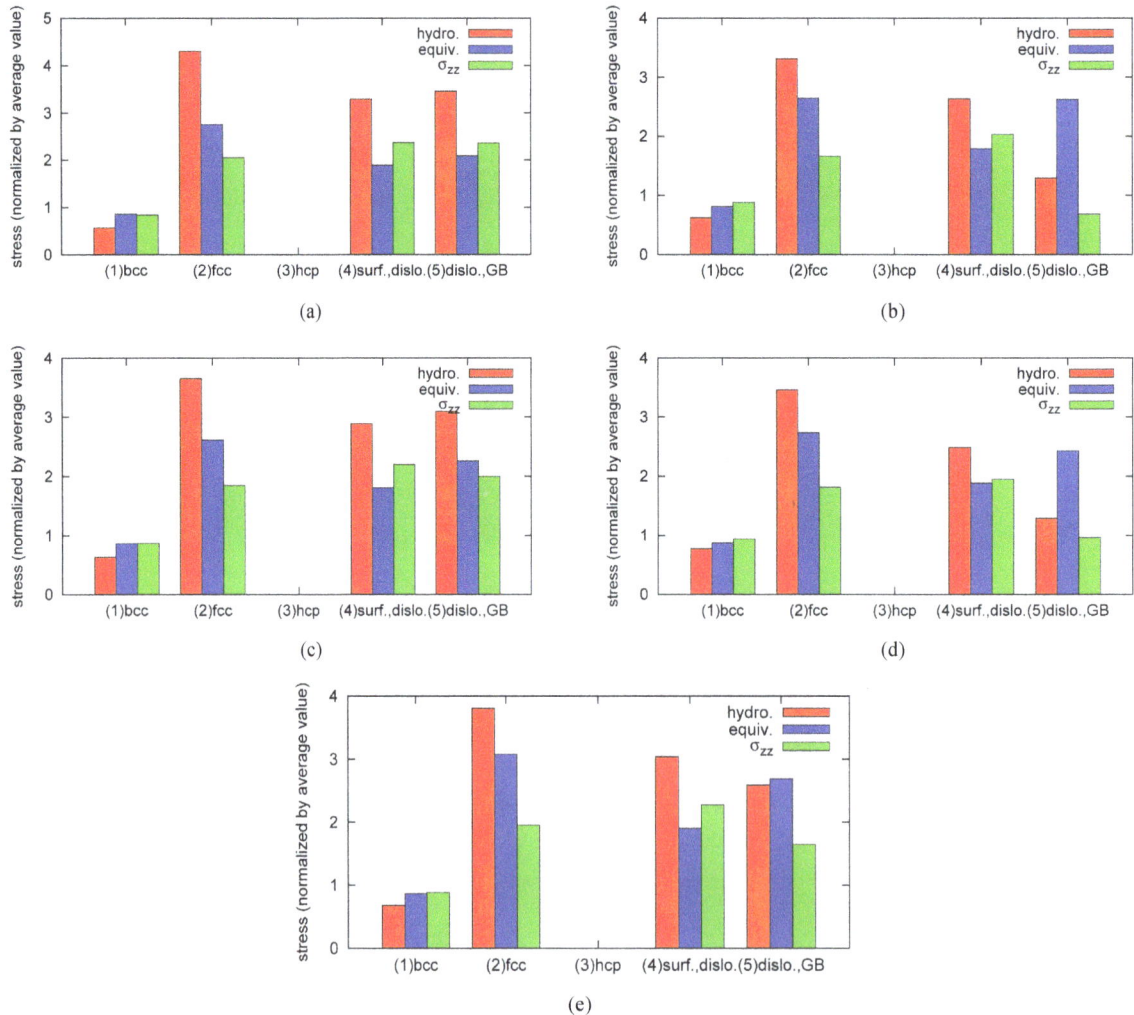

Figure 10. Relation between CNA categories and stress components ($v = 50$ m/s, the drawing in [100] direction). (a) 72.00 ps; (b) 76.00 ps; (c) 80.00 ps; (d) 84.00 ps; (e) 88.00 ps.

4. Conclusions

In order to assess the possibility of nano-sized wiredrawing, molecular dynamics (MD) modelling of α-iron is carried out. In the modelling, potential function, crystalline orientation and geometry of die etc. have to be considered so as to succeed in wiredrawing. In particular, introduction of friction model is inevitable. In this paper, to simplify the material behavior, the two-dimensional MD models with three different crystalline orientations are compared. Following results are obtained.

1) The interatomic function which is modified by the factor ω works well for reducing the friction between wire and die. The variety of ω values is compared. It is resulted that the smaller ω is, the smaller the resistance in drawing is obtained.

2) The invariant of atomic stress tensor is useful to recognize the status of atoms in the wire. Both hydrostatic stress σ_m and von Mises equivalent stress σ_{eq} are evaluated.

3) Phase transformation into fcc crystal occurs in wiredrawing of α-iron. It is caused by large hydrostatic stress σ_m in tensile regime (plus sign). The fcc phase is generally short-lived and is maintained only inside die region.

4) Dislocation structures are observed in wiredrawing. Atomistic equivalent stress clearly exhibits the core structure of a dislocation. A set of dislocation behaviors including nucleation, movement and annihilation at the wire surface completes in very small period.

Acknowledgements

This study is partly supported by the "Strategic Project to Support the Formation of Research Bases at Private Universities: Matching Fund Subsidy from MEXT (Ministry of Education, Culture, Sports, Science and Technology) (2012-2014)". The authors also acknowledge Nippon Steel & Sumitomo Metal Co. Ltd. and Kurimoto Ltd. for financial support.

References

[1]   Tang, Z. and Kotov, N.A. (2005) One Dimensional Assemblies of Nanoparticles: Preparation, Properties, and Promise. *Advanced Materials*, **17**, 951-962. http://dx.doi.org/10.1002/adma.200401593

[2]   Roters, F., Eisenlohr, P., Hantcherli, L., Tjahjanto, D.D., Bieler, T.R. and Raabe, D. (2010) Overview of Constitutive Laws, Kinematics, Homogenization and Multiscale Methods in Crystal Plasticity Finite-Element Modeling: Theory, Experiments, Applications. *Acta Materialia*, **58**, 1152-1211. http://dx.doi.org/10.1016/j.actamat.2009.10.058

[3]   Yoshida, T., Morita, Y. and Nakamachi, E. (2013) Crystal Texture Evolution Analyses in Drawing Process by Using Multi-Scale Finite Element Method. *Proceedings of JSME* (88*th Kansai Branch Annual Meeting*), **134-1**, (5-)23.

[4]   Nakatani, A. (2005) Plastic Deformation Analysis of Nanostructured Metal Using Molecular Dynamics. *Proceedings of JSME* (2005 *Annual Congress*), **8**, 470-471.

[5]   Chu, C.-Y. and Tan, C.-M. (2009) Deformation Analysis of Nanocutting Using Atomistic Model. *International Journal of Solids and Structures*, **46**, 1807-1814. http://dx.doi.org/10.1016/j.ijsolstr.2008.11.017

[6]   Karthikeyan, S., Agrawal, A. and Rigney, D.A. (2009) Molecular Dynamics Simulations of Sliding in an Fe-Cu Tribopair System. *Wear*, **267**, 1166-1176. http://dx.doi.org/10.1016/j.wear.2009.01.032

[7]   Ngan, A.H.W. and Wen, M. (2001) Dislocation Kink-Pair Energetics and Pencil Glide in Body-Centered-Cubic Crystals. *Physical Review Letters*, **87**, Article ID: 075505. http://dx.doi.org/10.1103/PhysRevLett.87.075505

[8]   Saitoh, K. and Liu, W.K. (2009) Molecular Dynamics Study of Surface Effect on Martensitic Cubic-to-Tetragonal Transformation in Ni-Al Alloy. *Computational Materials Science*, **46**, 531-544. http://dx.doi.org/10.1016/j.commatsci.2009.04.025

[9]   Dan, T. and Saitoh, K. (2012) Microstructure Evolution in Polycrystalline Metal under Severe Plastic Deformation by Strain-Controlled Molecular Dynamics. *Journal of Solid Mechanics and Materials Engineering*, **6**, 48-60. http://dx.doi.org/10.1299/jmmp.6.48

[10]   Pei, Q.X., Lu, C., Liu, Z.S. and Lam, K.Y. (2007) Molecular Dynamics Study on the Nanoimprint of Copper. *Journal of Physics D: Applied Physics*, **40**, 4928-4935. http://dx.doi.org/10.1088/0022-3727/40/16/026

[11]   Saitoh, K., Sameshima, Y., Takuma, M. and Takahashi, Y. (2014) Atomistic Simulation of Crystal Change and Carbon Diffusion in Nano-Sized Wiredrawing of Pearlitic Steel. *Technische Mechanik*, **32**, in press.

[12]   Nakagiri, A., Yamano, T., Konaka, M., Yoshida, K. and Asakawa, M. (2000) Chevron Crack and Optimum Drawing Condition in the Diagram of Mean Stress and Die-Wire Contact Length Ratio by FEM Simulation. 2000 *Conference*

Proceedings of the Wire Association International, Inc., 75-82.

[13] Atienza, J.M., Martinez-Perez, M.L., Ruiz-Hervias, J., Mompean, F., Garcia-Hernandez, M. and Elices, M. (2005) Residual Stresses in Cold Drawn Ferritic Rods. *Scripta Materialia*, **52**, 305-309.
http://dx.doi.org/10.1016/j.scriptamat.2004.10.010

[14] McAllen, P.J. and Phelan, P. (2007) Numerical Analysis of Axisymmetric Wire Drawing by Means of a Coupled Damage Model. *Journal of Materials Processing Technology*, **183**, 210-218.
http://dx.doi.org/10.1016/j.jmatprotec.2006.10.014

[15] Wright, R.N. (2009) Center Bursts—A Review of Criteria. *Wire Journal International*, **42**, 80-84.

[16] Yoshida, K. and Koyama, R. (2012) Reduction of Residual Stress of Drawn Wires. *Wire Journal International*, **45**, 56-60.

[17] Finnis, M.W. and Sinclair, J.E. (1984) A Simple Empirical N-Body Potential for Transition Metals. *Philosophical Magazine A*, **50**, 45-55. http://dx.doi.org/10.1080/01418618408244210
(Erratum: *ibid.*, 1986, **53**, 161. http://dx.doi.org/10.1080/01418618608242815)

[18] Saitoh, K., Daira, S. and Sameshima, Y. (2012) Nano-Scale Modelling and Simulation of Metal Wiredrawing by Using Molecular Dynamics Method. *Proceedings of International Conference on Materials Processing and Technology 2012 (MAPT2012)*, Hawaii, 28-29 June 2012, 211-216.

[19] Wright, R.N. (2010) Wire Technology "Process Engineering and Metallurgy". Butterworth-Heinemann, Oxford.

[20] Honeycutt, J.D. and Andersen, H.C. (1987) Molecular Dynamics Study of Melting and Freezing of Small Lennard-Jones Clusters. *The Journal of Physical Chemistry*, **19**, 4950-4963. http://dx.doi.org/10.1021/j100303a014

[21] Tsuzuki, H., Branicio, P.S. and Rino, J.P. (2007) Structural Characterization of Deformed Crystals by Analysis of Common Atomic Neighborhood. *Computer Physics Communications*, **177**, 518-523.
http://dx.doi.org/10.1016/j.cpc.2007.05.018

4

High Performance Polymer Light-Emitting Devices

Vivek Kant Jogi

School of Studies in Electronics, Pt. Ravi Shankar Shukla University, Raipur, India
Email: vivekkantjogi@gmail.com

Abstract

In order to improve the performance of polymer light-emitting devices, driving voltages, current efficiency, luminance and power efficiency of different cathode metals such as Ca/Al, CsF/Al, LiF/Al and LiF/Ca/Ag were compared. The results show that cathode metals CsF/Al contain the highest current efficiency, maximum luminance and power efficiency. Therefore, we can choose the CsF/Al to be the cathode for improving the performance of polymer light-emitting devices.

Keywords

PLED; Current Efficiency; Power Efficiency; Current Density

1. Introduction

A PLED consists of very thin layers of polymer films sandwiched by two electrodes [1]. In a typical PLED structure, there are two polymer layers, one of which functions as the hole transporting layer and the other functions as the light-emission layer as shown in **Figure 1(a)**. An indium tin oxide (ITO) layer is generally used as the transparent anode, which allows the light generated within the diode to leave the device [1]. The metal cathode is conveniently deposited on top of the polymer by thermal evaporation [1]. The devices' performance strongly depends on effective charge injection from the electrodes to the organic medium and charge transport in the organic materials.

In an ideal physical process in which the chemical reactions at the interface are ignored, the energy diagram can be represented as shown in **Figure 1(b)**. Here $\Delta\Phi h$ and $\Delta\Phi e$ respectively define the energy barriers for holes and electrons by the Mott-Schottky rule of vacuum level alignment [2]. Although this picture is simple and ideal for the interfacial effect on charge injection, it can nevertheless be successfully used to evaluate the metal electrode's work functions and the positions of the HOMO and the LUMO of the polymer.

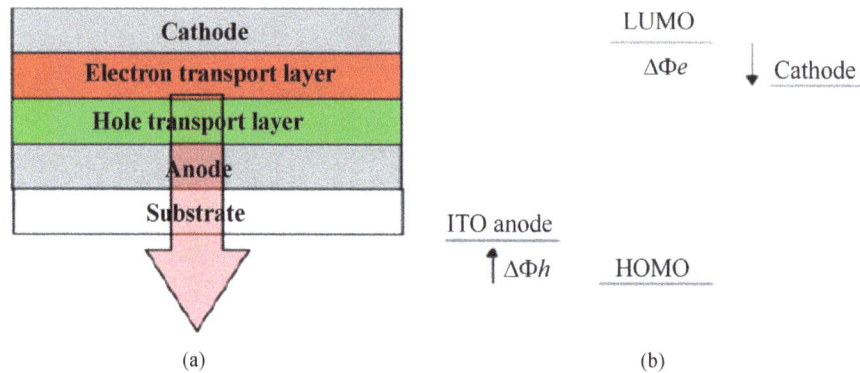

Figure 1. The basic structure (a) and the energy level for the operation (b) of a PLED.

2. Devices Fabrication Characteristic

Commercial available indium-tin oxide (ITO) from Sigma Aldrich is used as anode of the PLEDs because of its not only transparent but also high conductivity. Besides, its ionization potential is just about 4.8 eV that is easy to transport the holes into the PEDOT (HTL) layer. The ITO glass obtains 30 $\Omega/\square$ resistances. Moreover, it was rinsed by the de-ionized water for removing the dusts & the other particles on the surface and baked at about 120°C for an hour to evaporate the water vapor. After that, the UV-ozone treatment was used to enrich the oxygen and reduce the oxygen vacancies. That can improve the conductivity of the ITO layer for helping the transport of hole easily [3].

The poly (3,4-ethylene dioxythiophene), PEDOT solution was spin-coated onto the ITO layer at about 20 seconds. Control the spinning speed of the spin-coating machine is of importance. The first 5 seconds must use a slow spinning speed about 400 rpm after that the spinning speed wasturned to 2000 rpm to provide the homogeneous and high quality layer. Finally, the PEDOT layer was dried by the infrared lamp. The electron transport layer and emissive layer was spin coated by the poly (9,9-dioctylfluorene) (PFO) solution below 400 rpm spinning speed at the first 5 seconds to control the homogeneous layer. Then, the spinning speed of the spin-coating machine was increased to 1000 rpm for 40 seconds and baked by the infra-red lamp for the evaporating minimum 10 minutes.

For deposition of the metal cathode the thermal evaporation machine (Vacuum coating systems Model 12A4D) is used to deposit the thin-film metal layers on the dried PFO layer with the deposition rate of 1.5 nm/s. During the deposition, the pressure of the chamber must need to be controlled to about 5×10^{-6} mbar for the successful thermal evaporation. Besides, the thickness of thin metal layers is 100 nm each (involving the protective metal layer).The different fabricated specimens are

    Device A: ITO/PEDOT/PFO/Ca/Al,
    Device B: ITO/PEDOT/PFO/CsF/Al,
    Device C: ITO/PEDOT/PFO/LiF/Al,
    Device D: ITO/PEDOT/PFO/LiF/Ca/Ag.

The current, voltage, luminance and electroluminescence (EL) of the devices are measured using spectrophotometer.

3. Results and Discussion

PFO layer was used to be the emissive layer so the EL spectrums of these four devices have two blue emission peaks that are 440 nm and 460 nm. **Figures 2-5** show the comparison of luminance vs. voltage, Current Efficiency, Power Efficiency and Current Density of Device A, B, C and D respectively.

From **Figure 2**, Device A (Blue) like to tend in steady that mean it would not have a higher amount luminance appear even though the voltage was increased. Device C (Purple) initial luminance occurred at the about 3.6 V which is higher than other specimens as well as it also tendentious to steady so it maximum luminance would not be higher and its power consumption is the largest. The Device B (Red) is the best performance compared with the other curves. Device B (Red) achieved the higher luminance at the lower voltages. For example, if the designed luminance is 700 cd/m^2, the driving voltage of CsF/Al just 6.5 V. Comparing with the other

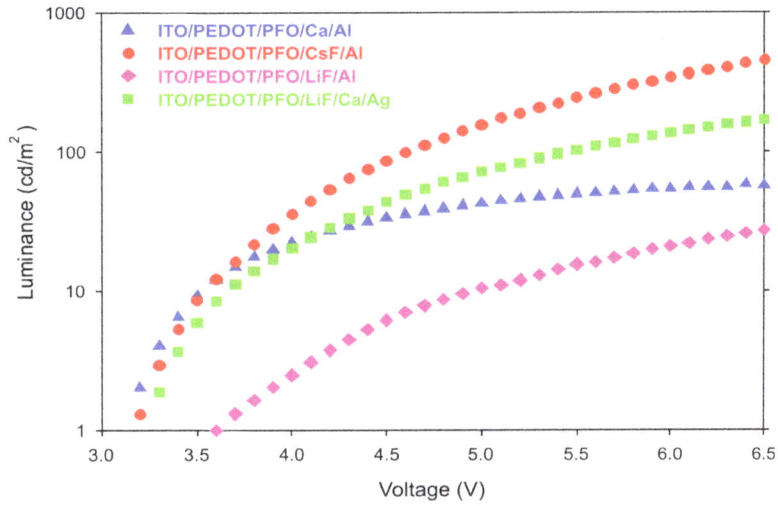

Figure 2. Comparison of luminance vs. voltage.

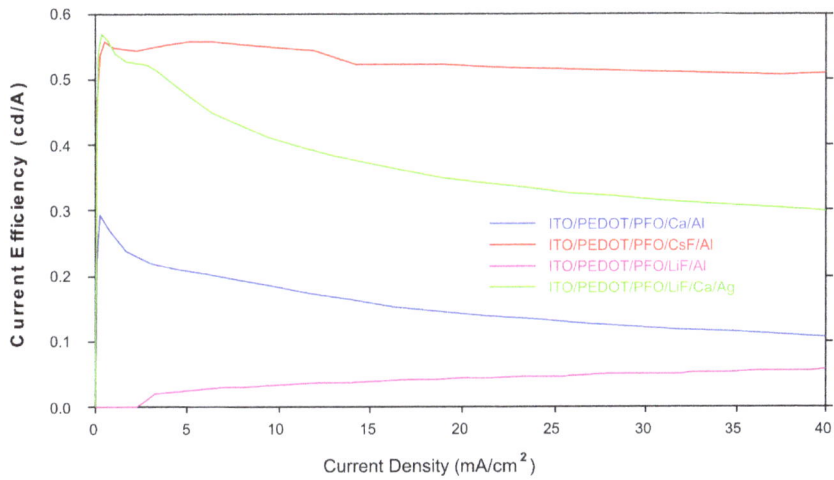

Figure 3. Comparison of current efficiency.

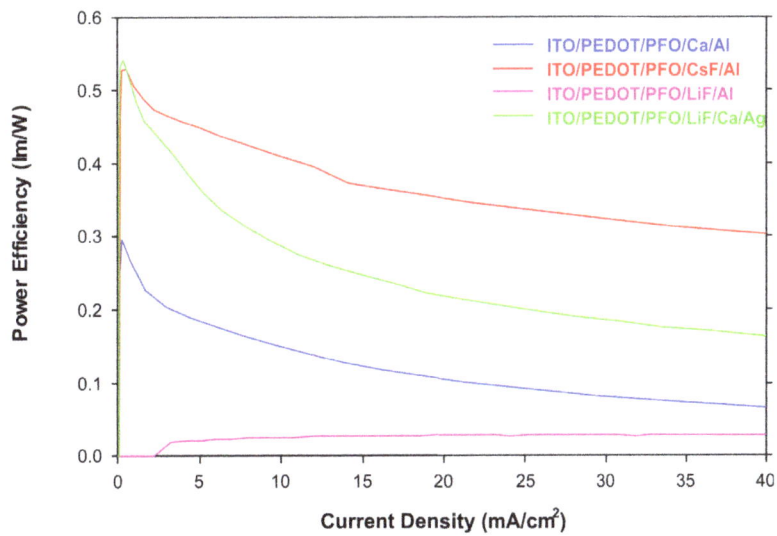

Figure 4. Comparison of power efficiency.

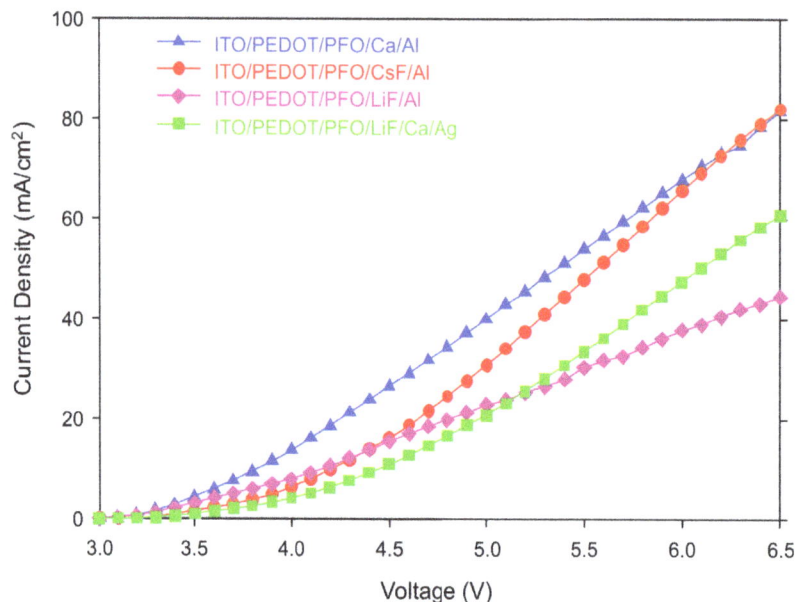

Figure 5. Comparison of current density.

specimens, the maximum luminance at the same voltage of CsF/Al is the highest. Therefore, we can reduce the driving voltage and increase the luminance through the changing the cathode materials such as LiF/Ca/Ag or CsF/Al. From **Figure 3**, only the Device B (Red) (CsF/Al) keeps steady, which is the most stable curve out of these four specimens. Furthermore, CsF/Al also is the highest curve of the graph that means it has the highest current efficiency. For example, if the current density is 40 mA/cm^2, the current efficiency of CsF/Al is 0.51 cd/A which is the highest one mean it has lowest power consumption. Therefore CsF/Al and LiF/Al are the lowest and highest power consumption respectively. From **Figure 4**, when the current density is 40 mA/cm^2, the power efficiency of CsF/Al is 0.3 lm/W which is highest. The power efficiency of CsF/Al and LiF/Ca/Ag are the better choice of the cathode for providing the higher power efficiency and lower power consumption of PLEDs. From **Figure 5**, Device B (Red) CsF/Al has the most of rapid increases, achieved the highest level of current density. For example, when the applied voltage is 6.5 V, the CsF/Al curve can achieve the highest current density 8.2 mA/cm^2. It expresses that CsF/Al can generate the highest brightness when the driving voltage is 6.5 V. Besides, its curve increases rapidly that mean it is easy to achieve the designed brightness level. Therefore, it is high sensitive and low reaction time lap. Comparison of the total performance of these four specimens from **Table 1**, we got the results that CsF/Al is the best materials because it's performance such as the best of current efficiency, power efficiency, current density and maximum luminance. That means, it does not only provide the higher luminance with lower driving voltage but also reduce the power consumption. Furthermore, CsF/Al is easy to achieve the turn on voltage to provide the luminance quickly due to its properties of current density (low reaction time lap). By the consideration of the above factors, LiF/Ca/Ag also can be used to replace the Ca/Al as a cathode of PLEDs though its performance is worse than CsF/Al. Compared with the normal materials (Ca/Al); LiF/Ca/Ag really has a large improvement. Besides, the LiF/Al is the worst specimens out of the other three samples.

$$PERFORMANCE : \mathbf{CsF/Al} > LiF/Ca/Ag > Ca/Al > LiF/Al$$

4. Conclusion

In order to improve PLED performance, multi-layer structures were studied. Four types of cathode materials were tested and compared. Those are Ca/Al, CsF/Al, LiF/Al and LiF/Ca/Ag. Lower work function is good for the electron injection from the cathode to emissive layer, so the cathode should be low work function. By comparison of those four specimens, LiF/Al has the worst performance, and CsF/Al has the best performance out of the other specimens. The cathode materials can be replaced by the CsF/Al to achieve lower power consumption, driving voltage and turn-on voltage. Moreover, CsF/Al does not only increase the devices' luminance

Table 1. Total performance of the devices.

Device	Luminance	Current Efficiency	Power Efficiency	Current Density
	[cd/m^2] at 6.5 V	[cd/A] at 40 mA/cm^2	[lm/w] at 40 mA/cm^2	[mA/cm^2] at 6.5 V
Ca/Al	60	0.15	0.1	82
CsF/Al	700	0.51	0.36	82
LiF/Al	40	0.1	0.05	45
LiF/CsF/Ag	200	0.35	0.53	60

but also reduce the response time lap. As the supporting results of this research, the total performance of the PLEDs can be improved by selecting the appropriate cathode materials as CsF/Al.

References

[1]   Deng, X.-Y. (2011) Light-Emitting Devices with Conjugated Polymers. *International Journal of Molecular Sciences*, **12**, 1575-1594. http://dx.doi.org/10.3390/ijms12031575

[2]   Pope, M. and Swenberg, C.E. (1982) Electronic Processes in Organic Crystals and Polymers. Oxford University Press, New York.

[3]   Bernius, M.T., Inbasekaran, M., O'Brien, J. and Wu, W.S. (2000) Progress with Light-Emitting Polymers. *Advanced Materials*, **12**, 1737-1750. http://dx.doi.org/10.1109/9.402235

5

Unusual Spectral Change Due to a Cyanine Dye Adsorbed on an Inorganic Layered Material upon Photoirradiation

Mari Ishihara*, Ryuji Hirase, Hideki Yoshioka

Hyogo Prefectural Institute of Technology, Kobe, Japan
Email: *mari@hyogo-kg.jp

Academic Editor: Yarub Al-Douri, University Malaysia Perlis, Malaysia

Abstract

Photoinduced spectral change can be utilized for various optical devices. The photoinduced spectral change due to an organic dye was demonstrated for the organic-inorganic hybrid film without the aid of photochromism with a simple preparation method for the first time. By the hybridization of a cyanine dye of 2-[5-(1,3-dihydro-3,3-dimethyl-1-octadecyl-2H-indol-2-ylidene)-1,3-pentadienyl]-3,3-dimethyl-1-octadecyl-3H-indolium perchlorate (NK3175) with an inorganic layered material of cation-exchangeable clay, smectite (SWN), a spectral change attributed to NK3175 was generated upon the irradiation of UV light. This result might serve as useful information on the methodology to produce optically controlled function for photoresponsive systems. Furthermore, the hybrid film of SWN and NK3175 was characterized by the use of XRD and FT-IR measurements. NK3175 molecules adsorbed onto external surfaces of SWN were confined by oriented SWN. It was suggested that the enhanced intermolecular interaction between NK3175 molecules caused by the hybridization with SWN resulted in the change of the aggregation state of NK3175 upon the UV light irradiation, which accounts for the spectral change of NK3175.

Keywords

Photoinduced Spectral Change, Cyanine Dye, Clay, Hybrid Material

1. Introduction

Organic dyes have attracted much attention from both viewpoints of fundamental science and practical applica-

*Corresponding author.

tions such as memories, switching devices, electroluminescent devices, solar cells, optical filters, etc. Change of physical properties such as optical reflectance, refractivity, and fluorescence caused by photoirradiation for photo-functional organic dyes has been utilized for various optical devices. Inorganic layered materials such as clay minerals are widely utilized as host materials for functional organic-inorganic hybrid systems due to their characteristic properties of intercalation and adsorption of organic guest species [1]-[6]. Photo-functional organic-inorganic hybrid materials with an ordered structure have been constructed by utilizing intercalation and adsorption of photo-functional organic guest species into or on clay layers [1] [2] [7]. In such cases, distances between the photo-functional organic dyes become shorter. Therefore, it is expected that intermolecular interactions between these compounds are enhanced [1] [2] [8]. In fact, the intercalation and adsorption of the photo-functional organic dyes have enabled aggregation and alignment control of the organic dyes [1] [2] [8]-[14] and have affected photochemical properties, such as fluorescence behaviors [4] [5] [7] [8] [15]-[21].

The cooperative control of physical properties of the other organic material caused by an influence of photochromism is one of the active ways to create novel photo-functional systems [22]-[30]. Photochromic compounds have been extensively studied due to their possible application in various optical devices, such as memories and switches [31]. For photochromic memories one of the most pressing problems is on the method for non-destructive readout [31]. Monitoring change in ultraviolet (UV)-visible absorption is not a feasible method for non-destructive readout of photochromic memories; therefore, several alternative methods for non-destructive readout have been demonstrated [31]. On the other hand, photochromic compounds and organic dyes can be selected, so as to that the absorption of an organic dye and two isomers of a photochromic compound are in different spectral regions for the cooperative photoresponsive systems including these compounds. Accordingly there might be a possibility that monitoring change in the UV-visible absorption without an occurrence of a reverse photo isomerization reaction can be utilized for a non-destructive readout in this system. We proposed to monitor the absorption in the UV-visible region as a non-destructive readout method for a cooperative photoresponsive system and performed the photoinduced spectral change due to a cyanine dye by the use of a simple preparation method [30]. In order to amplify a cooperative effect of photochromism, we made use of an inorganic layered material together with an organic dye and a photochromic molecule to successfully prepare the hybrid film as a novel photo-functional organic-inorganic hybrid material [30]. It was shown that an electronic absorption spectrum ascribed to the cyanine dye was significantly changed by UV light irradiation for the hybrid film of the cyanine dye, a photochromic molecule of a diarylethene, and an inorganic clay mineral, smectite, [30]. However, the further mechanism of the photoinduced spectral change due to the cyanine dye for the above hybrid film is not elucidated. For the above hybrid film, smectite as well as the diarylethene is considered to play an important role in bringing about the photoinduced spectral change due to the cyanine dye. The interaction between the cyanine dye and smectite might affect the microstructure and the photoinduced spectral behavior for the above hybrid film.

In this study, we investigated the interaction between a cyanine dye and smectite and the photoinduced spectral behavior for the hybrid film of a cyanine dye and smectite. We demonstrate the interesting phenomena that without the aid of photochromism, a cyanine dye is induced to exhibit the photoresponsive property by hybridizing with an inorganic clay mineral, although a cyanine dye itself does not show any photoresponsive properties. This study is expected to give new information on the methodology to introduce optically controlled function for photoresponsive systems.

2. Experimental

2.1. Materials

A cyanine dye of 2-[5-(1,3-dihydro-3,3-dimethyl-1-octadecyl-2H-indol-2-ylidene)-1,3-pentadienyl]-3,3-dimethyl-1-octadecyl-3H-indolium perchlorate (NK3175; Hayashibara Biochemical Labs., Inc.) was used without further purification (**Figure 1**).

Cation-exchangeable clay, SWN (CO-OP Chemical Co., Inc.), was used as a host material. It is a type of hydrophilic smectite, and has good transparency in the visible region. The composition is [(Mg$_{2.67}$Li$_{0.33}$)(Si$_4$O$_{10}$)(OH)$_2$]Na$_{0.33}$, and the cation exchangeable capacity (CEC) is c.a. 0.6 meq·g^{-1}. Distilled and deionized water (conductivity $< 4 \times 10^{-6}$ S·cm^{-1}) and ethanol (Wako Pure Chemical Industries Co. Inc., spectroscopic grade) were used as solvents for the preparation of the organic-inorganic hybrid films.

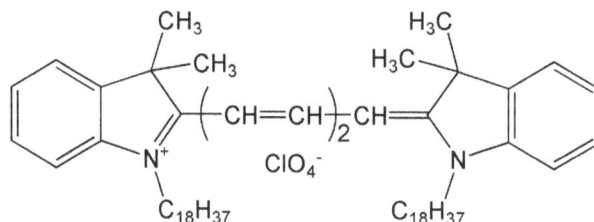

Figure 1. Chemical structure of a cyanine dye, NK3175.

2.2. Sample Preparation

Solution of cyanine dye of NK3175 (1.5×10^{-6} mol) in ethanol (10 mL) added to an aqueous suspension of SWN (0.050 g/5mL) and stirred for 27 h at 60°C in the dark. Obtained viscous solution was then casted on fused silica plates or Au-coated glass plates and dried at room temperature and transparent blue films were prepared. For FT-IR measurements, Au-coated glass plates were used. The obtained films are, hereafter, referred to as the hybrid films of SWN and NK3175. The ethanol solution of NK3175 (2.5×10^{-6} M) was also prepared for comparison.

2.3. Measurement

Electronic absorption spectra were recorded on a Shimadzu SolidSpec-3700 spectrophotometer. X-ray diffraction (XRD) patterns were measured with a RIGAKU RINT 2500 or a RIGAKU Smart Lab diffractometer using CuKα radiation source operating at 40 kV and 200 mA as the applied voltage and current, respectively. FT-IR spectra were measured on a Nicolet is 50 FTIR at 4 cm^{-1} resolution. FT-IR reflection spectra of films were taken using a variable angle reflection accessory at an incident angle of 45 degrees. In the case of the measurement of polarized FT-IR spectra, a ZnSe polarizer was used. An attenuated total reflection (ATR) FT-IR spectrum of NK3175 powder was taken using Smart iTR accessory. UV light of around 254 nm (c.a. 1100 μW·cm^{-2}) was irradiated for 5 minutes using a handy UV lamp.

3. Results and Discussion

3.1. Spectral Change of the Hybrid Films of SWN and NK3175 upon Photoirradiation

Figure 2 shows electronic absorption spectra of the hybrid film of SWN and NK3175 and, for reference, ethanol solution of NK3175 before and after irradiation of UV light at 254 nm. For the solution of NK3175, the bands at 598 and 647 nm were observed (**Figure 2(a)**). These peaks were not changed by UV light irradiation at 254 nm. These bands could be ascribed to H-dimers and monomers of NK3175, respectively, according to the previous reports [8] [16] [32]. Before the UV light irradiation, the absorption bands at 603 and 652 nm ascribed to H-dimers and monomers of NK3175, respectively, were lightly red-shifted from those for the solution of NK3175, which were caused by adsorption of cationic NK3175 on negatively charged SWN [16]-[18] [33]. In addition, the relative intensity of the band at 603 nm to the band at 652 nm in the hybrid film in comparison with that in the solution of NK3175 increased, indicating that hybridization with SWN promotes the formation of H-dimers of NK3175.

On the other hand, it was interestingly found that in the hybrid film of SWN and NK3175 the spectral change due to NK3175 was occurred upon UV light irradiation at 254 nm. Upon the irradiation of UV light, the bands at 412 and 531 nm were clearly appeared, whereas the bands at 603 and 652 nm significantly decreased (**Figure 2(c)**). This change was essentially in good accord with our previous report which demonstrate a spectral change in the hybrid films of SWN, NK3175 and a diarylethene [30]. The change was somewhat smaller in this study probably because of the absence of the movement accompanied by a photochromic reaction [30]. The irradiated spectrum gradually returned to the one before UV light irradiation and reached a stationary state on keeping this film for about 23 h in the dark at room temperature, which indicates the observed spectral change do not arise from the photodecomposition of NK3175. Upon repeated irradiation of UV light, the similar spectral change took place as in the first time again. According to the previous report [8] [33], we presumed that the bands at 412 and 531 nm were ascribed to higher-order H-aggregates of NK3175 resulted from disaggregated H-dimers and monomers of NK3175. After UV light irradiation, after keeping this film in the dark the slow spectral change occurred for hours. It was reported that the aggregation of a cationic dye solution was instantly formed in the dispersions after mixing a cationic dye with a clay mineral, and then disaggregation of a cationic dye proceed slowly for hours and days [16] [34]. We assume that a similar situation occurs here. We have found that the similar spectral change due to other cyanine dyes, such as 2-[7-(1,3-dihydro-3,3-dimethyl-1-octadecyl-2H-indol-2-ylidene)-1,3,5-heptatrienyl]-3,3-dimethyl-1-octadecyl-3H-indolium perchlorate and 3-ethyl-2-[5-(3-ethyl-2(3-H)-benzoxazolylidene)-1,3-pentadienyl] benzoxazolium iodide also occurred upon UV light irradiation for the hybrid films with SWN.

3.2. Spectral Change of the Hybrid Film of SWN and NK3175 upon Heating

Figure 3 shows the electronic absorption spectra of the hybrid film of SWN and NK3175 on heating. Upon heating this film at 80°C for 1 h, the bands at 652 nm slightly shifted to 648 nm, which implies a change of a polar environment around NK3175 molecules [18] [35]. The spectral change such as the appeared bands at 531 and 412 nm and the decreased bands at 652 and 603 nm observed upon the UV light irradiation were not recognized upon heating. From these results, it was confirmed that the spectral change due to NK3175 upon UV light irradiation was caused not by heating but by photoirradiation.

3.3. X-Ray Diffraction and FT-IR Reflection Spectra of the Hybrid Film of SWN and NK3175

Figure 4 shows XRD patterns of the hybrid film of SWN and NK3175 and a casted film of SWN which was obtained from the suspension of SWN (0.051 g) in distilled and deionized water (5 mL) and ethanol (25 mL) stirred for 30 h at 60°C. The 100 reflections were observed at d = 14.7 Å (6.02 degree) and 14.2 Å (6.21 degree),

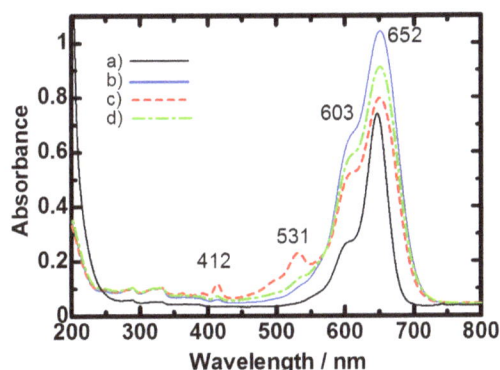

Figure 2. Electronic absorption spectra of a) the ethanol solution of NK3175 and the hybrid film of SWN and NK3175, b) before irradiation, c) after UV light (λ = 254 nm) irradiation, and d) after keeping for 23 h in the dark at room temperature.

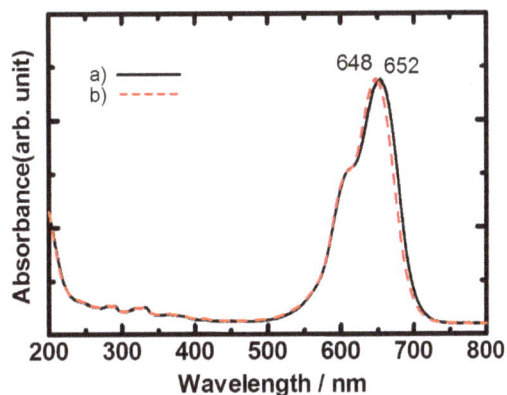

Figure 3. Electronic absorption spectra of the hybrid film of SWN and NK3175: a) before heating, b) after heating at 80°C for 60 min.

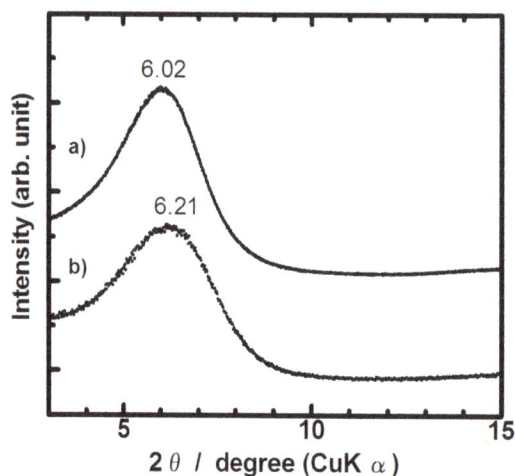

Figure 4. XRD patterns of a) the hybrid film of SWN and NK3175 and b) the casted film of SWN.

respectively. The slight enhancement of the d_{001} values indicates that NK3175 molecules were not intercalated into SWN layers.

FT-IR spectroscopy revealed the adsorption of a cationic dye of NK3175 on the anionic surface of SWN which appreciably oriented parallelism to the surface of the hybrid film of SWN and NK3175. **Figure 5** and **Figure 6** show the FT-IR spectra of the hybrid film of SWN and NK3175. In **Figure 5(a)**, the bands at 1491, 1481, and 1456 cm^{-1} are assignable to the stretching vibration of the central conjugated system of NK3175. The bands at 1481 and 1456 cm^{-1} might contain a contribution from the in-plane skeletal vibration of the phenyl ring and a contribution from the CH_3 asymmetric deformation vibration of NK3175, respectively [36]-[38]. The bands at 1491, 1481, and 1456 cm^{-1} were shifted and changed in relative intensity compared to those for NK3175 powder (**Figure 5(b)**). Moreover, the broad band around 1388 cm^{-1} and the band at 1335 cm^{-1} for the hybrid film (**Figure 5(a)**) were relatively smaller than those in NK3175 powder (**Figure 5(b)**). These bands are attributable to atomic motions in the methine chain of NK3175 and the band around 1388 cm^{-1} might contain a contribution from the CH_3 symmetric deformation mode of NK3175 [36]-[38]. These were attributable to the adsorption of the cationic dye of NK3175 on the negatively charged external surface of SWN, which corresponds to the results from the electronic absorption spectra. **Figure 6** shows the polarization dependency on the bands ascribed to the Si-O vibrations of SWN. In the p-polarized spectrum the bands at 1090 cm^{-1} and 977 cm^{-1} are assignable to the out-of-plane Si-O stretching and the in-plane Si-O stretching vibration of SWN, respectively. The band at 965 cm^{-1} observed in the s-polarized spectrum is attributable to the in-plane Si-O stretching vibration of SWN [33] [39]. These suggest that tetrahedral layers of SWN considerably oriented parallel to the surface of the hybrid film of SWN and NK3175 [30] [33] [39].

Taking into accounts the above results, one of the possible mechanism of the spectral change due to NK3175 induced by photoirradiation for the hybrid film of SWN and NK3175 is as follows; NK3175 molecules adsorbed onto the external surface of SWN are confined by the oriented SWN to approach each other. This situation is considered to enhance the intermolecular interactions between NK3175 molecules. Computational study for cyanine dyes previously reported that upon photoexcitation, the electron density on the methine chain increased, leading to the rotation of the methine chain [40]. For the hybrid film of SWN and NK3175, in a similar manner, upon photoirradiation the electron density on the methine chain of NK3175 somewhat increases, giving rise to the rotation of the methine chain, that is, the conformational change of NK3175. In the present case, NK3175 molecules approach each other so that they could strongly interact to form unique aggregation such as higher-order H-aggregates of NK3175 upon UV light irradiation. Consequently, the spectral change due to NK3175 is induced by photoirradiation for the hybrid film of SWN and NK3175. The spectral change due to NK3175 upon photoirradiation was enhanced in the hybrid film of SWN, NK3175 and the diarylethene [30] in comparison with that in the hybrid film of SWN and NK3175. It is probably because not only the above effect but also photochromism of the diarylethene which affected the state of aggregation of NK3175 have contributed to the enhanced spectral change upon photoirradiation in the hybrid film of SWN, NK3175 and the diarylethene [30].

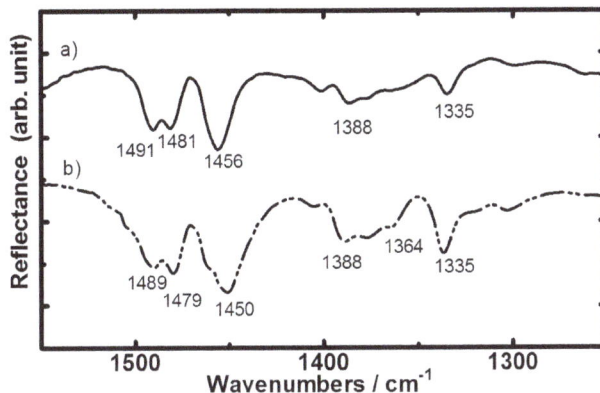

Figure 5. Unpolarized FT-IR spectra of a) the hybrid film of SWN and NK3175 and b) NK3175 powder.

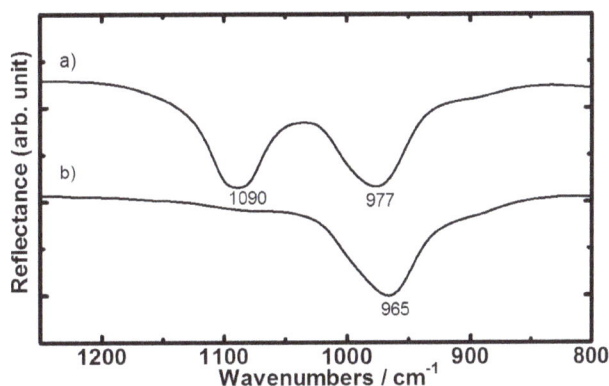

Figure 6. Polarized FT-IR spectra in the Si-O stretching vibration region of the hybrid film of SWN and NK3175 with a) *p*-polarization and b) *s*-polarization.

4. Conclusion

The unusual spectral change ascribed to a cyanine dye was induced upon the irradiation of UV light without the help of photochromism, for the first time, by hybridizing a cyanine dye with an inorganic clay mineral with a simple preparation method. This result is expected to provide new information on the methodology to produce optically controlled function for photoresponsive systems. In addition, for the hybrid film of smectite of SWN and a cyanine dye of NK3175, it was suggested that NK3175 molecules adsorbed onto the external surface of SWN were confined by oriented SWN, enhancing the intermolecular interaction of NK3175 molecules, which induced the change of the state of aggregation of NK3175 upon UV light irradiation.

Acknowledgements

We would like to thank Dr. Nobutaka Tanigaki and Dr. Keiko Tawa of National Institute for Advanced Industrial Science and Technology and Professor Hideyuki Nakano of Muroran Institute of Technology for their valuable discussions and advices. This work was partly supported by Research for Promoting Technological Seeds type A and Adaptable and Seamless Technology Transfer Program through Target-driven R & D from the Japan Science and Technology Agency (JST).

References

[1] Ogawa, M. and Kuroda, K. (1995) Photofunctions of Intercalation Compounds. *Chemical Reviews*, **95**, 399-438. http://dx.doi.org/10.1021/cr00034a005

[2] Shichi, T. and Takagi, K. (2000) Clay Minerals as Photochemical Reaction Fields. *Journal of Photochemistry and*

Photobiology C: Photochemistry Reviews, **1**, 113-130. http://dx.doi.org/10.1016/S1389-5567(00)00008-3

[3]   Takagi, S., Shimada, T., Ishida, Y., Fujimura, T., Masui, D., Tachibana, H., Eguchi, M. and Inoue, H. (2013) Size-Matching Effect on Inorganic Nanosheets: Control of Distance, Alignment and Orientation of Molecular Adsorption as a Bottom-Up Methodology for Nanomaterials. *Langmuir*, **29**, 2108-2119. http://dx.doi.org/10.1021/la3034808

[4]   Zhou, C.H., Shen, Z.F., Liu, L.H. and Liu, S.M. (2011) Preparation and Functionality of Clay-Containing Films. *Journal of Material Chemistry*, **21**, 15132-15153. http://dx.doi.org/10.1039/c1jm11479d

[5]   Ras, R.H.A., Umemura, Y., Johnston, C.T., Yamagishi, A. and Schoonheydt, R.A. (2007) Ultrathin Hybrid Films of Clay Minerals. *Physical Chemistry Chemical Physics*, **9**, 918-932. http://dx.doi.org/10.1039/b610698f

[6]   Ogawa, M., Ishii, I., Miyamoto, N. and Kuroda, K. (2003) Intercalation of a Cationic Azobenzene into Montmorillonite. *Applied Clay Science*, **22**, 179-185. http://dx.doi.org/10.1016/S0169-1317(02)00157-6

[7]   Tsukamoto, T., Shimada, T. and Takagi, S. (2013) Unique Photochemical Properties of *p*-Substituted Cationic Triphenylbenzene Derivatives on a Clay Layer Surface. *The Journal of Physical Chemistry C*, **117**, 2774-2779. http://dx.doi.org/10.1021/jp3092144

[8]   Bujdak, J. (2006) Effect of the Layer Charge of Clay Minerals on Optical Properties of Organic Dyes. *Applied Clay Science*, **34**, 58-73. http://dx.doi.org/10.1016/j.clay.2006.02.011

[9]   Estevez, M.J.T, Arbeloa, F.L. and Arbeloa, T.L. (1994) On the Monomeric and Dimeric States of Rhodamine 6G Adsorbed on Laponite B Surfaces. *Journal of Colloid and Interface Science*, **162**, 412-417. http://dx.doi.org/10.1006/jcis.1994.1055

[10]   Yariv, S., Nasser, A. and Baron, P. (1990) Metachromasy in Clay Minerals. Spectroscopic Study of the Adsorption of Crystal Violet by Laponite. *Journal of the Chemical Society, Faraday Transactions*, **86**, 1593-1598. http://dx.doi.org/10.1039/ft9908601593

[11]   Lucia, L.A., Yui, T., Sasai, R., Yoshida, H., Takagi, S., Takagi, K., Whitten, D.G. and Inoue, H. (2003) Enhanced Aggregation Behavior Antimony (V) Porphyrins in Polyfluorinated Surfactant/Clay Hybrid Microenvironment. *The Journal of Physical Chemistry B*, **107**, 3789-3797. http://dx.doi.org/10.1021/jp026648a

[12]   Takagi, K., Kurematsu, T. and Sawaki, Y. (1991) Intercalation and Photochromism of Spiropyrans on Clay Interlayeres. *Journal of the Chemical Society, Perkin Transactions*, **2**, 1517-1522. http://dx.doi.org/10.1039/p29910001517

[13]   Sasai, R., Ogiso, H., Shindachi, I., Shichi, T. and Takagi, K. (2000) Photochromism in Oriented Thin Films Prepared by the Hybridization of Diarylethenes in Clay Interlayers. *Tetrahedron*, **56**, 6979-6984. http://dx.doi.org/10.1016/S0040-4020(00)00519-6

[14]   Iyi, N., Kurashima, K. and Fujita, T. (2002) Orientation of an Organic Anion and Second-Staging Structure in Layered Double-Hydroxide Intercalates. *Chemistry of Materials*, **14**, 583-589. http://dx.doi.org/10.1021/cm0105211

[15]   Bujdak, J., Iyi, N. and Sasai, R. (2004) Spectral Properties, Formation of Dye Molecular Aggregates and Reactions in Rhodamine 6G/Layered Silicates Dispersions. *The Journal of Physical Chemistry B*, **108**, 4470-4477. http://dx.doi.org/10.1021/jp037607x

[16]   Mishra, A., Behera, R.K., Behera, P.K., Mishra, B.K. and Behera, G.B. (2000) Cyanines during the 1990s: A Review. *Chemical Reviews*, **100**, 1973-2012. http://dx.doi.org/10.1021/cr990402t

[17]   Ogawa, M., Kawai, R. and Kuroda, K. (1996) Adsorption and Aggregation of a Cationic Cyanine Dye on Smectites. *The Journal of Physical Chemistry*, **100**, 16218-16221. http://dx.doi.org/10.1021/jp960261o

[18]   Bujdak, J., Martinez, V.M., Arbeloa, F.L. and Iyi, N. (2007) Spectral Properties of Rhodamine 3B Adsorbed on the Surface of Montmorillonites with Variable Layer Charge. *Langmuir*, **23**, 1851-1859. http://dx.doi.org/10.1021/la062437b

[19]   Sasai, R., Iyi, N., Fujita, T., Arbeloa, F.L., Martinez, V., Takagi, K. and Itoh, H. (2004) Luminescence Properties of Rhodamine 6G Intercalated in Surfactant/Clay Hybrid Thin Solid Films. *Langmuir*, **20**, 4715-4719. http://dx.doi.org/10.1021/la049584z

[20]   Suzuki, Y., Tenma, Y., Nishioka, Y., Kamada, K., Ohta, K. and Kawamata, J. (2011) Efficient Two-Photon Absorption Materials Consisting of Cationic Dyes and Clay Minerals. *The Journal of Physical Chemistry C*, **115**, 20653-20661. http://dx.doi.org/10.1021/jp203809b

[21]   Nakato, T., Kusunoki, K., Yoshizawa, K., Kuroda, K. and Kaneko, M. (1995) Photoluminescence of Tris(2,2'-bipyridine)ruthenium(II) Ions Intercalated in Layered Niobates and Titanates: Effect of Interlayer Structure on Host-Guest and Guest-Guest Interactions. *The Journal of Physical Chemistry*, **99**, 17896-17905. http://dx.doi.org/10.1021/j100051a015

[22]   Ichimura, K. (2000) Photoalignment of Liquid-Crystal Systems. *Chemical Reviews*, **100**, 1847-1874. http://dx.doi.org/10.1021/cr980079e

[23]   Ruslim, C., Hashimoto, M., Matsunaga, D., Tamaki, T. and Ichimura, K. (2004) Optical and Surface Morphological

Properties of Polarizing Films Fabricated from a Chromonic Dye by the Photoalignment Technique. *Langmuir*, **20**, 95-100. http://dx.doi.org/10.1021/la035366e

[24] Seki, T. (2007) Smart Photoresponsive Polymer Systems Organized in Two Dimensions. *Bulletin of the Chemical Society of Japan*, **80**, 2084-2109. http://dx.doi.org/10.1246/bcsj.80.2084

[25] O'Neill, M. and Kelly, S.M. (2011) Ordered Materials for Organic Electronics and Photonics. *Advanced Materials*, **23**, 566-584. http://dx.doi.org/10.1002/adma.201002884

[26] Kawatsuki, N. (2011) Photoalignment and Photoinduced Molecular Reorientation of Photosensitive Materials. *Chemistry Letters*, **40**, 548-554. http://dx.doi.org/10.1246/cl.2011.548

[27] Kurihara, S., Nomitama, S. and Nonaka, T. (2001) Photochemical Control of the Macrostructure of Cholesteric Liquid Crystals by Means of Photoisomerization of Chiral Azobenzene Molecules. *Chemistry of Materials*, **13**, 1992-1997. http://dx.doi.org/10.1021/cm0007555

[28] Moriyama, M., Mizoshita, N., Yokota, T., Kashimoto, K. and Kato, T. (2003) Photoresponsive Anisotropic Soft Solids: Liquid-Crystalline Physical Gels Based on a Chiral Photochromic Gelator. *Advanced Materials*, **15**, 1335-1338. http://dx.doi.org/10.1002/adma.200305056

[29] Matsumoto, M., Tachibana, H., Sato, F. and Terrettaz, S. (1997) Photoinduced Self-Organization in Langmuir-Blodgett Films. *The Journal of Physical Chemistry B*, **101**, 702-704. http://dx.doi.org/10.1021/jp9629093

[30] Ishihara, M., Hirase, R., Mori, M., Yoshioka, H. and Ueda, Y. (2009) Photoinduced Spectral Changes in Hybrid Thin Films of Functional Dyes and Inorganic Layered Material. *Thin Solid Films*, **518**, 857-860. http://dx.doi.org/10.1016/j.tsf.2009.07.103

[31] Irie, M. (2000) Diarylethenes for Memories and Switches. *Chemical Reviews*, **100**, 1685-1716. http://dx.doi.org/10.1021/cr980069d

[32] Bujdak, J. and Iyi, N. (2006) Molecular Aggregation of Rhodamine Dyes in Dispersions of Layered Silicates: Influence of Dye Molecular Structure and Silicate Properties. *The Journal of Physical Chemistry B*, **110**, 2180-2186. http://dx.doi.org/10.1021/jp0553378

[33] Iyi, N., Sasai, R., Fujita, T., Deguchi, T., Sota, T., Arbeloa, F.L. and Kitamura, K. (2002) Orientation and Aggregation of Cationic Laser Dyes in a Fluoromica: Polarized Spectrometry Studies. *Applied Clay Science*, **22**, 125-136. http://dx.doi.org/10.1016/S0169-1317(02)00144-8

[34] Bujdak, J., Iyi, N. and Fujita, T. (2002) The Aggregation of Methylene Blue in Montmorillonite Dispersions. *Clay Minerals*, **37**, 121-133. http://dx.doi.org/10.1180/0009855023710022

[35] Neumann, M.G., Schmidt, C.C. and Gessner, F. (1996) Time-Dependent Spectrophotometric Study of the Interaction of Basic Dyes with Clays II: Thionine on Natural and Synthetic Montmorillonites and Hectorites. *Journal of Colloid and Interface Science*, **177**, 495-501. http://dx.doi.org/10.1006/jcis.1996.0063

[36] Fujimoto, Y., Katayama, N., Ozaki, Y., Yasui, S. and Iriyama, K. (1992) Spectroscopic Studies of Thiatri-, Penta- and Heptamethine Cyanine Dyes II. Infrared and Resonance Raman Spectra of Thiatri-, Penta- and Heptamethine Cyanine Dyes. *Journal of Molecular Structure*, **274**, 183-195. http://dx.doi.org/10.1016/0022-2860(92)80157-D

[37] Sato, H., Kawasaki, M., Kasatani, K. and Katsumata, M. (1988) Raman Spectra of Some Indo-, Thia- and Selena-Carbocyanine Dyes. *Journal of Raman Spectroscopy*, **19**, 129-132. http://dx.doi.org/10.1002/jrs.1250190210

[38] Yang, J.P. and Callender, R.H. (1985) The Resonance Raman Spectra of Some Cyanine Dyes. *Journal of Raman Spectroscopy*, **16**, 319-321. http://dx.doi.org/10.1002/jrs.1250160507

[39] Ras, R.H.A., Johnston, C.T., Franses, E.I., Ramaekers, R., Maes, G., Foubert, P., De Schryver, F.C. and Schoonheydt, R.A. (2003) Polarized Infrared Study of Hybrid Langmuir-Blodgett Monolayer Containing Clay Mineral Nanoparticles. *Langmuir*, **19**, 4295-4302. http://dx.doi.org/10.1021/la026786r

[40] Cao, J.F., Wu, T., Hu, C., Liu, T., Sun, W., Fan, J.L. and Peng, X.J. (2012) The Nature of the Different Environmental Sensitivity of Symmetrical and Unsymmetrical Cyanine Dyes: An Experimental and Theoretical Study. *Physical Chemistry Chemical Physics*, **14**, 13702-13708. http://dx.doi.org/10.1039/c2cp42122d

6

The Effect of Carbon Rod—Specimens Distance on the Structural and Electrical Properties of Carbon Nanotube

Mohammad M. Uonis, Bassam M. Mustafa, Anwar M. Ezzat

Department of Physics, College of Science, Mosul University, Mosul, Iraq
Email: Bassam_alemam@yahoo.com

Abstract

The research studies the effect of the distance between the sample and the plasma sputtering source on the properties of the junction (silicon wafer-carbon nanotubes). The silicon wafer is fixed at (near, medium and far distances from the plasma source which is in the form of high purity graphite rod heated electrically). For the three cases, thickness of the sample is constant (20 nm). The samples were studied by scanning electron (SEM) and atomic force microscopes (AFM), X-ray and Raman spectra. For optimum distances the carbon layer is in the form of multi wall carbon nanotube (MWCNT). SEM images shows no formation of CNT on the Si wafer for near distance, which is consistent with the AFM images, X-ray and Raman spectrograms and no existence of characteristics (002) peaks whereas it appears for medium and longer distances, and by experience the optimum distance was found. This means that at closer distance high energy and high intensity plasma particles prevent the formation of CNT. This effect decreases with increasing distance of substrate from the graphite rod.

Keywords

Carbon Nanotubes, Si-CNT Junction, Plasma Sputtering of CNT

1. Introduction

There are many applications of carbon nanotubes in various scientific fields such as electronics, materials, medical science and others. Researchers were interested in finding different methods to produce these nanotubes and to study different variables affecting these methods to gain best properties for these tubes: their directions of growth, lengths and densities. Different methods are used in the production of CNTS some of them involve very high temperature (arc discharge: 5000°C - 20,000°C, laser vaporization: 4000°C - 5000°C) and others are performed at low temperatures (chemical vapor deposition CVD: below 1000°C) [1].

In this work preparation of Si-CNT junction without catalyst was done using plasma-sputtering system. The effects of distance variation between the carbon rod and the samples on nanotubes productions and their optical properties have to be examined through the SEM & AFM images, X-ray and Raman spectra.

2. Experimental Method

Our research involves the production of Si-C junctions using plasma-sputtering method. The main procedure is through the conventional sputter chamber which is evacuated with a rotary pump until the vacuum level (10^{-2} tor), then Introducing Argon gas into the chamber to allow the vacuum to recover then. Apply the lowest voltage that will allow a plasma to strike the carbon rod [2].

The carbon layers (20 nm thickness) were directly deposited on silicon wafer from carbon rods without any kind of catalysts [3]. The current flow through carbon rods is about 70 A. The effect of distance variation between the carbon rod and the samples on nanotubes productions and their properties was examined through the SEM, AFM images, X-ray and Raman spectroscopy.

The system for synthesis of carbon nanotubes on silicon wafers without catalyst by plasma sputtering method is shown in **Figure 1**. A high voltage is applied to the electrode causes an ionization of the gases, resulting in plasma formation. This method allows the control of growth parameters and as a result affecting the growth rate.

The most two important parameters in this technique are distance of the sample from the carbon rod thickness and the current flow through the carbon rod.

Measurements of the properties of the samples prepared by the above method are:

The surface microstructure obtained by a cold field scanning electron microscope (SEM) (JEUM-JSM-6756 F) operating at a voltage of 10 keV. Raman spectra measurements using spectrometer (GM SER No 87120) (Germany) which consists of a laser excitation beam that is focused through a microscope on the sample surface. The back scattered light intensity is measured as a function of its frequency shift. These shifts induced by the inelastic energy exchange between photons and vibration modes. The spectra obtained give information on the bonding environment in the sample. X-ray diffraction is done using the diffractometer (XRD, Bruker/D8-a with Cu Kα radiation λ = 1.54178 Å). The current-voltage characteristics obtained using the circuit shown in **Figure 2** revealed a standard pattern as shown later.

3. Results & Discussion

Figure 3 shows scanning electron microscope images for 20 nm carbon layer deposited on silicon wafers. These samples were located at three different distance from carbon rod with 3 mm increment. For the three cases,

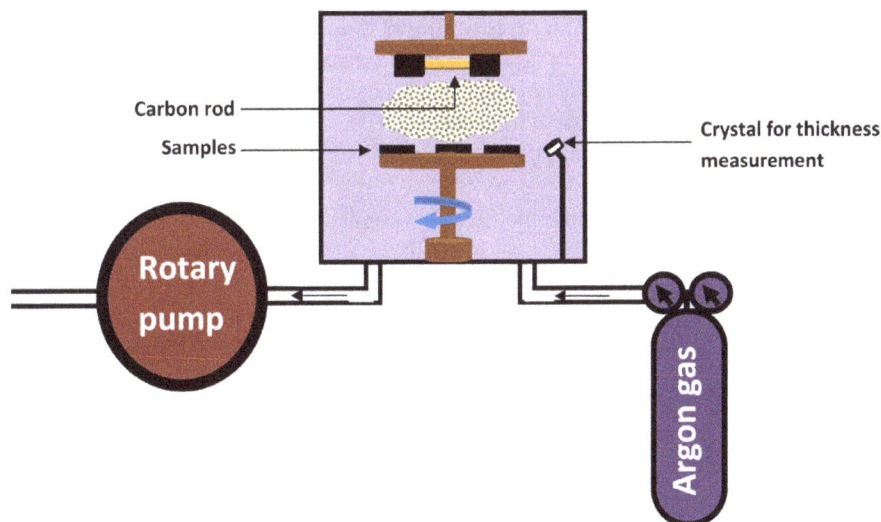

Figure 1. Schematic of plasma sputtering system.

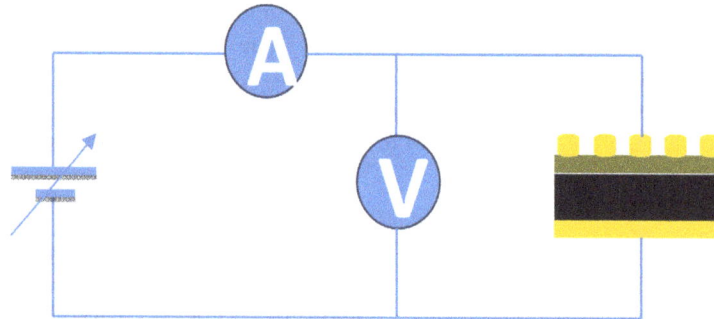

Figure 2. The circuit used for measuring I-V characteristics.

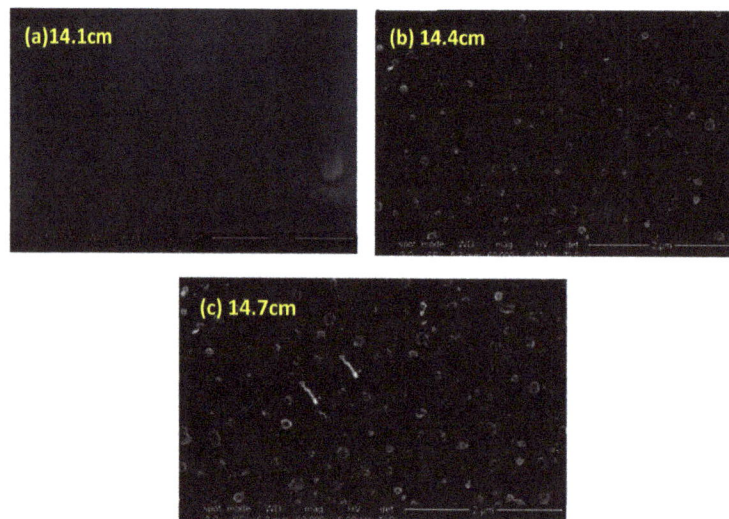

Figure 3. Scanning electron microscope images for the samples at three distance (a) 14.1 cm, (b) 14.4 cm and (c) 14.7 cm.

thickness of the sample is constant (20 nm) .the samples were studied by scanning electron (SEM) and atomic force microscopes (AFM), X-ray and Raman spectra. For optimum distances the carbon layer is in the form of multi wall carbon nanotube (MWCNT).

Figure 3(a) show the SEM image of the nearest sample to carbon rod (1.41 cm), SEM images shows no formation of CNT on the Si wafer for near distance, this is consistent with the AFM images, X-ray and Raman spectrograms where no existence of characteristics (002) peaks are noticed whereas **Figure 3(b)** & **Figure 3(c)** for the medium and longer distances shows formation of CNT. This means that at closer distance high energy and high intensity plasma particles prevent the formation of CNT, this effect decreased with increasing distance of substrate-graphite rod [4]-[6].

Raman spectra for the samples revealed: the G and D bands appear for the smallest distances at 1610 - 1620 cm^{-1} and at 1360 - 1370 cm^{-1} respectively, also intensity of Raman peaks increase with distance by comparing (b) and (c) in **Figure 4**, while for the smallest distance (a) in the same figure, the peaks (D & G) disappeared [6]-[10]. The most noticeable features seen in **Figure 4** are the increase in disorder induced D mode with distance increase, which ascribed to prevent the formation of the both phases (order and disorder modes) in the deposited carbon layers.

X-ray diffractogram shows the characteristic peak (002) of the graphite layers which indicate surly the existence of multiwall Carbon nanotube, intensity of the peak for smaller distance is lower than that of high distance [11]. From **Figure 5**, we can observe that some peaks, which belong to SiC, and disappeared with decreasing distance. This is fully compatible with the SEM images and Raman spectrum.

Figure 6 shows the Atomic force microscope images for the samples at three distances. **Figure 6(c)** at near distances from the graphite rod the center of the plasma will lead to a distortion of the samples surfaces, increase

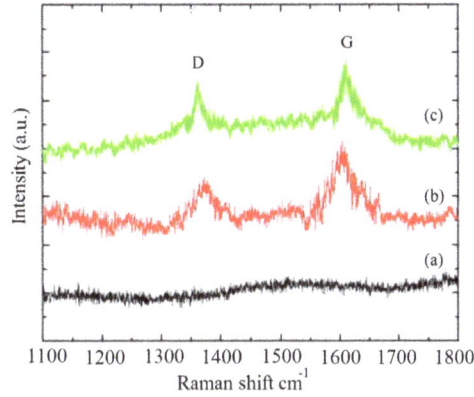

Figure 4. Raman spectra for the samples with distances from the Carbon rod (a) 14.1 cm, (b) 14.4 cm and (c) 14.7 cm.

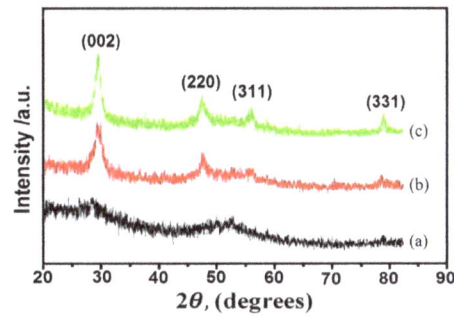

Figure 5. X-ray spectrum for the samples at three distance (a) 14.1 cm, (b) 14.4 cm and (c) 14.7 cm.

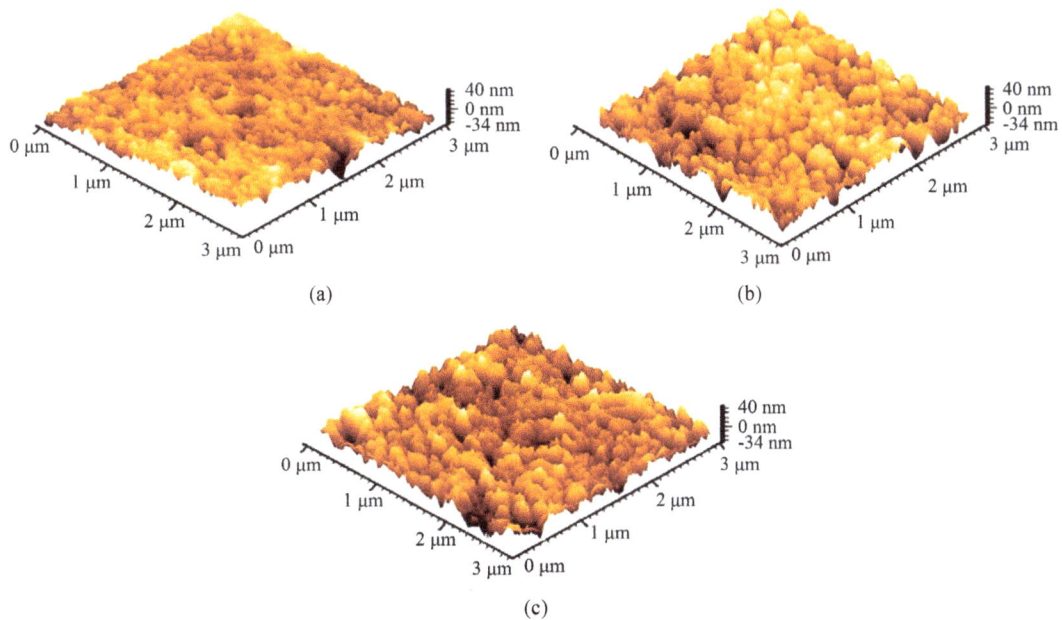

Figure 6. Atomic force microscope images for the samples at three (a) 14.1 cm, (b) 14.4 cm and (c) 14.7 cm.

in surface roughness and preventing the formation of grains and carbon nanotubes, whereas grains are well defined for the medium and far distances Which we will consider as optimum **Figure 6(a)** & **Figure 6(b)** [11] [12].

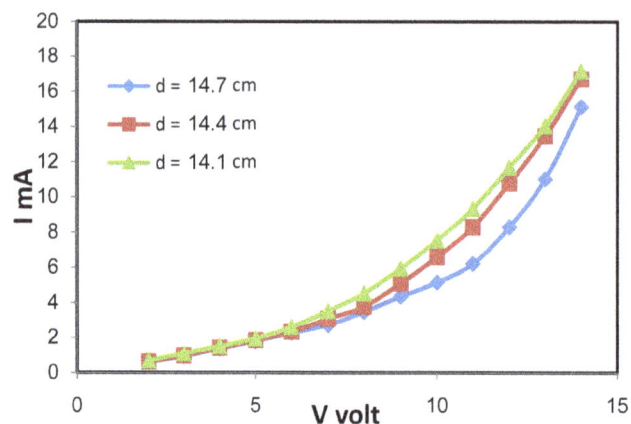

Figure 7. The I-V characteristics for Si-C junction with gold electrodes.

The I-V characteristics for Si-C junction with gold electrodes on the front and backsides of the junction are shown in **Figure 7**. It is very clear that the junction has a semiconducting behavior, which is a characteristic property of the zigzag carbon nanotubes I-V characteristics support our deduction and best curve is with distance 1.47 cm.

The study by scanning electron (SEM) and atomic force microscopes (AFM), X-ray and Raman spectra all shows that distance between sample and source of plasma sputtering source affecting the formation of CNT on the Si wafer .For near distance no formation of CNT , whereas it appears for medium and longer distances. This means that at closer distance high energy and high intensity plasma particles prevent the formation of CNTi.eit prevent the formation of the hexagonal C loops which is the essential unit of the graphite layer that forms the CNT, this effect decreased with increasing distance of substrate from the graphite rod which means that the distance between sample and source is a critical factor.

4. Conclusion

Distance between sample and source of plasma sputtering source affect the formation of CNT on the Si wafer. For near distance there is no formation of CNT, whereas it appears for medium and longer distances. This means that at closer distance high energy and high intensity plasma particles prevent the formation of CNT. This effect decreases with increasing distance of substrate from the graphite rod.

References

[1]   Poole Jr., C.P. and Owens, F.J. (2003) Introduction to Nanotechnology. John Wiley & Sons, Inc., Hoboken.

[2]   Seshan, K. (2001) Handbook of Thin-Film Deposition Processes and Techniques, Principles, Methods, Equipment and Applications. Noyes Publications/William Andrew Publishing, New York.

[3]   Rümmeli, M.H., Bachmatiuk, A., Börrnert, F., Schäffel, F., Ibrahim, I., Cendrowski, K., Simha-Martynkova, G., Plachá, D., Borowiak-Palen, E., Cuniberti, G. and Büchner, B. (2011) Synthesis of Carbon Nanotubes with and without Catalyst Particles. *Nanoscale Research Letters*, **6**, 303. http://dx.doi.org/10.1186/1556-276X-6-303

[4]   Hofmann, S., Kleinsorge, B., Ducati, C., Ferrari, A.C. and Robertson, J. (2004) Low-Temperature Plasma Enhanced Chemical Vapor Deposition of Carbon Nanotubes. *Diamond and Related Materials*, **13**, 1171-1176. http://dx.doi.org/10.1016/j.diamond.2003.11.046

[5]   Gan, K.J., Chang, C.H., Lu, J.J., Lin, C.L., Su, Y.K., Li, B.J. and Yeh, W.K. (2011) Growth of Carbon Nanotube Using Microwave Plasma Chemical Vapor Deposition and Its Application to Thermal Dissipation of High-Brightness Light Emitting Diode. *WCE* 2011, London, 6-8 July 2011.

[6]   Choi, Y.C., Bae, D.J., Lee, Y.H. and Lee, B.S. (2000) Growth of Carbon Nanotubes by Microwave Plasma-Enhanced Chemical Vapor Deposition at Low Temperature. *Journal of Vacuum Science & Technology A*, **18**, 1864. http://dx.doi.org/10.1116/1.582437

[7]   Jorio, A., Pimenta, M.A., Souza Filho, A.G., Saito, R., Dresselhaus, G. and Dresselhaus, M.S. (2003) Characterizing Carbon Nanotube Samples with Resonance Raman Scattering. *New Journal of Physics*, **5**, 139.1-139.17.

[8]   Bokova, S.N., Obraztsova, E.D., Grebenyukov, V.V., Elumeeva, K.V., Ishchenko, A.V. and Kuznetsov, V.L. (2010) Raman Diagnostics of Multi-Wall Carbon Nanotubes with a Small Wall Number. *Physica Status Solidi (B)*, **247**, 2827-2830. http://dx.doi.org/10.1002/pssb.201000237

[9]   Lehman, J.H., Terrones, M., Mansfield, E., Hurst, K.E. and Meunier, V. (2011) Evaluating the Characteristics of Multiwall Carbonnanotubes. *Carbon*, **49**, 2581-2602. http://dx.doi.org/10.1016/j.carbon.2011.03.028

[10] Oddershede, J., Nielsen, K. and Stahl, K. (2007) Using X-Ray Powder Diffraction and Principal Component Analysis to Determine Structural Properties for Bulk Samples of Multiwall Carbon Nanotubes. *Zeitschrift für Kristallographie*, **222**, 186-192.

[11] Zdrojek, M., Gebicki, W., Jastrzebski, C., Melin, T. and Huczko, A. (2004) Studies of Multiwall Carbon Nanotubes Using Raman Spectroscopy and Atomic Force Microscopy. *Solid State Phenomena*, **99-100**, 265-268. http://dx.doi.org/10.4028/www.scientific.net/SSP.99-100.265

[12] Bellucci, S., Gaggiotti, G., Marchetti, M., Micciulla, F., Mucciato, R. and Regi, M. (2007) Atomic Force Microscopy Characterization of Carbon Nanotubes. *Journal of Physics: Conference Series*, **61**, 99-104.

7

The Role of Sputtering Current on the Optical and Electrical Properties of Si-C Junction

Mohammad M. Uonis, Bassam M. Mustafa, Anwar M. Ezzat*

Department of Physics, College of Science, Mosul University, Mosul, Iraq
Email: *prof.dr.anwar@uomosul.edu.iq

Abstract

The effect of sputtering current that flow in a carbon rod on the structural and transport properties of Si-C junction is studied. Si-C junction is fabricated by plasma sputtering in Argon gas atmosphere without catalysts with thickness of 20, 40 and 60 nm. Images of the specimen by scanning electron microscope (SEM) and atomic force microscope (AFM) show that the carbon layer is as carbon nanotubes with diameters about 20 - 30 nm. X-ray and Raman spectrums show peak characteristics of the carbon nanotubes, the G and D bands appear for all thicknesses indicating free of defect carbon nanotubes. Two parameters about the thickness of the carbon layer and the sputtering current for different thicknesses and currents were studied. Nanotubes evidence was clear. We noticed that the sputtering current and thickness of layers affect the structure of CNT layer leading to the formation of grains. Increasing plasma current led to decrease grain formation however increasing thickness ends to increase grain size; moreover it led to amorphous structure formation and this was proved through X-ray, Raman spectra and AFM images.

Keywords

Carbon Nanotubes, SEM, AFM and X-Ray Multi Wall Images, Plasma Sputtering and Si-C Junction

1. Introduction

Carbon Nanotubes (CNT's) are cylindrical structures of nanoscale diameter; it composed of carbon atoms in a hexagonal arrangement. They were discovered in 1991 by Japanese scientist Sumio Iijima [2]. The unique properties of CNT's, such as tremendous strength and excellent thermal and electrical conductivity, have caused this

*Corresponding author.

material to become the focus of intense research by many groups [1]-[3].

There are three commonly used methods of CNT synthesis. Arc-discharge method employs evaporation of graphite electrodes in electric arcs that involve very high temperatures (~4000°C). Laser-vaporization technique employs evaporation of high-purity graphite target by high-power lasers in conjunction with high-temperature furnaces. Chemical vapor deposition (CVD) required incorporating catalyst-assisted by thermal decomposition of hydrocarbons [1]-[5].

All the above methods need catalysts such as Ni, Fe …etc. In our research, we use plasma scattering of carbon rod to prepare Si-C junction, which was studied earlier thoroughly in our published paper [6].

In this paper, we want to find the effect of changing the sputtering current and the thickness effect on the structural and electrical properties of the CNT layer in the Si-C junction.

2. Experimental Method

We report the synthesis of carbon nanotubes on silicon wafers without catalyst of large aspect ratios by plasma sputtering method as shown in **Figure 1**. The carbon-containing and reacting gases introduced into the chamber through a network with mass flow controllers allowing to regulate the flow rate and gas composition of the mixture. A high voltage applied to the electrode above the sample causes an ionization of the gases, resulting in plasma formation. This method allows the control of growth parameters to influence growth rateand diameter.

The two most important features of this technique are carbon layer thickness and current flow through the carbon rod, which studied using scanning electron microscopy (SEM), Atomic force microscope (AFM), X-ray diffraction (XRD) and Raman spectroscopy.

The surface microstructure obtained by a cold field scanning electron microscope (SEM) (JEUM-JSM-6756 F) operating at a voltage of 10 keV. Scanning electron microscope (Jeol; JDS-7391LV) was used to record Energy dispersive scattering (EDS).

Raman spectra measurements (GM SER No 87,120) in (Germany made). The current voltage characteristics obtained using A WXS140s Wacom sun simulator.

Micro-Raman spectroscopy consists of a laser excitation beam that focused through a microscope on the sample surface. The back scattered light intensity is measured as a function of its frequency shift. These shifts induced by the inelastic energy exchange between photons and vibration modes. The spectra obtained give information on the bonding environment in the sample. X-ray diffraction (XRD, Bruker/D8-advance with Cu Kα radiation ($\lambda = 1.54178$ Å), the diffraction angle was scanned from 20± to 100± at the scanning speed of 0:02± per second.

3. The Results

Many measurements were done include Scanning electron microscopy (SEM), Atomic force microscope (AFM), X-ray diffraction (XRD) and Raman spectra are taken.

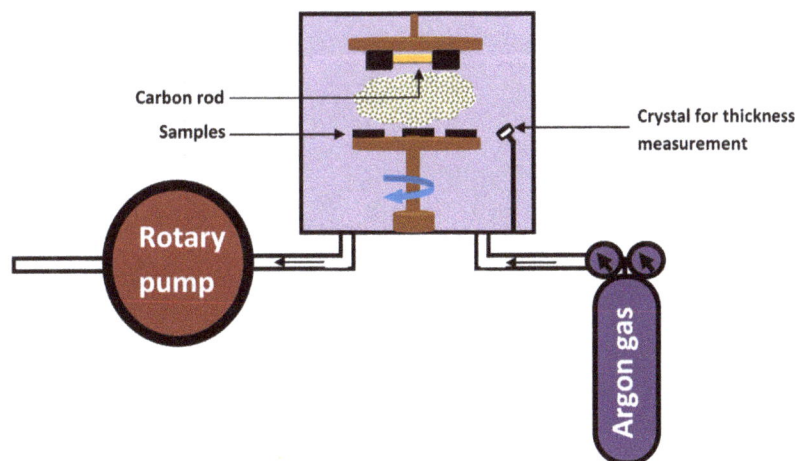

Figure 1. Schematic of plasma sputtering system.

Two parameters were studied, the thickness of the carbon layer and the sputtering current. For all thicknesses and all currents included in the study, the effect of plasma sputtering current is opposite to increasing of layers thickness. Increasing plasma current will decrease grain formation and increasing thickness will increases grain size and moreover it lead to amorphous structure formation, this were proved by X-ray, Raman spectra and AFM images.

Carbon layers deposited using 70 A current in the carbon rod are shown in **Figures 2(a)-(c)**. From **Figure 2(a)**, it is noticed that the sputtering process produces small grains and these grains grows with thickness increase as shown in **Figure 2(b)** and **Figure 2(c)** for 40 and 60 nm respectively.

In **Figure 2(c)**, clearly shown bigger grains due to cohesion between the small grains as a result of the continued growth in their dimensions with increasing the thickness of the carbon layer.

In **Figure 2(a)**, **Figure 2(d)** and **Figure 2(g)**, represents SEM images due to sputtering currents in the carbon layer with values 70, 80 and 90 A with constant thickness (20 nm).The increase in current flow in the carbon rod produces further excitation(increase in the energy of plasma) in plasma which tries to obstruct the formation of grains and this will lead to produce more amorphous phase at the expense of carbon nanotubes formation (ordered phase) in the deposited carbon layer.

As a conclusion, the increase of thickness will produce grains that grown slowly with high currents. This agree with the results of X-ray and Raman spectra, which will be discussed later.

Raman spectra revealed the G and D bands clearly for all thicknesses and just for 70 A and 80 A. However, for high current 90 A, the G and D bands distorted as shown in **Figure 3**.

From **Figure 3(a)** and **Figure 3(b)**, the most important features seen are the disorder induced D band at 1340 - 1350 cm^{-1} and the tangential G band at 1550 - 1580 cm^{-1}, which is related to the graphite tangential Raman active mode where the two atoms in graphene unit cell are vibrating tangentially one against the other. A presence in the Raman spectra of D bands with the frequency 1350 cm^{-1} is ascribed to a presence of amorphous Carbon in the sample, while the presence of G bands with a frequency 1580 cm^{-1} give us information about the existence of ordered Carbon structure [7]-[12].

We note that the specific peaks shown for 70 A and 80 A are clear but they decrease when increasing current to 90 A. The Raman spectrum distorted due to the increase in plasma energy by increasing current.

The ratio of I_D/I_G bands (about 74% for 70 A and 77% for 80 A) give indication that the more probable phase is the ordered mode, but for 90 A power, the amorphous mode increase as a result of high sputtering power, so that we conclude that the amorphous mode will increase with increasing of flow current, this conclusion matches with conclusion that drown from the SEM images [10].

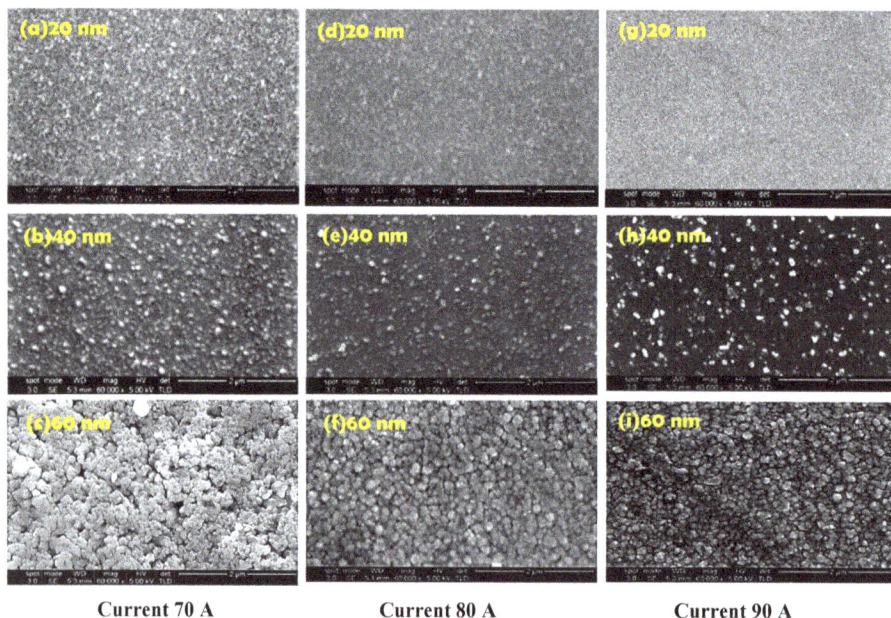

Current 70 A   Current 80 A   Current 90 A

Figure 2. SEM images of Carbon layers for a range of currents 70 A, 80 A and 90 A.

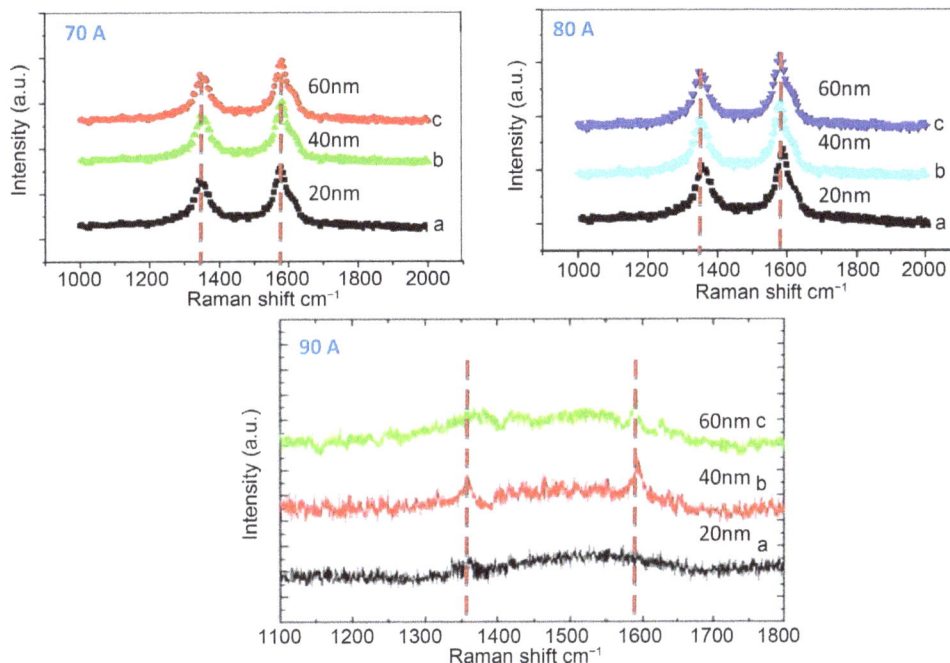

Figure 3. The Raman spectra of Carbon layers for a range of plasma currents 70 A, 80 A and 90 A.

The Raman features associated with the radial breathing mode (RBM) from the large diameter tubes is usually too poor to be observable. The D band activated in the first order scattering process of sp2 Carbons by the presence of in plane substitution heteroatom vacancies, grain boundaries, or other defects and by finite size effects, all of which lower the crystalline symmetry of the quasi-infinite lattice. Therefore, the D mode could be used as a diagnostic of disruptions in the hexagonal framework of MWNTs and is induced by double resonance process. This result also observed elsewhere [10].

The shape of the G Raman peak gives possibility to distinguish between semiconducting or metallic nanotubes. Here, we think, that the nanotubes are associated with semiconducting type of conductivity as mentioned in ref. [12].

In purified sample G band is narrower than in non-purified, so the spectra shows that our samples were well purified. At low thickness the intensity is somewhat lower than that of high thickness, this indicate that the diameter of the nanotube is increasing, this is clear through comparing (a) with (b) and (c) spectra in **Figure 2**.

X-ray measurements for Carbon nanotubes is somewhat different from that of crystalline or polycrystalline samples. X-ray diffraction of CNT gives indication of the presence of graphite layers. X-ray diffraction pattern shows clearly the characteristic peak (002) of the graphite layers which indicate surly the existence of multiwall Carbon nanotube. From **Figure 4**, it can be observed that the peaks for power 70 A Can be specified clearly for all thicknesses and they are distorted with increasing sputtering current due to increase in amorphous phase, In addition, the increase in thickness is inversely proportional to spectrum distortion for 80 A and 90 A currents, also it is noted that some peaks which belong to Si-C are disappeared in high current spectrums. XRD spectrum also shows characteristic peaks of Si, which belong to the substrate, and Si-C, which is normally formed at the junction but the intensity, is low indicating small quantity of Si-C.

Figure 5 shows the atomic force microscope AFM images of the carbon layer on silicon wafer for different thicknesses and powers. It can be observed that the surface roughness of the carbon layers increases with thickness for all currents, this roughness is due to the formation of grains and their growth, so that the amorphous mode (grains) and short CNTS were clearly shown in this figure which increases with thickness.

In **Figure 6**, the three dimensional images of the samples for 90 A current are shown. It is clear that there is an enlargement in the grain size with increasing thickness.

Figures 7(a)-(c) show the I-V characteristics for the Si-CNT junction for all currents and all thicknesses under study, it is noticed that the current decreases with sample thickness for the same current in the carbon rod. Moreover for the same thickness, the current in I-V characteristics increases with the current flows in the carbon

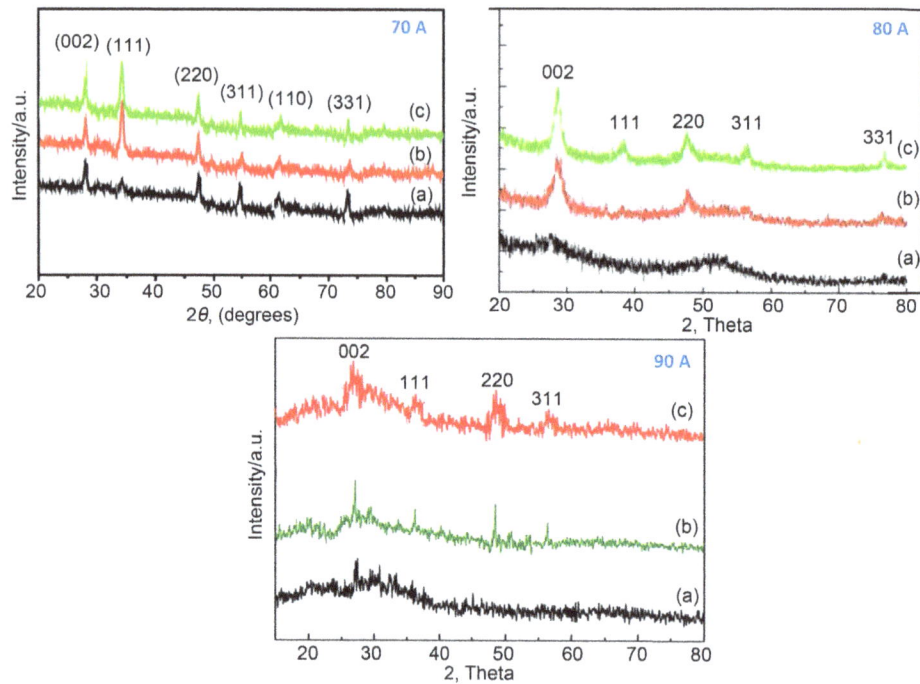

Figure 4. X-ray spectrum for Carbon layers (a) 20 nm thickness, (b) 40 nm thickness and (c) 60 nm thickness.

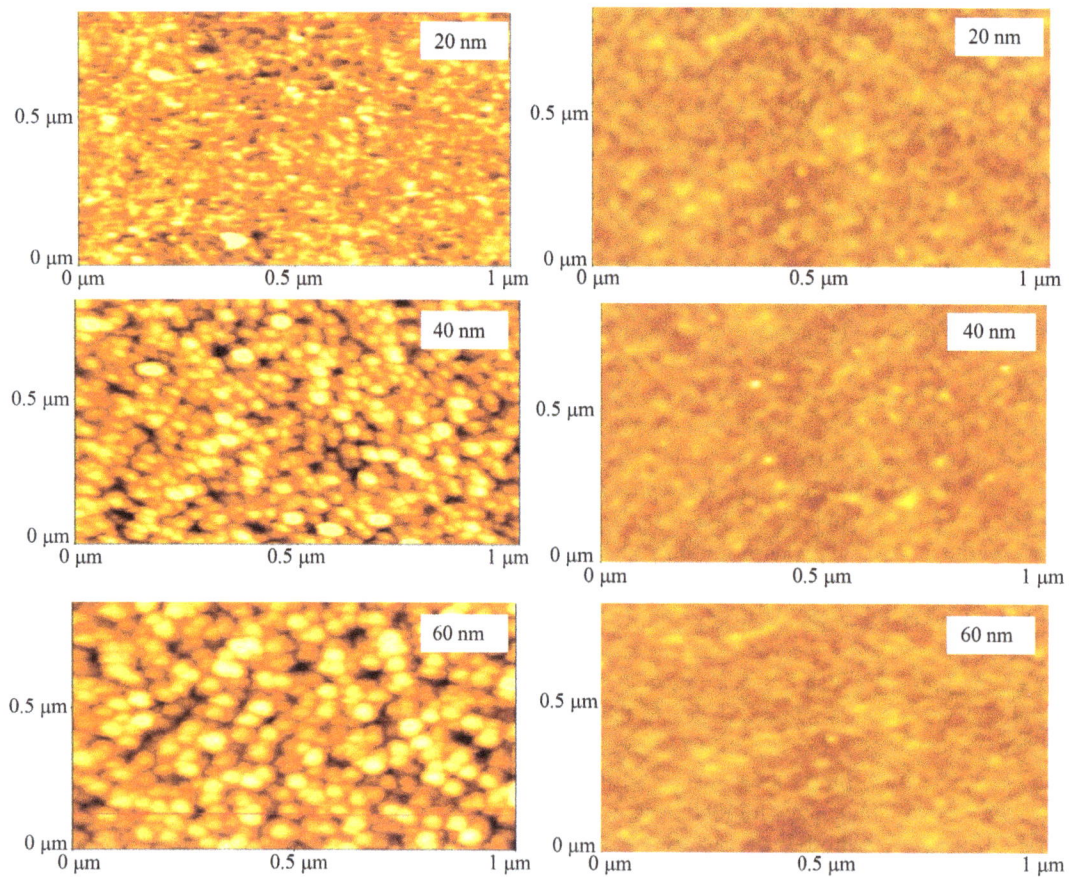

Figure 5. AFM images of Carbon layers for a range of currents 70 A, 80 A and 90 A.

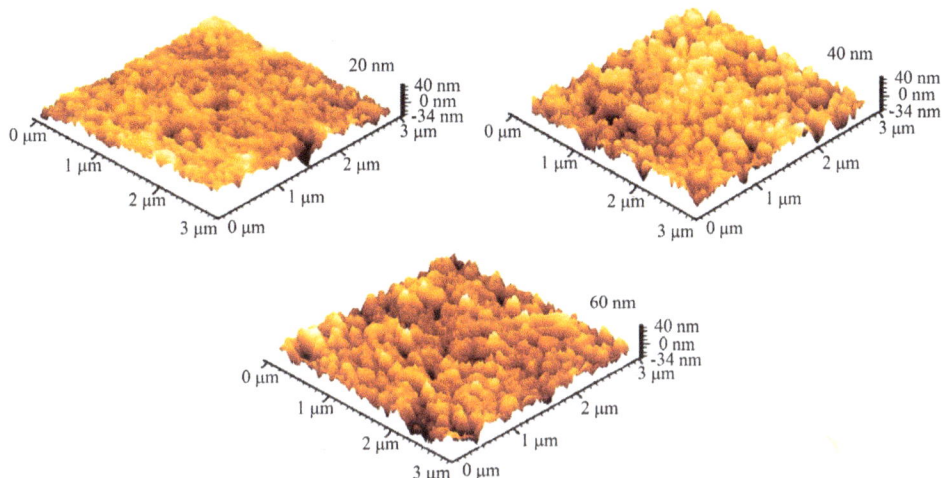

Figure 6. AFM images of Carbon layers for plasma current 90 A.

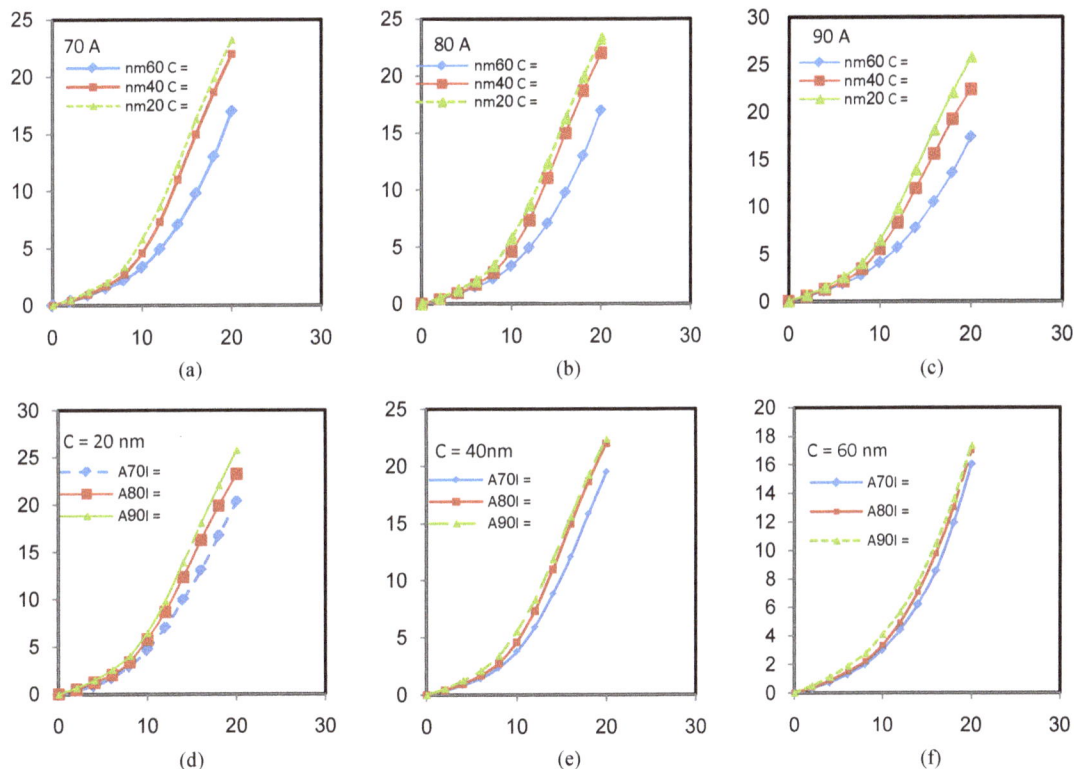

Figure 7. The I-V characteristics of Si-CNT junction for different plasma current and thicknesses.

rod as in **Figures 7(a)-(c)**. As a general behavior of the junction, I-V characteristics show semiconducting behavior.

For constant thickness, the increase in the plasma current flowing in the carbon rod will increase the energy of the plasma, which pervert the formation of the grains in the deposited carbon layer, reduces the amorphous phase, and thus reduces distortions occurring in this layer, which in turn is reflected in the increase in I-V characteristics current.

4. Conclusions

1) The joined effect of increasing thickness of the CNT layer in Si-CNT junction and the sputtering current in

the carbon rod led to the formation of grains.

2) At certain sputtering current increasing thickness of the layer increases the grain size.

3) Increasing sputtering current that flows in the carbon rod led to increase the plasma energy, which lead to increase percentage of amorphous phase in the CNT layer.

4) I-V characteristic shows semiconducting behavior and the current decreases with increasing thickness for the same sputtering current in the carbon rod, whereas the current increases with increasing sputtering current in the carbon rod.

Acknowledgements

The Authors would like to thank the College of Science, administration at Mosul University for supporting this work. Also thanks to Abid Al Karem M Muhammed (Sumysyate University/Department of Nano/Ukraine) for his great help in samples measurements.

References

[1]   Valentini, L., Armentano, I., Kenny, J.M., Lozzi, L. and Santucci, S. (2003) Pulsed Plasma-Induced Alignment of Carbon Nanotubes. *Materials Letters*, **57**, 3699-3704. http://dx.doi.org/10.1016/S0167-577X(03)00166-6

[2]   Lehman, J.H., Terrones, M., Mansfield, E., Hurst, K.E. and Meunier, V. (2011) Evaluating the Characteristics of Multiwall Carbon Nanotubes. *Carbon*, **49**, 2581-2602. http://dx.doi.org/10.1016/j.carbon.2011.03.028

[3]   Hofmann, S., Kleinsorge, B., Ducati, C., Ferrari, A.C. and Robertson, J. (2004) Low-Temperature Plasma Enhanced Chemical Vapor Deposition of Carbon Nanotubes. *Diamond and Related Materials*, **13**, 1171-1176. http://dx.doi.org/10.1016/j.diamond.2003.11.046

[4]   Choi, Y.C., Bae, D.J., Lee, Y.H. and Lee, B.S. (2000) Growth of Carbon Nanotubes by Microwave Plasma-Enhanced Chemical Vapor Deposition at Low Temperature. *Journal of Vacuum Science & Technology A*, **18**, 1864-1868.

[5]   Muratore, C., Reed, A.N., Bultman, J.E., Ganguli, S., Cola, B.A. and Voevodin, A.A. (2013) Nanoparticle Decoration of Carbon Nanotubes by Sputtering. *Carbon*, **57**, 274-281. http://dx.doi.org/10.1016/j.carbon.2013.01.074

[6]   Ezzat, A.M., Mustafa, B.M. and Uonis, M.M. (2014) Fabrication of Si-CNT Junction by Plasma Sputtering of Graphite Rods on Silicon Wafers. *International Journal of Advanced Research*, **2**, 108-113.

[7]   Costa, S., Borowiak-Palen, E., Kruszyńska, M., Bachmatiuk, A. and Kaleńczuk, R.J. (2008) Characterization of Carbon Nanotubes by Raman Spectroscopy. *Materials Science-Poland*, **26**, 433-441.

[8]   Bokova, S.N., Obraztsova, E.D., Grebenyukov, V.V., Elumeeva, K.V., Ishchenko, A.V. and Kuznetsov, V.L. (2010) Raman Diagnostics of Multi-Wall Carbon Nanotubes with a Small Wall Number. *Physica Status Solidi* (*B*), **247**, 2827-2830. http://dx.doi.org/10.1002/pssb.201000237

[9]   Jorio, A., Pimenta, M.A., Souza Filho, A.G., Saito, R., Dresselhaus, G. and Dresselhaus, M.S. (2003) Characterizing Carbon Nanotube Samples with Resonance Raman Scattering. *New Journal of Physics*, **5**, 139.1-139.17.

[10] Murphy, H., Papakonstantinou, P. and Okpalugo, T.I.T (2006) Raman Study of Multiwalled Carbon Nanotubes Functionalized with Oxygen Groups. *Journal of Vacuum Science & Technology B*, **24**, 715.

[11] Zdrojek, M., Gebicki, W., Jastrzebski, C., Melin, T. and Huczko, A. (2004) Studies of Multiwall Carbon Nanotubes Using Raman Spectroscopy and Atomic Force Microscopy. *Solide State Phenomena*, **99**, 265.

[12] Bellucci, S., Gaggiotti, G., Marchetti, M., Micciulla, F., Mucciato, R. and Regi, M. (2007) Atomic Force Microscopy Characterization of Carbon Nanotubes. *Journal of Physics*: *Conference Series*, **61**, 99-104.

8

Application of the Level Set Method in Three-Dimensional Simulation of the Roughening and Smoothing of Substrates in Nanotechnologies

Branislav Radjenović, Marija Radmilović-Radjenović

Institute of Physics, University of Belgrade, Zemun, Serbia
Email: marija@ipb.ac.rs

Abstract

This paper contains results of the comprehensive studies of the effect of the isotropic etching mode on roughening of the nanocomposite materials and on smoothing of the roughed nanostructure made of homogeneous materials. Three-dimensional simulation results obtained illustrate the influence of the isotropic etch process on dynamics of the roughening and smoothing of the surfaces, indicating the opposite effects of the same etch process on the surfaces made of different materials. It was shown that root mean square roughness obeys simple scaling laws during both roughening and smoothing processes. The exponential time dependences of the rms roughness have been determined.

Keywords

Nanocomposite, Roughness, Smoothing, Isotropic Etching

1. Introduction

The surface evolution induced by the ion-bombardment has been subjects of a numerous experimental and theoretical studies in recent years. It was found that the ion sputtering represents one of the crucial processes in producing various nanostructured surfaces or interfaces in nanotechnology [1]-[3]. On the other hand, ion beam-based finishing technologies are very important for the surface shape corrections and low-energy ion beams can be beneficially used to tailor the microscopic surface roughness of solid surfaces on a micron and sub-micron scales [4]-[6].

The control of plasma induced roughness or perhaps control of surface roughness by plasma etching is recognized as one of the limiting and key issues in applications of plasma etching in new generations of plasma technologies [7] [8]. Control of the roughness has become a fast growing area of intensive experimental and theoretical studies. In the majority of such studies, the etching of homogeneous materials with a constant etch rate along the whole volumes has been investigated. However, etched materials, such as nanocomposite materials, usually do not have the same etch rates [7] [8]. The presence of two phases with different etch rates (the ratio of the two etch rates is s and the abundance of one phase is p) strongly affects the formation of the surface roughness and that the etch rate is higher during the isotropic process as compared to the anisotropic process for [7].

The energetic ions that hit the solids can cause different modifications of the surface topography via opposing effects like surface roughening and surface smoothing [9] [10]. The evolution of the surface caused by the ion bombardment can be described as the interplay between the dynamics of the surface roughening due to sputtering and smoothing due to material transport during surface diffusion. These processes are competitive and responsible for the creation of quasiperiodic ripples [10] and self-affine topographies [11]-[13]. Although there is a large number of studies devoted to ripple formation there are only a few studies on the scaling of the surfaces evolving under ion bombardment [12]-[14]. Among them, [3] [6] [7] [14] deal with surface roughening and [3] [15] involve surface smoothing.

In this paper we present our three-dimensional simulation (3D) results based on the level set method with sparse field method for solving differential equations. We have simulated both the roughness formation during isotropic etching of nanocomposite materials and smoothing of the homogeneous materials.

2. Method

Isotropic etching can be regarded as a non-directional removal of material from a substrate via a chemical process using a corrosive liquid or a plasma. Etchants that are currently used show significant surface roughening, as well as dependence of etching rates on feature density, size, electrical connectivity, and location on the wafer. Smooth etching of silicon applying plasmas is highly desirable for some integrated circuit applications and for manufacturing devices such as microstructures, microsensors, and electro-optic devices. Smooth etching is also advantageous to produce surfaces that will be bonded together, since stronger bonds are formed between flat silicon substrate surfaces having minimum roughness.

Although roughness is usually undesirable, it represents complicated and very expensive to control in manufacturing. Developing integral modeling systems for realistic geometries has evolved into one of the most efficient methods for optimizing plasma processing equipment [16]-[18]. The level set methods are powerful techniques for analyzing and computing moving fronts in a variety of different cases. Level set method has advantages as compared to the other simulation methods since it provides results much faster and with multi scale approach may give local roughness for unlimited resolution. The method have a wide range of applications, including problems in fluid mechanics, combustion, manufacturing of computer chips, computer animation, image processing and many other fields [16]-[18].

The smoothing process enhanced by the isotropic etching has been modelled for a substrate made of homogenous materials. The obtained simulation results show how properties of material on different scales affect the roughness and smoothing. Analyzing the obtained simulation results, the time dependence of the roughness exponent has been determined for both processes.

In this paper we discuss the application of the level set method for the three-dimensional (3D) modelling of the roughening of the nano-composite surface which characteristics depend not only on the properties of their individual constituents but also on their morphology and interfacial characteristics [7] [8]. We assumed that the nanocomposite materials consist of two phases (polymer and graphite nanoparticles) characterized by the ratio of the two etch rates $s = 2$ and concentration of the easily etched material $p = 0.5$. The materials have been randomly distributed and represented by 3D cubic lattice.

3. Simulation Technique and Procedure

The basic idea behind the level set method is to represent the surface in question at a certain time t as the zero level set (with respect to the space variables) of a certain function $\varphi(t,x)$, the so called level set function shown in **Figure 1**. The initial surface is given by $\{x|\varphi(0,x) = 0\}$. The evolution of the surface in time is caused by the surface processes in the case of the etching [19]. The velocity of the point on the surface normal to the surface

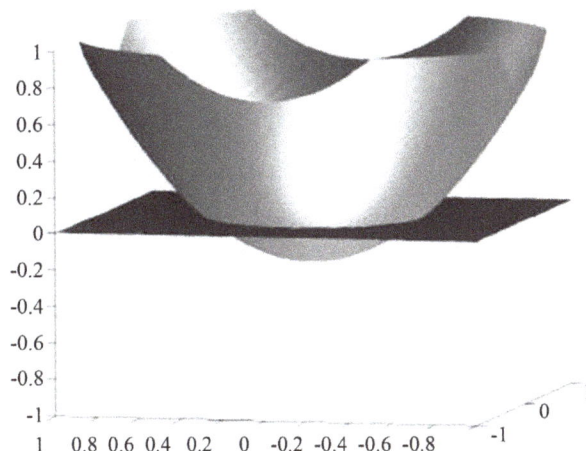

Figure 1. Level set function.

will be denoted by $R(t,x)$, and is called velocity function. The velocity function generally depends on the time and space variables and we assume that it is defined on the whole simulation domain. At a later time $t > 0$, the surface is as well the zero level set of the function $\varphi(t,x)$. Namely, it can be defined as a set of points $\{x|\varphi(t,x) = 0\}$. This leads to the level set equation in Hamilton-Jacobi form:

$$\frac{\partial \varphi}{\partial t} + H\left(\nabla \varphi(t,x)\right) = 0 \tag{1}$$

in the unknown function $\varphi(t,x)$, where Hamiltonian function is given by $H = R(t,x)|\nabla \varphi(t,x)|$ and where $\varphi(0,x) = 0$ determines the initial surface. Several approaches for solving level set equations exist which increase accuracy while decreasing computational effort. The most important are narrow band level set method, widely used in etching process modeling tools, and recently developed sparse-filed method [3]. The sparse-field method use an approximation to the level set function that makes it feasible to recompute the neighborhood of the zero level set at each time step. As a result, the number of computations increases with the size of the surface, rather than with the resolution of the grid.

We say that Hamiltonian function H is convex if the following condition is fulfilled:

$$\frac{\partial^2 H}{\partial \phi_{x_i} \partial \phi_{x_j}} \geq 0 \tag{2}$$

where ϕ_{x_i} is a partial derivative of $\varphi(t,x)$ with respect of x_i. The non-convex Hamiltonians are characteristic for plasma etching and deposition simulations. During these processes the etching (deposition) rate, that defines the surface velocity function $R(t,x)$, depends on the geometric characteristics of the profile surface itself, or more precisely, on the angle of the incidence of the incoming particles. In the cases under study here we shall consider an etching beam coming down in the vertical direction. These conditions are characteristic for ion milling technology, but angular dependence of the etching rates appears, more or less, in all etching processes. The simplest finite difference scheme that can be applied in these cases is the Lax-Friedrichs scheme, one that relies on the central difference approximation to the numerical flux function, and preserves monotonicity through a second-order linear smoothing term. It is shown [4] that it is possible to use the Lax-Friedrichs scheme in conjunction with the sparse field method, and to preserve sharp interfaces and corners by optimizing the amount of smoothing in it. This is of special importance in the simulations of the etching processes in which spatially localized effects appear, like notching and microtrenching.

4. Results

The evolution of the surface morphology of nanocomposite material (characterized by the etch rate ratio $s = 2$ and the abundance of one phase $p = 0.5$) during isotropic etching is shown in **Figure 2**. The roughness of a surface exposed to the isotropic etch process increases with processing time evolving from a flat structure to a highly rough structure.

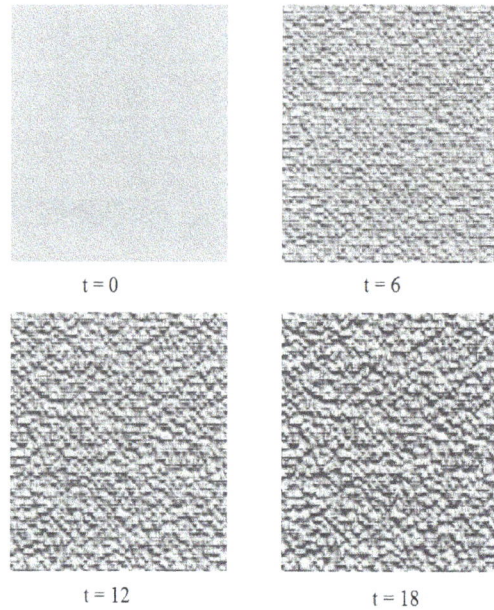

Figure 2. The evolution of the surface roughness during the etching time caused by isotropic etching of a nanocomposite material composed of two phases with the ratio of the etch rates is $s = 2$ and the abundance of one phase is $p = 0.5$.

The opposite process involves smoothing of a rough surface as illustrated in **Figure 3**. Obviously, application of the isotropic etching leads to decreasing the surface roughness over time. As can be noticed, an ideal smooth surface can not be obtained by employing isotropic etch process only.

The rms roughness R_q quantifies the surface roughness and is defined as a standard deviation of the height fluctuations along a profile. **Figure 4(a)** displays the time evolution of the rms roughness R_q illustrating the roughness dynamics. The surface was intentionally roughened by isotropic plasma etching to reach rms value of 2.5. For the roughening, rms roughness R_q changes with the time t according to β, where $\beta = 0.7945$. At the same time, the mean height of the surface during roughening decreases almost monotonically as shown in **Figure 4(b)**. The etching rates can be established from the slope of the curves.

During smoothing process (as can be observed form **Figure 5(a)** however, the rms roughness R_q decreases from 2.5 to 1.0 indicating that such procedure does not lead to an ideal smooth surface. In that case, the slope of the curve is different and rms roughness R_q also follows β scaling with $\beta = -1.4855$ describing decreasing of the rms roughness R_q. **Figure 5(b)** shows decreasing of the mean height of the homogenous surface during smoothing processes.

5. Conclusions

In this paper we have demonstrated applications of the level set method in performing three-dimensional simulations of the surface topology evolution. Both surface roughening and smoothing processes are simulated. The obtained simulation results indicate that the isotropic etch process has different influences on the surface composed of different materials. The evolution of surface roughness appeared during isotropic etching of material consisting of two randomly distributed phases with different etch rates. On the other hand, surface roughness could be reduced by the isotropic etching.

Explanation can be attributed to the facts that the isotropic etching will attack the substrate made of homogeneous or isotropic materials uniformly in all directions. Concerning the roughness kinetics, the rms R_q dramatically increases in time during roughening. Otherwise rms roughness R_q decreases in the case of smoothing process. Based on the simulation results, the time dependence of the rms R_q has been determined. Simulation results, presented here, apart from their theoretical relevance, have practical implications for surface treatments of various materials.

Figure 3. Images of smoothing of a roughed substrate made of homogeneous materials over the equidistant time intervals during isotropic etch process.

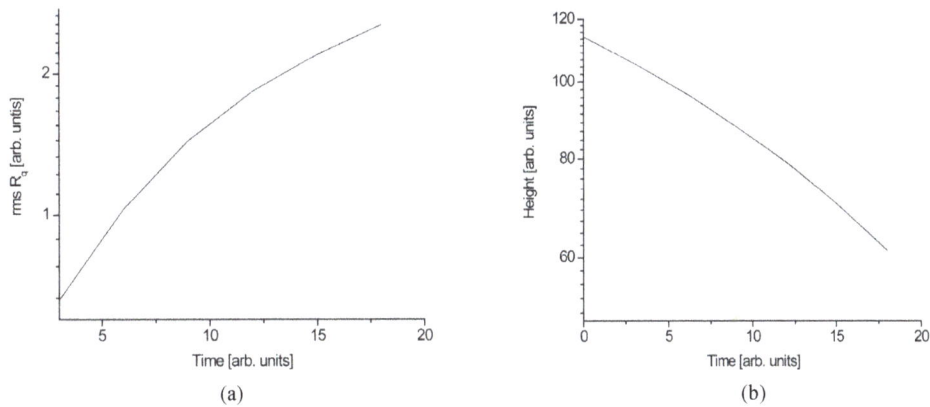

(a)   (b)

Figure 4. Dependence of: (a) the rms roughness Ropq and (b) the mean height of tophe surface on etching time of a nanocomposite material characterized by the relative concentration of two phases $p = 0.5$ and the etch rate ratio $s = 2$.

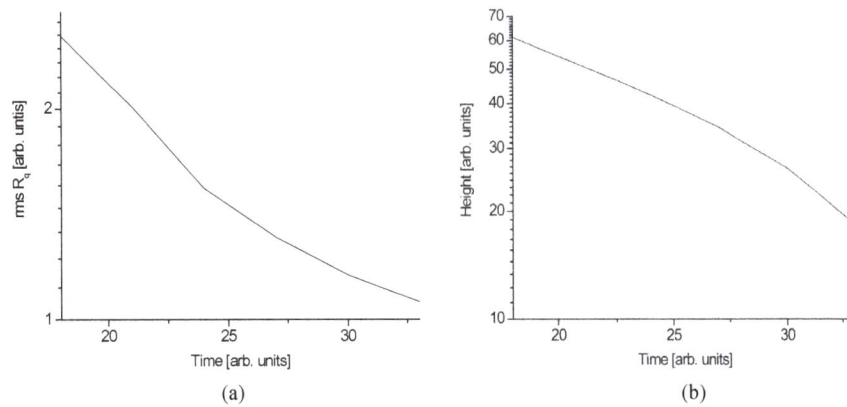

(a)   (b)

Figure 5. The time evolution of: a) the rms roughness R_q and b) the mean height of the roughed homogenous surface during smoothing process induced by isotropic etching.

Acknowledgements

This study was supported by Grant No. MNTRS141025 from the Ministry of Science of Serbia.

References

[1]   Bianconi, G. and Barabási, A.-L. (2001) Competition and Multiscaling in Evolving Networks. *Europhysics Letters*, **54**, 436-442. http://dx.doi.org/10.1209/epl/i2001-00260-6

[2]   Frost, F., Fechner, R., Ziberi, B., Flamm, D. and Schinler, A. (2004) Large Area Smoothing of Optical Surfaces by Low-Energy Ion Beams. *Thin Solid Films*, **459**, 100-105. http://dx.doi.org/10.1016/j.tsf.2003.12.107

[3]   Roy, A., Bhattacharjee, K., Lenka, H.P., Mahapatra, D.P. and Dev, B.N. (2008) Surface Roughness of Ion-Bombarded Si(100) Surfaces: Roughening and Smoothing with the Same Roughness Exponent. *Nuclear Instruments and Methods in Physics Research Section B: Beam Interactions with Materials and Atoms*, **266**, 1276-1280. http://dx.doi.org/10.1016/j.nimb.2007.10.045

[4]   Bhattacharjee, K., Bera, S., Goswami, D.K. and Dev, B.N. (2005) Nanoscale Self-Affine Surface Smoothing: Dependence on Ion Fluence and Initial Surface Roughness. *Nuclear Instruments and Methods in Physics Research Section B: Beam Interactions with Materials and Atoms*, **230**, 524-532. http://dx.doi.org/10.1016/j.nimb.2004.12.095

[5]   Yarin, A.L., Megairids, C.M., Mattia, D. and Gogotsi, Y. (2008) Smoothing of Nanoscale Roughness Based on the Kelvin Effect. *Nanotechnology*, **19**, Article ID: 365702. http://dx.doi.org/10.1088/0957-4484/19/36/365702

[6]   Guo, W. and Sawin, H.H. (2009) Review of Profile and Roughening Simulation in Microelectronics Plasma Etching. *Journal of Physics D: Applied Physics*, **42**, Article ID: 194014 http://dx.doi.org/10.1088/0022-3727/42/19/194014

[7]   Radmilović-Radjenović, M., Radjenović, B. and Petrović, Z.L.J. (2009) Application of Level Set Method in Simulation of Surface Roughness in Nanotechnologies. *Thin Solid Films*, **517**, 3954-3957. http://dx.doi.org/10.1016/j.tsf.2009.01.123

[8]   Twardowski, T.E. (2007) Introduction to Nanocomposite Materials: Processing Characterization. DesTech Publications, Lancaster.

[9]   Makabe, T. and Petrović, Z.L.J. (2006) Plasma Electronics. Taylor and Francis, New York.

[10] Chason, E., Mayer, T.M., Kellerman, B.K., McIlroy, D.T. and Howard, A.J. (1994) Roughening Instability and Evolution of the Ge(001) Surface during Ion Sputtering. *Physical Review Letters*, **72**, 3040-3043. http://dx.doi.org/10.1103/PhysRevLett.72.3040

[11] Eklund, E.A., Bruinsma, R., Rudnick, J. and Williams, R.S. (1991) Submicron-Scale Surface Roughening Induced by Ion Bombardment. *Physical Review Letters*, **67**, 1759-1762. http://dx.doi.org/10.1103/PhysRevLett.67.1759

[12] Krim, J., Heyvart, I., Haesendonck, D.V. and Bruynseraede, Y. (1993) Scanning Tunneling Microscopy Observation of Self-Affine Fractal Roughness in Ion-Bombarded Film Surfaces. *Physical Review Letters*, **70**, 57-60. http://dx.doi.org/10.1103/PhysRevLett.70.57

[13] Goswami, D.K. and Dev, B.N. (2003) Nanoscale Self-Affine Surface Smoothing by Ion Bombardment. *Physical Review B*, **68**, Article ID: 033401. http://dx.doi.org/10.1103/PhysRevB.68.033401

[14] Habenicht, S., Bolse, W., Lieb, K.P., Reimann, K. and Geyer, U. (1999) Nanometer Ripple Formation and Self-Affine Roughening of Ion-Beam-Eroded Graphite Surfaces. *Physical Review B*, **60**, Article ID: R2200. http://dx.doi.org/10.1103/PhysRevB.60.R2200

[15] Radjenović, B. and Radmilović-Radjenović, M. (2012) The Effects of Isotropic Etching on Roughening and Smoothing of Nanostructure. *Electronic Materials Letters*, **8**, 491-494. http://dx.doi.org/10.1007/s13391-012-2063-5

[16] Radjenović, B., Lee, J.K. and Radmilović-Radjenović, M. (2006) Sparse Field Level Set Method for Non-Convex Hamiltonians in 3D Plasma Etching Profile Simulations. *Computer Physics Communications*, **174**, 127-132. http://dx.doi.org/10.1016/j.cpc.2005.09.010

[17] Radjenović, B. and Radmilović-Radjenović, M. (2012) The Effect of Different Etching Modes on the Smoothing of the Rough Surfaces. *Materials Letters*, **86**, 165-167. http://dx.doi.org/10.1016/j.matlet.2012.07.068

[18] Radjenović, B. and Radmilović-Radjenović, M. (2013) Three-Dimensional Simulations with Fields and Particles in Software and Inflector Designs. *Journal of Software Engineering and Applications*, **6**, 390-395. http://dx.doi.org/10.4236/jsea.2013.68048

[19] Radjenović, B. and Radmilović-Radjenović, M. (2014) The Implementation of the Surface Charging Effects in Three-Dimensional Simulations of SiO_2 Etching Profile Evolution. *Engineering*, **6**, 1-6.

9

Neural Network Based on SET Inverter Structures: Neuro-Inspired Memory

Bilel Hafsi[1,2]*, Rabii Elmissaoui[3], Adel Kalboussi[1]

[1]IEMN Laboratory, University of Lille1, Avenue Poincaré, 59652 Villeneuve d'Ascq Cedex, France
[2]Microelectronics and Instrumentation Laboratory, Faculty of Sciences of Monastir, University of Monastir, Monastir, Tunisia
[3]Research Unit on Study of Industrial Systems and Renewable Energies, National Engineering School of Monastir, Monastir, Tunisia
Email: *Bilel.Hafsi@iemn.univ-lille1.fr

Academic Editor: Yarub Al-Douri, University Malaysia Perlis, Malaysia

Abstract

This paper presents a basic block for building large-scale single-electron neural networks. This macro block is completely composed of SET inverter circuits. We present and discuss the basic parts of this device. The full design and simulation results were done using MATLAB and SIMON, which are a single-electron tunnel device and circuit simulator based on a Monte Carlo method. Special measures had to be taken in order to simulate this circuit correctly in SIMON and compare results with those of SPICE simulation done before. Moreover, we study part of the network as a memory cell with the idea of combining the extremely low-power properties of the SET and the compact design.

Keywords

Single-Electron, Neuron, Synapse, Inverter, Neural Network, Single-Electron Memory, Perceptron, MATLAB, SIMON

1. Introduction

Single-Electron Devices (SEDs) have attracted much attention since the 1980s when it was that they could be used to fabricate memory devices, low-power logic devices and high-performance sensors. Among SEDs, Single-Electron Transistors (SETs) have been the most studied device, because it was seen as a potential successor for

---

*Corresponding author.

present metal-oxide-semiconductor MOS transistor [1].

This device is made of an island connected through two tunneling junctions to a drain and a source electrode. The distance between the two conductors is in order of few nanometers, and electron can possibly tunnel through the insulator [2], when the absolute voltage difference across the junction is decreased due to the event [3].

The idea of combining Single-Electron Transistors (SETs) in neural networks architectures has raised considerable interest over recent years because of its potentially unique functionalities. The rationale behind neural networks and SET devices is the possibility of taking advantages of both technologies [3] [4]: the high gain, speed of the SETs, learning capacity and problem-solving abilities of the Artificial Neural Networks (ANNs), in order to lead to Single-Electron Memories (SEMs) with low-power consumption, high density of the SETs and brain behavior qualities.

In this paper, we present simulations of the two main parts of an elementary neural network based on SET (Perceptron) using SIMON [5] simulator and MATLAB [6]. The purpose of our work was therefore the presentation of the advantages to simulate such an elementary neural network by a Monte Carlo simulator [5] which features a graphical circuit editor embedded in a graphical user interface as well as the simulation of co-tunnel events and a single-step interactive analyses mode. The idea then was to present this neural network as a model of a "smart" SET/SEM device.

2. Single-Electron Devices

2.1. Single-Electron Transistor

Since the 1980s, developments in both semiconductor technology and theory have lead to a completely new field of research focusing on devices whose operation is based on the discrete nature of electrons tunneling through thin potential barriers. These devices which exhibit charging effects including Coulomb blockade are the single-electron devices [7].

The single-electron tunneling technology is the most promising future technology generations to meet the required increase in density and performance and decrease in power dissipation [8]. The main component of SET is the tunnel junction that can be implemented using silicon or metal-insulator-metal structures, GaAs quantum dots.

The tunnel junction can be thought of as a leaky capacitor then electrons are blockaded, the blockade can be overcome and a current flows if the junction voltage V is above the threshold given by the charging energy. Then the junction behaves like resistor.

Recent research in SET gives new ideas which are going to revolutionize the random access memory and digital data storage technologies.

2.2. Single-Electron Memories

During the last few years the development of new memory devices has attracted much attention These single-electron memories provides a great potential for SETs to be used in the design of nonvolatile Random Access Memory (RAMs), for example, for mobile computer and communication applications [9].

Several single-electron memory cells have been proposed in the literature such us the single-electron flip-flop, the Multiple Tunnel Junction (MTJ) proposed by Nakazato and Ahmed [10], the Q0-independent memory and others.

Each memory differs from the other by such properties as the complexity of the architecture, the dependence of background charges and the operating temperature.

Neural networks are suitable architectures for SET devices and the combination of the two technologies has many advantages. Because they are too small SET devises can yield to a powerful neural network with extremely low power dissipation.

3. Neural Network

3.1. The Neural Biological Model

The human brain contains more than a billion of neuron cells, each cell works like a simple processor. The massive interaction between all cells and their parallel processing only makes the brain's abilities possible [11].

By definition, "neurons are basic signaling units of the nervous being in which each neuron is a discrete cell whose several processes are from its cell body" [12].

Figure 1 shows the biological model of the neuron which has four main regions to its structure: The cell body, the dendrites, the axon and the heart of the cell "the cell body" which contains the nucleolus and maintains protein synthesis.

The dendrites collect synaptic potentials from other neurons and transmit them to the soma. The soma performs a non-linear processing step (called fire model): if the total synaptic potential exceeds a certain threshold then a spike is generated [13]. A spike is transmitted to other neurons via synapses, the points of connection between two neurons.

Synapses dominate the architecture of the brain and are responsible for structural plasticity [14], robustness of the brain and electrochemical communication takes place at these junctions.

3.2. Artificial Neural Model

3.2.1. Perceptron

The most well-known neural processing element for artificial analog neural networks is one of the Perceptron models of Rosenblatt (1962) [15]. This model is an extension to the older neural model of McCulloch and Pitts (1943) [16].

The perceptron forms the basic neural processing element for the artificial analog neural networks, a single layer neural network with weights and biases trained to produce a correct target vector [17]. It represents the combination of a neuron and n number of synapses. **Figure 2** shows the complete perceptron.

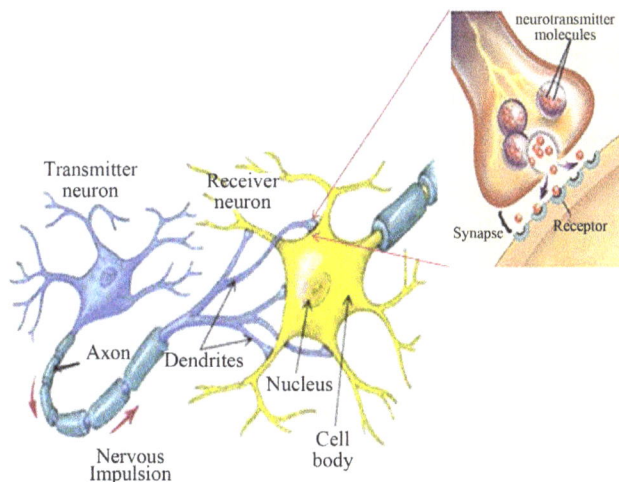

Figure 1. Biological neuron model.

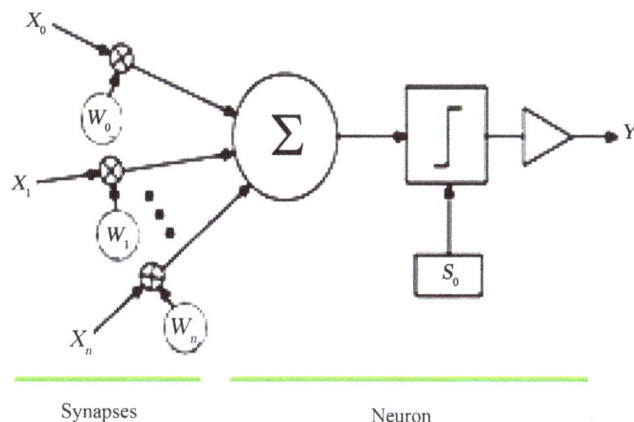

Figure 2. The complete perceptron, consisting of multiple synapses and a binary neuron.

In the literature, many types of neural networks hardware based on single-electron transistor can be found. Our work is based on a complete two-input perceptron implemented in SET technology by Van de Haar and Hoekstra [18], which is composed of two important stage, a SET inverter performs the function of the synapse while another one perform the activation function of the neuron.

SIMON simulations are shown using an input test-pattern that enters all possible states of the neural network. We should know that the various coupling capacitances between device elements, such as the capacitance of tunnel junction, are extracted from SIMON simulator after several trials using stationary simulation. In fact, for every such time step "even number tunnel events are simulated and averaged".

3.2.2. The Synapse

The synapse is the point of connection between neurons, it allows to store the analog weight value W_i and to multiply them by digital input signal X_i.

The basic principle of a synapse is to perform the following two functions.
- Modify connection strength.
- Store weight.

In the SET inverter, the analog values of W_i and X_i are represented by the voltages V_{Wi} and V_{Xi}. **Figure 3** shows the SET inverter structure based synapse where V_{W0} is Weight value.

The digital input signals V_{X0} are set by the supervisor or by the output of another perceptron. The analog weights W_i and the digital inputs signals X_i are normalized between 0 and 1. In order to translate these digital values to voltages in the SET inverter structure, we simulate the transfer function using SIMON simulator with the parameters set as chosen in **Figure 3**.

The transfer function of the SET inverter based synapse is shown in **Figure 4**. V_{X0} is set to 0 and 6 mV in order to obtain this simulation result.

The transfer function is approximately linear when the input voltage V_{W0} evolves in the region 3.9 mV $< V_{W0} <$ 5.8 mV. In this case the structure has an inversion operation, as a consequence the analog weight value $W_i = 1$ is mapped to 3.9 mV and $W_i = 0$ is mapped to 5.8 mV.

The translation **Table 1** shows the translation of parameter W_i into voltage V_{Wi}. Note that parameter X_i can only yield 0 or 1 in other words V_{Xi} can be 0 or 6 mV.

3.2.3. The Neuron

The task of the neuron is to sum all weighted analog input signals X_i and classify the summed signals by means of an activation function. The activation function is a limiting function and different activation functions are found, in fact, the most commonly used functions, are the hard-limiter and the sigmoid function [17].

The neuron should perform the following two functions:

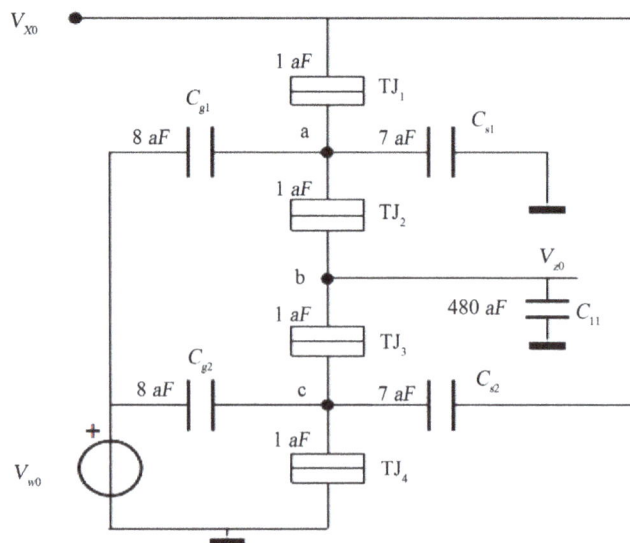

Figure 3. Synapse based on the SET inverter structure.

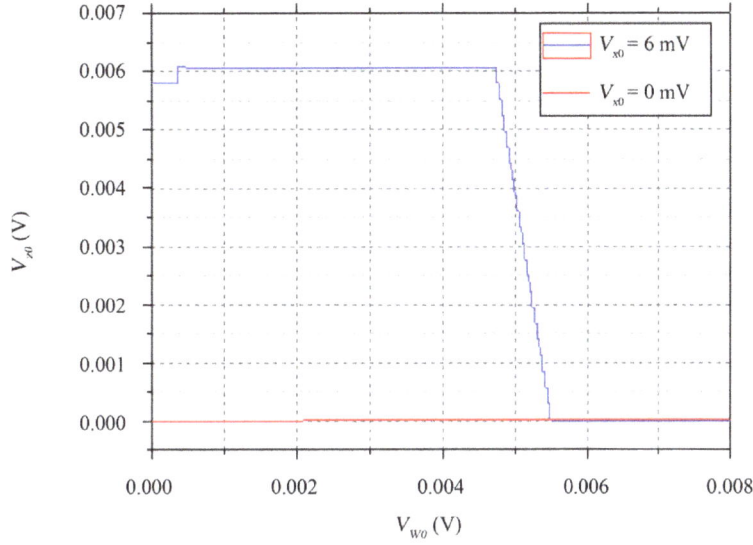

Figure 4. Transfer function of the SET inverter based synapse.

Table 1. Translate table W_i on voltages V_{wi}.

W_i	0	0.1	0.2	0.3	0.4	0.5	0.6	0.7	0.8	0.9	1
V_{wi} (mV)	5.5	5.42	5.34	5.26	5.18	5.1	5.02	4.94	4.86	4.78	4.7

- Gather input signals.
- Classify the results.

In **Figure 5**, a two-input neuron based on a SET inverter structure is shown. The coupling capacitors C_{C0} and C_{C1} perform the summation function and the activation function is carried out by the SET inverter structure [17].

The parameter set for the neuron is scaled with a factor 0.8 with respect to the parameter set of the synapse, in order to obtain output signals in the same voltage range.

Two important parameters can be distinguished. The first parameter is the SET junction capacitance value C_T, this parameter describes the behavior of a SET circuit operating in the single-electron transport regime. The second parameter is the tunnel resistance R_T, which is needed to describe the behavior of a SET circuit in the high current regime.

The neuron's activation function is a hard-limiter function, as it is shown in **Figure 6** the threshold voltage, which is the input voltage V_g for which the output voltage changes, is set to 3.3 mV approximately in this configuration. Which correspond to the threshold parameter $S_0 = 0.55$.

3.2.4. Simulation of the Perceptron

The perceptron developed by Rosenblatt (1957) is an artificial neuron [15]-[17], whose activation function is a step function, it consists of multiple synapses and a neuron which is designed using the SET inverter structure. If the adder output is equal or greater than the threshold Z, then the neuron fires. If the adder output is smaller than the threshold Z then there is no neuron output.

In Equation (1), the output signal Y is expressed in terms of the input variables X_i and W_i for a general case:

$$Z = \sum_{i=1}^{n} w_i \cdot x_i - \theta \tag{1}$$

$$Y = \begin{cases} 1 & \text{if } Z \geq 0 \\ 0 & \text{if } Z < 0 \end{cases}$$

Figure 7 shows a complete two-input perceptron, which is based on SET inverter structures. This perceptron was first published at the ICES2003 [17] [18] and simulated using P-Spice. Therefore, in our work special

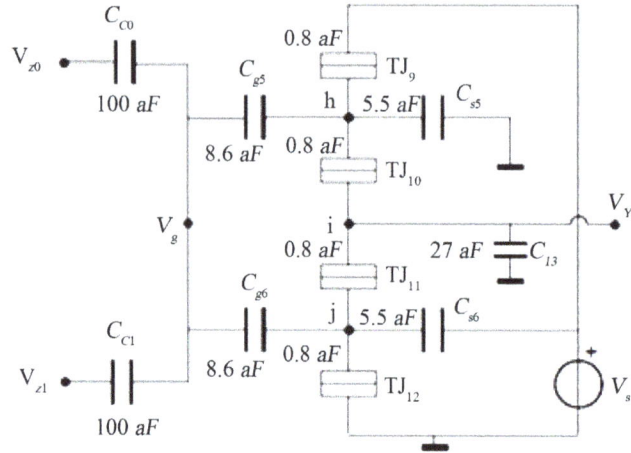

Figure 5. Two-input SET inverter based neuron.

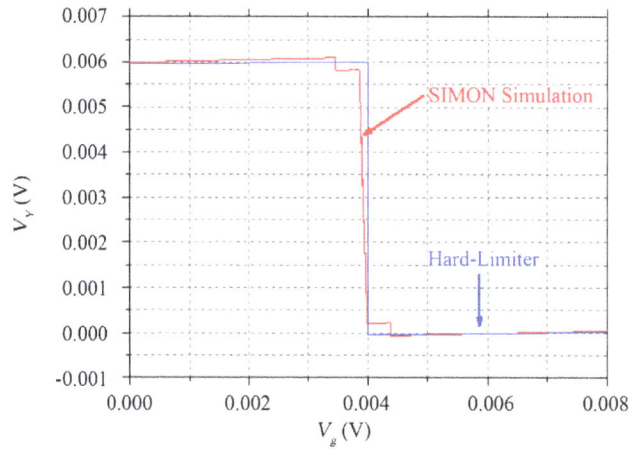

Figure 6. Transfer function of the SET inverter structure based neuron.

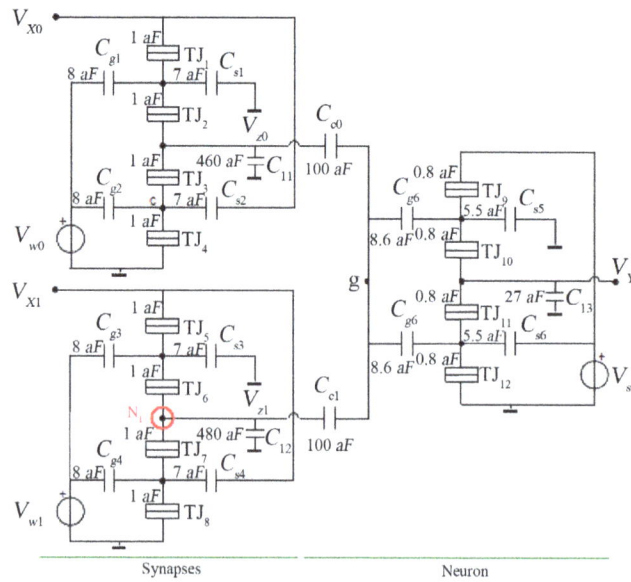

Figure 7. Two-input perceptron, based on SET inverter structures.

measures had to be taken in order to simulate this circuit correctly in SIMON.

Before this perceptron can be simulated, the desired output has to be defined. In order to check if the output signal is the desired signal, a good test-set of input signals has to be created using MATLAB taking into account Equation (1). In **Table 2** a test-set of input signals and the desired output of a two-input perceptron is shown.

The threshold level S_0 is assumed to be 0.35 (which is an arbitrary choice). Then the operation of the perceptron is fully tested using SIMON simulator and results are presented in **Table 3**. Comparing with the results obtained with P-Spice in [17] the output signal V_Y agrees with the desired output signals obtained from the test patterns of **Table 2**. However, these outputs are inverted with respect to the outputs obtained using MATLAB. In order to obtain a non-inverted output signal V_Y, an extra inversion stage (identical to the neuron stage of **Figure 5**) can be added to the perceptron of **Figure 7**.

A very interesting idea that brings multiple benefits to nanotechnology is to have a multi-level memory to a single electron [19]. In this context, we have tried to highlight the discrete of the charge of the electron node memory N_1.

The shapes of control signals (the weight V_{Wi} & the input V_{Xi}) are very important to observe the charge evolution "Q" versus time. We have considered in this work that the architecture works in the same way as a memory.

Figure 8 shows the shape of the charge $Q(t)$ in the island which allows us to count the number of electrons stored in the memory node (node 1) and to know the instant t when writing or erasing.

At first, the input voltage V_{X1} is zero and there is no electron at the quantum dot N_1. By applying a negative input voltage $V_{X1} = -6$ mV, a certain number of electrons are stored inside the quantum dot and no more electrons will be transported out of there. Because of the Coulomb blockade effect and the potential barrier, the electron cannot flow back to the ground.

This result shows that the number of electrons in the memory node (the red curve for $V_{w1} = 9$ mV) after the charging increases when the input voltages increase until it reaches 5 electrons.

After returning V_{X1} to zero electrons are not lost, this property provides a great potential for the network to be used in the design of nonvolatile RAMs.

The weight V_{W1} plays the role of parameters which are adjusted at the training process. As we can see in **Figure 8** the number of charge stored in node N_1 depends of the value of V_{W1}. When the value of V_{W1} is equal to 9 mV five electrons are stored in the node N_1 but when we decrease the value of the weight the number of the writing electron increase until it reach $\approx$19 electron for $V_{w1} = 1$ mV. It can be seen that this memory can be con-

Table 2. Test-set of input signals and the desired output.

X_0	X_1	W_0	W_1	Y
0	1	1	0	0
0	1	0.3	0.8	1
1	0	0.3	0.8	0
1	1	0.4	0.4	1

Table 3. A test-set of input signals.

V_{X0}	V_{W0}	V_{X1}	V_{W1}	P-Spice Simulation V_Y	SIMON Simulation V_Y
0 mV	4.7 mV	0 mV	4.7 mV	6.13 mV	5.07 mV
0	1	0	1	0	0
0 mV	5.26 mV	6 mV	4.86 mV	148 μV	150 μV
0	0.3	1	0.8	1	1
6 mV	5.26 mV	0 mV	4.86 mV	6.01mV	4.9 mV
1	0.3	0	0.8	0	0
6 mV	5.18 mV	6 mV	5.18 mV	198 μV	196 μV
1	0.4	1	0.4	1	1

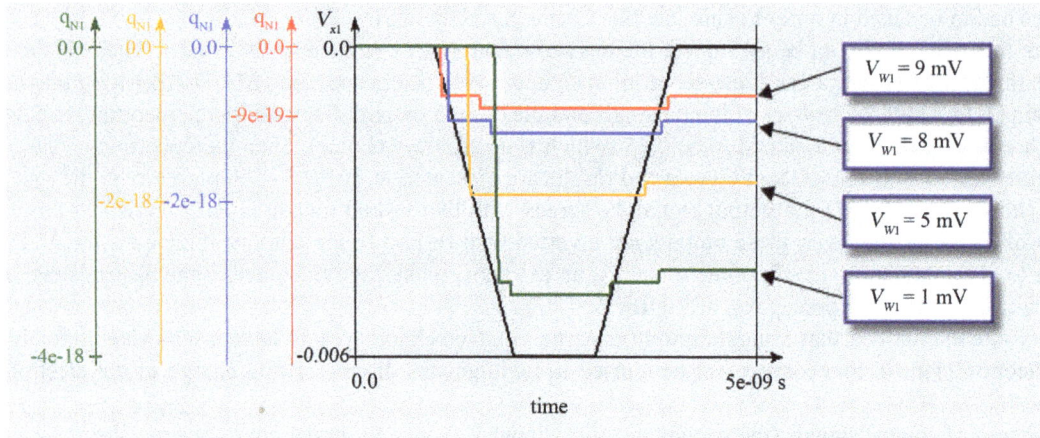

Figure 8. Simulation results of the charge and influence of the value of the weight V_{W1} and the input V_{X1}.

sidered as a smart memory because we can manipulate the number of charges stored in the memory by a learning algorithm.

This learning algorithm is used to train the network for a particular purpose, in our case achieve extraordinary storage densities at an extremely low power consumption.

4. Conclusions

In this paper a complete two-input perceptron, which is based on SET inverter structures, was presented, discussed and simulated using SIMON simulator and MATLAB. It was also proved that this architecture works as a random access memory. Two features are important in this memory: the first is its low-power consumption and the second is its dependence of the charge stored to the variation of the weight.

The advantages of such Monte Carlo simulator "SIMON" which is designed to solve capacitance systems that contain tunnel junctions are the possibility to investigate the evolution of charge versus time by the main of simulation process based on orthodox theory.

This work opens the prospect of smart memory in which one can manipulate the number of bits stored by the mean of learning algorithm.

References

[1]   Beaumont, A. (2009) Room Temperature Single-Electron Transistor Featuring Gate-Enhanced On-State Current. *IEEE Electron Device Letters*, **31**, 249.

[2]   Mandal, S. (2013) Single Electron Transistor. *International Journal of Innovations in Engineering and Technology (IJIET)*, **2**.

[3]   Van de Haar, R., *et al.* (2003) Simulation of a Neural Node Using SET Technology. Springer-Verlag, Berlin, Heidelberg, 377-386.

[4]   Guo, L., Leobandung, E. and Chou, S.Y. (1997) A Silicon Single Electron Transistor Memory Operating at Room Temperature. *Science*, **275**, 649-651. http://dx.doi.org/10.1126/science.275.5300.649

[5]   Wasshuber, C. (1998) SIMON2.0. Institute for Micro Electronics, TU, Vienna.

[6]   MATLAB R2009b. http://www.mathworks.com

[7]   Scholze, A. (2000) Simulation of Single-Electron Devices. Ph.D. Thesis, Swiss Federal Institute of Technology, Zurich.

[8]   Paulthurai, A. and Dharmaraj, B. (2012) Single Electron 2-Bits Multiplier. *International Journal of Computer Applications*, **42**, 17-20. http://dx.doi.org/10.5120/5680-7719

[9]   Boubaker, A., Krout, I. and Kalboussi, A. (2011) Study and Modelling Hybrid MTJ/Ring Memory Using Simon Simulator.

[10] Nakazato, K. and Ahmed, A. (1995) The Multiple-Tunnel Junction and Its Application to Single-Electron Memory and Logic Circuits. *Japanese Journal of Applied Physics*, **34**, 700-706. http://dx.doi.org/10.1143/JJAP.34.700

[11] Hafsi, B., Boubaker, A., Krout, I. and Kalboussi, A. (2013) Simulation of Single Electron Transistor Inverter Neuron:

Memory Application. *International Journal of Information and Computer Sciences (IJICS)*, **2**, 8-15.

[12] Bajpai, S., Jain, K. and Jain, N. (2011) Artificial Neural Networks. *International Journal of Soft Computing and Engineering (IJSCE)*, **1**, 28.

[13] Gürcan, Ö., Bernon, C. and Türker, K.S. (2012) Towards a Self-Organized Agent-Based Simulation Model for Exploration of Human Synaptic Connections. arXiv:1207.3760.

[14] Kuzum, D., Jeyasingh, R.G.D., Lee, B. and Philip Wong, H.S. (2011) Nanoelectronic Programmable Synapses Based on Phase Change Materials for Brain-Inspired Computing. *Nano Letters*, **12**, 2179-2186.

[15] Rosenblatt, F. (1962) Principles of Neurodynamics: Perceptrons and the Theory of Brain Mechanisms. Spartan, New York.

[16] McCulloch, W.S. and Pitts, W.H. (1943) A Logical Calculus of the Ideas Immanent in Nervous Activity. *Bulletin of Mathematical Biophysics*, **5**, 115-133. http://dx.doi.org/10.1007/BF02478259

[17] Van de Haar, R. (2004) Simulation of Single-Electron Tunnelling Circuits Using SPICE. Ph.D. Dissertation, Delft University of Technology, Delft.

[18] Van de Haar, R. and Hoekstra, J. (2003) Simulation of a Neural Node Using SET Technology. *Proceeding ICES'03, Proceedings of the 5th International Conference on Evolvable System: From Biology to Hardware*, Springer-Verlag, Berlin, Heidelberg, 377-386.

[19] Boubaker, A., Sghaier, N., Souifi, A. and Kalboussi, A. (2010) Simulation and Modeling of the Write/Erase Kinetics and the Retention Time of Single Electron Memory at Room Temperature. *Journal of Semiconductor Technology and Science*, **10**, 143-151.

10

Research Advances in Photocatalysis of Inorganic Hollow Spheres

Ting Tian[1], Jing Hu[1,2]*, Zuobing Xiao[1,2]

[1]School of Perfume and Aroma Technology, Shanghai Institute of Technology, Shanghai, China
[2]Shanghai Research Institute of Fragrance & Flavor Industry, Shanghai, China
Email: *hujing@sit.edu.cn

Academic Editor: Yarub Al-Douri, University Malaysia Perlis, Malaysia

Abstract

Inorganic hollow spheres have shown their superiority in photocatalytic area due to the large specific surface area, controllable structure and their own special optical, electrical, magnetic properties. According to the classification of inorganic hollow spheres as photocatalysts, recent research progress and application status have been summarized in this paper. At last, the future developments of inorganic hollow spheres in photocatalytic field have been discussed.

Keywords

Inorganic Hollow Spheres, Photocatalysis, Application, Progress

1. Introduction

Hollow spheres, also known as 0D nanomaterial, were spheroidal aggregates with hollow structure assembled by their 1D nanomaterial. Due to their characteristics of controlled morphology, uniform size, large specific surface area and low density, extensive attention has been paid so far in material area [1]-[5]. Particularly, inorganic hollow spheres with special optical, electrical, magnetic, mechanical and catalytic properties have been widely used in bio-pharmaceutical, catalyst, carrier, controlled release, photonic crystal, electrochemistry and environmental protection [6]-[11]. Under the urgent situation of environmental pollution, inorganic hollow spheres as photocatalyst turned into a way to solve environmental problems. Inorganic hollow spheres have peculiar hierarchical porous structure, which makes reactant molecules easily transfer to the active sites of porous wall to improve their photocatalytic efficiency. Moreover, the hollow structure allows multiple reflections of ul-

---

*Corresponding author.

traviolet and visible light in inner hole to enhance utilization of light [12]. In this paper, according to the classification of photocatalytic inorganic hollow spheres, recent research progress has been introduced and application status has been concluded. The future developments of inorganic hollow spheres in photocatalytic field have been discussed.

2. Metal Oxide Hollow Spheres

2.1. Titanium Dioxide (TiO_2)

The valence and conduction band of TiO_2 is composed of the filled 2p orbitals of oxygen and 3d, 4s and 4p orbitals of titanium, respectively. Its energy gap is about 3.2 eV, which makes TiO_2 being the most promising photocatalyst. The specific band potential gives its strong oxidizing property, high chemical durability and photoelectric conversion efficiency, as well as price superiority. Furthermore, the large specific surface area and low density of hollow structure can effectively enhance photocatalytic activity of TiO_2 [13]. The TiO_2 hollow spheres obtained by Li and co-workers can completely remove azo-dye Rhodamine B (Rh B) under visible light irradiation in 240 min, which is only 50% for DeGussa P25 under the same experimental conditions [14]. However, TiO_2 can be excited only under ultraviolet light (less than 5% of the full solar spectrum). Besides, easily recombination of photogenerated electrons and holes and low interfacial charge transfer rate into TiO_2 lead to reduction of photocatalytic efficiency. Therefore, three principal methods have been used to enhance photocatalytic activity of TiO_2: ameliorate crystal composition, enlarge specific surface area and modify surface of TiO_2 hollow sphere. In the process of crystal growth, the more exposed of {001} and {110} facets which have high interface energy means higher photocatalytic activity [15]-[17]. Wang et al. [18] prepared TiO_2 hollow sphere with sixty percent exposure of {001} facets, which has higher photocatalytic activity than that of commercial photocatalyst P25. **Figure 1** shows XRD pattern , SEM images and TEM images of anatase TiO_2 hollow microspheres with exposed {001} facets. Jiao et al. [19] discovered that the interface energy of {116} facets was similar to that of {110} facets through X-ray diffraction analysis. The photocatalytic property of TiO_2 hollow sphere consisting of highly active {116} plane-oriented crystallites exceed that of common TiO_2 hollow sphere, because curved {116} facets can multi-reflect incident light to enhance capture rate. The TiO_2 hollow spheres with special morphology were obtained to add specific surface area thus improve photocatalytic performance such as multi-shell structure [20] [21]. Tao et al. [22] prepared flower-like TiO_2 hollow spheres assembled by nanosheet, which has a higher surface area (65 m^2/g) than that value of commercial TiO_2 (7 m^2/g). Therefore, Degradation performance of methyl orange (MO) by the flower-like TiO_2 hollow sphere is better than that of commercial TiO_2.

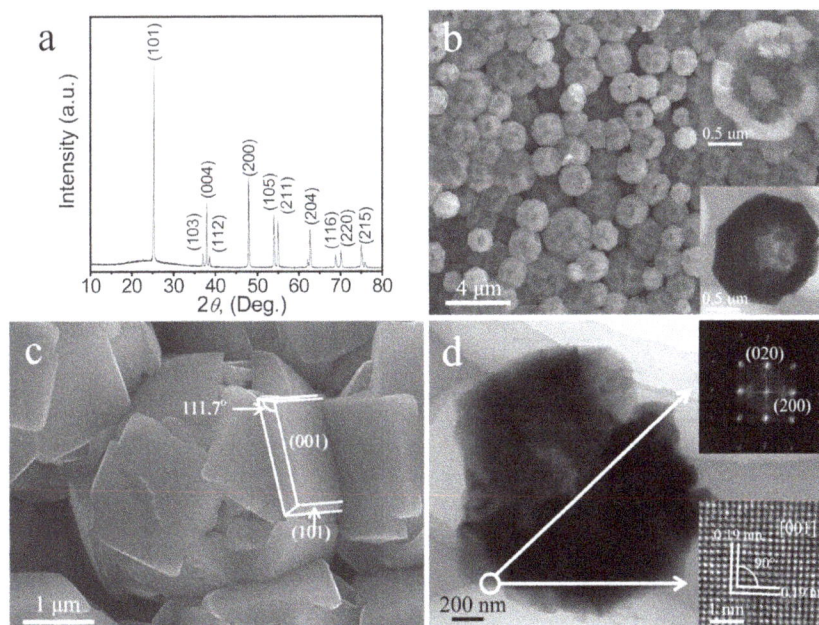

Figure 1. XRD pattern (a), SEM images (b) (c) and TEM images of anatase TiO_2 hollow microspheres with exposed {001} facets (d) [18].

In recent years, surface modification of TiO_2 hollow spheres has been widely researched to improve their photocatalytic efficiency. 1) Doping metal ion into TiO_2 lattice to create defect position can capture electron from conduction band, reduce recombination of photogenerated electrons and holes in photocatalytic process and accelerate interface charge-transfer to raise photocatalytic activity. Transition metal, noble metal and rare-earth metal can be incorporate elements, such as vanadium, platinum, cerium, neodymium, etc. Vanadium-doped TiO_2 hollow spheres prepared by Liu et al. [23] can quickly degrade methylene blue under visible-light irradiation to show their excellent photocatalytic activity and renewability. Feng and coworkers [24] prepared platinum doped TiO_2 hollow spheres using carbon spheres as templates through hydrothermal precipitation method and adding hydrazine hydrate to reduce chlomplatinie acid. When platinum-doped content is 2.0%, the decolorization rate of obtained TiO_2 hollow spheres can achieve 100% under 2 hour ultraviolet light irradiation. Wang and coworkers [25] [26] use carbon spheres to prepare cerium and neodymium doped TiO_2 hollow spheres with dope concentration of 4% and 3.9% and apply them to dye decomposition. Results show that the apparent rate constant of doped TiO_2 hollow spheres in degrading dye is 31 and 9 times as that of P25. 2) Non-metal elements also can be doped, such as boron [27], nitrogen [28] [29], carbon [30] [31], fluorine [32], etc. Non-metal ion possess relatively high energy orbit. Once doped it can replace some part of oxygen in the TiO_2 lattice, thus bring in new level to lower energy gap and extend absorption wavelength. Yu et al. [33] prepared trifluoroacetic acid modified TiO_2 hollow spheres by one-pot hydrothermal treatment using titanium sulfate as titanium source, the photocatalytic activity of which manifested 2 times higher than that of P25. In order to further improve photocatalytic activity of TiO_2 hollow spheres, two or more kinds of ion can be doped to simultaneously provide electrons and holes trap to effectively suppress the recombination. Wang et al. [34] prepared cerium and nitrogen co-doped TiO_2 hollow spheres owning enhanced visible light photocatalytic performance in dye X-3B decolorization because nitrogen decreased the energy gap and photogenerated electron of TiO_2 valence band can transfer to the 4f orbit of cerium to inhibiting recombination. 3) Noble metal particles deposited onto the surface of TiO_2 hollow spheres can take advantage of their surface plasmon resonance to extend absorption wavelength to visible light region, therefore enhance photocatalytic performance [35] [36]. Xiang et al. [37] combine microwave and hydrothermal treatment to prepare TiO_2 hollow spheres with silver (Ag) nanoparticles deposited onto the surface, which have obvious absorption in the wavelength region from 400 to 600 nm and show better photocatalytic activity than pure TiO_2 and commercial Degussa P25 powders. The single-crystalline anatase TiO_2 hollow nano-hemispheres with bimetallic Ag/Pt nanoparticles uniformly loaded on both interior and exterior of the nano-hemispheres prepared by Jiang and co-workers exhibited excellent photocatalytic ability in the degradation of Rh B/ciprofloxacin (RhB/CIP) and hydrogen generation [38]. **Figure 2** shows H_2 production rates of TiO_2, $Ag@TiO_2$, $Pt@TiO_2$ and $Ag/Pt@TiO_2$ photocatalysts with CH_3OH as the sacrificial reagent.

2.2. Other Metal Oxide

Oxide of d region elements also own good photocatalytic activity, such as vanadium pentoxide [39], manganese dioxide [40], nickel oxide [41], tantalum oxide [42], ferric oxide (Fe_2O_3) [43], copper (I) oxide (Cu_2O) [44], tungsten oxide (WO_3) [45] [46], zinc oxide (ZnO) [47] [48], etc. Patrinoiu et al. [49] impregnated carbonaceous

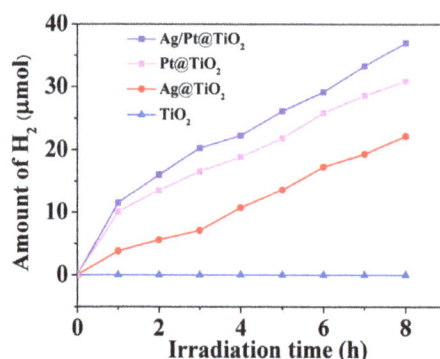

Figure 2. H_2 production rates of TiO_2, $Ag@TiO_2$, of and $Ag/Pt@TiO_2$ photocatalysts with CH_3OH as sacrificial reagent [38].

spheres templates with hydrated zinc acetate and then removed templates by thermal treatment to obtain zinc oxide hollow spheres. Under ultraviolet light irradiation, the phenol mineralization rate by zinc oxide hollow spheres can exceed 80%, which notably superior to commercial zinc oxide (Merck). **Figure 3** shows effects of ZnO hollow sphere (HS) and commercial product (Merck) nature on degradation of phenol under same irradiation conditions. Cao and co-workers [50] prepared hexagonal α-Fe$_2$O$_3$ hollow spheres assembled by nanosheets through a microwave-assisted solvothermal route using ferric trichloride hexahydrate, sodium hydroxide and sodium dodecyl benzene sulfonate as raw materials and ethylene glycol as solvent. Degradation percentage of salicylic acid by the obtained hollow spheres under ultraviolet light in one hour can reach 60% and overtop 40% after using 2 times. Li *et al.* [51] prepared Nb$_2$O$_5$ hollow nanospheres with high surface energy (001) planes via ostwald ripening process, which owned high thermal stability, strong intensity of blue emission and efficiently split water under visible light irradiation.

Moreover, oxide of some p region elements also have photocatalytic activity, such as tin oxide (SnO$_2$), indium oxide (In$_2$O$_3$), etc. Manjula *et al.* [52] utilize glucose as structure-directing agent to prepare porous tin oxide hollow spheres, which not only can photodegrade dye but also can be reused. The indium oxide hollow spheres gained by Li *et al.* [53] using emulsion vesicles as templates can photodecompose Rh B under ultraviolet light irradiation.

Surface modification still can be employed to enhance photocatalytic property of the aforementioned metal oxide hollow spheres. Ma *et al.* [54] doped tungsten oxide hollow spheres with silver-silver chloride to further improve their photocatalytic activity. Rahimi *et al.* [55] successfully prepared Ba-Cd-Sr-Ti doped ferroferric oxide (Fe$_3$O$_4$) nanohollow spheres via a simple solvothermal method without any templates. The photocatalytic degradation rate of congo red solution by this nanohollow spheres under visible light irradiation can reach to 99.5% when pH is 6.

2.3. Composite Metal Oxide

When two or more kinds of semiconductor form a compound system with fixed microstructure, it can restrain the recombination of photogenerated electrons and holes to significantly improve the photocatalytic performance. Hence, composite metal oxide hollow spheres owned a more excellent photocatalytic activity than that of single metal oxide spheres. Composite metal oxide can be divided into metal oxide modification and non-metal oxide modification. 1) Two kinds of metal oxide can be mutually compounded, such as ZnO [56], Fe$_3$O$_4$ [57], zirconium oxide (ZrO$_2$) [58] composited with TiO$_2$; In$_2$O$_3$ [59], SnO$_2$ [60], CuO [61], GeO$_2$ [62] composited with ZnO; Fe$_2$O$_3$ composited with MnO [63]; bismuth oxide (Bi$_2$O$_3$) composited with WO$_3$ [64], V$_2$O$_5$ [65] [66], etc. When WO$_3$ which has a low band gap combined with TiO$_2$, it suppresses recombina-

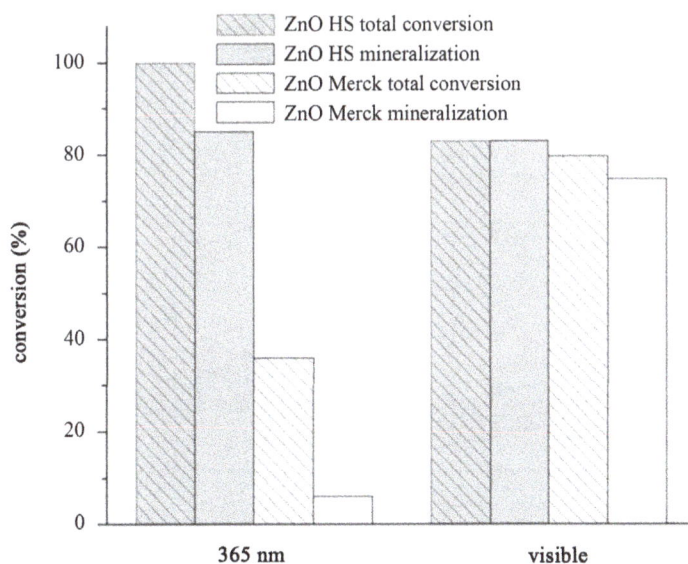

Figure 3. Effects of ZnO hollow sphere (HS) and commercial product (Merck) nature on degradation of phenol under same irradiation conditions [49].

tion of photogenerated electrons and holes on hollow spheres surface as well as broadens photo-responsive scope to enhance utilization rate for optical energy. Tian *et al.* [67] prepared hierarchical flower-like Bi_2MoO_6 hollow spheres via a solvothermal process in the presence of ethylene glycol, which can remove over 95% Rh B within 2 h under visible-light irradiation. Li *et al.* [68] prepared bismuth tungstate (Bi_2WO_6) hollow spheres using polystyrene particles as the template, with Bi_2O_3 deposited on their surface and sequentially calcination to obtain double-shell Bi_2O_3-Bi_2WO_6 hollow spheres. The internal electric field of p-n junction formed by close contact between p-type Bi_2O_3 and n-type Bi_2WO_6 can speed up separation of photogenerated charge to boost photocatalytic performance, which makes this composite hollow spheres can completely decompose Rh B in 3 hours under visible light irradiation. An Ag_2ZnGeO_4 photocatalyst was obtained by Zhang and co-workers [69] via an ion-exchange reaction between amorphous Zn_2GeO_4 suspension and Ag ions solutions. The Ag_2ZnGeO_4 hollow sphere obtained through ostwald ripening process shows superior photocatalytic activity. 2) Few research has been focused on non-metal oxide modification. The composite between TiO_2 and silicon oxide (SiO_2) is the relatively familiar one. Li *et al.* [70] prepared composite SiO_2-TiO_2 hollow spheres by coating colloid carbon microspheres template with Si-doped TiO_2 layer in a one-pot hydrothermal approach and sequentially calcination. The photodegradation rate of methylene blue solution under ultraviolet irradiation in 2 hours by the composite SiO_2-TiO_2 hollow spheres (80%) exceeds that of TiO_2 hollow spheres (54%). In order to further improve photocatalytic activity of the above composite hollow spheres, measures such as surface noble metal deposition and element doping can be taken. Zhao *et al.* [71] prepared Ag modified hollow SiO_2/TiO_2 hybrid spheres through successively coating polystyrene spheres (PS) with SiO_2 and TiO_2 layer, evenly loading Ag nanoparticles by reducing silver nitrate onto surface of TiO_2 layer, and then carrying out a calcination process to eliminate the template. The modified hollow spheres can efficiently degrade Rh B under both ultraviolet and visible light irradiation. **Figure 4** shows UV-vis spectra of Rh B with Ag modified hollow SiO_2/TiO_2 hybrid

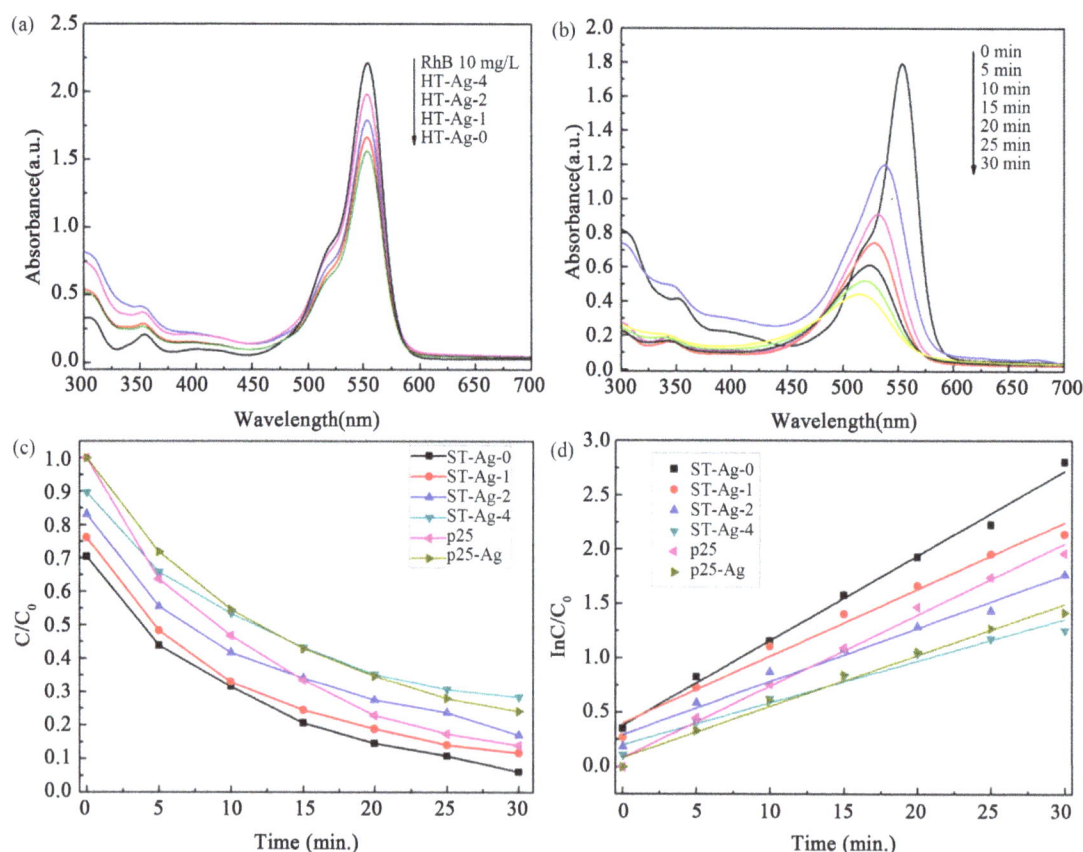

Figure 4. (a) UV-vis spectra of Rh B with Ag modified hollow SiO_2/TiO_2 hybrid spheres after 1 h dark absorption, photocatalytic degradation of Rh B under UV irradiation; (b) UV-vis spectra of the aqueous solutions of Rh B dye; (c) Photodegradation of the Rh B dye; (d) The corresponding pseudo-first-order kinetic rate plot [71].

spheres after 1 h dark absorption, photocatalytic degradation of Rh B under UV irradiation, UV-vis spectra of the aqueous solutions of Rh B dye, photodegradation plot of the Rh B dye and the corresponding pseudo-first-order kinetic rate plot. Zhang *et al.* [72] prepared hollow cobalt, nitrogen co-doped TiO_2/SiO_2 microspheres using PS microspheres as templates, tetraethylorthosilicate and tetrabutyltitanate as precursors. The compound hollow spheres have a wide absorption wavelength to 600 nm and their photodegradation rate for Rh B under visible light irradiation in 40 minute can reach to 98%.

3. Metal Sulfide Hollow Spheres

3.1. Single Metal Sulfide

Metal sulfide as photocatalysts have attracted researchers' general attention in recent years. Cadmium sulfide (CdS), a typical semiconductor of II-VI group with the band gap at 2.42 eV, owns outstanding photocatalytic property. Li *et al.* [73] prepared hollow CdS nanospheres with a diameter of about 130 nm and controllable shell thickness through 1-butyl-3-methylimidazolium-bis(trifluoro-methylsulfonyl)-imide ionic liquids as the templates, polyvinylpyrrolidone adjusting the formation of spheres and hexamethylenetetramine regulating the size and shell thickness. The photodegradation rate for methylene blue by the hollow CdS spheres can achieve 81% under ultraviolet irradiation in 80 minutes. Elements doping can effectively overcome its instability for light. Luo and co-workers [74] prepared nickel ion (Ni^{2+}) doped CdS hollow spheres via a template-free one-pot method. The obtained 1.2 mol% Ni-doped CdS hollow spheres can completely decompose Rh B under visible light ($\lambda > 420$ nm) in 35 minutes and the degradation rate keeps over 98% after using 4 times. **Figure 5** shows the photogagradation plot for Rh B under visible light of Ni-doped CdS hollow spheres.

Besides, zinc sulfide (ZnS), indium sulfide (In_2S_3) and copper sulfide (CuS) also have a certain degree of photocatalytic property. Yu and co-workers [75] prepared hexagonal wurtzite ZnS hollow spheres via one-pot template-free hydrothermal route using zinc acetate dehydrate and thiourea as raw materials. The ZnS hollow sphere shows strong absorption less than 365 nm. Zhang *et al.* [76] successfully synthesized Bi-doped ZnS hollow spheres (BZ) with enhanced ultraviolet and visible-light photocatalytic activity because of the formation of an isolated state originating from Bi 6s above the top of the valence band of ZnS and the electron excitation from Bi 6s state to the conduction band occurred under visible light irradiation. Under the optimal content of Bi dopant (R = 0.3, which is defined as the nominal atomic ratio of Bi to Zn), that is 0.3 at%, the hydrogen production rate is 1030 and 134 $\mu mol \cdot h^{-1} \cdot g^{-1}$ under UV and visible-light irradiation, respectively.

Rengaraj *et al.* [77] obtained tetragonal porous In_2S_3 hollow spheres composed of two-dimensional nanosheets and nanorods via a one-step solvothermal method using thiosemic arbazide as both sulfur source and capping ligand. The above-mentioned hollow spheres can photodegrade 30% percentage of methyl blue solution

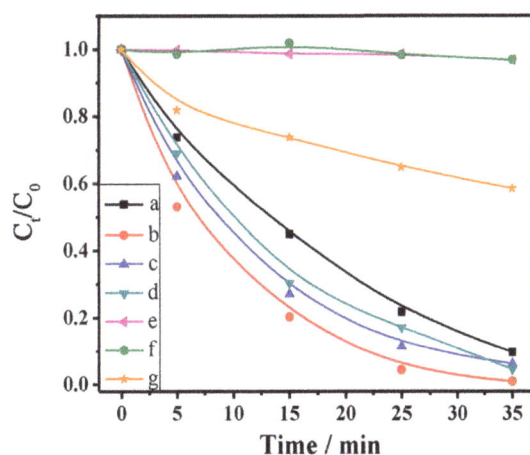

Figure 5. C_t/C_0 versus time curves of Rh B photodegradation under visible-light ($\lambda > 420$ nm) irradiation over (a) CdS; (b) 1.2 mol% Ni-doped CdS; (c) 3 mol% Ni-doped CdS; (d) 5 mol% Ni-doped CdS; (e) Catalyst free; (f) Light off; and (g) Degussa P25 [74].

under visible light irradiation in 3 hours. Meng *et al.* [78] prepared hierarchical flower-like CuS hollow nanospheres via a solvothermal approach. The CuS hollow nanospheres obtained can photodegrade Rh B and 2,4-dichlorophenol aqueous solution because of the synergistic effect of surface hierarchical structure with large surface area, porous hollow sphere structure and high visible light utilization.

3.2. Composite Metal Sulfide

Compared to single metal sulfide, composite metal sulfide has more superior photocatalytic activity and broaden light responsive scope. However, until now few studies of composite metal sulfide hollow spheres concentrated on phohocatalytic realm. The way to get composite metal sulfide hollow spheres is mainly by simple coupling and doping. Zhu *et al.* [79] prepared composite (Ag, Cu)$_2$S hollow spheres with cation exchange method using spherical aggregates of CuS nanoparticles as templates. The composite (Ag, Cu)$_2$S hollow spheres have strong absorption at ultraviolet light scope from 200 nm to 500 nm and infrared light scope from 1000 nm to 2500 nm, but weak absorption at visible light scope from 500 nm to 800 nm, which makes selective light absorption come true. Yu *et al.* [80] prepared composite CuS-ZnS hollow nanospheres by an ion-exchange method using ZnS solid spheres as a precursor and copper nitrate as raw materials. The photodegradation rate for Rh B by the composite hollow nanospheres is obviously higher than that of CuS and ZnS solid spheres. Bhirud *et al.* [81] prepared CdIn$_2$S$_4$ hollow spheres assembled by nanoparticles with a flower like morphology using cadmium nitrate tetrahydrate, indium nitrate trihydrate, thiourea and cetyl trimethyl ammonium bromide as raw materials. The rate of hydrogen production from H$_2$S photodecomposition for CdIn$_2$S$_4$ hollow spheres is 3171 μmol·h^{-1}, which is almost three fold enhancement than the highest rate of hydrogen production from normal bulk CdS (847 μmol·h^{-1}). Xing and co-workers [82] prepared a type of ternary Ag/Ag$_2$S/Ag$_3$CuS$_2$ hollow microspheres with Cu$_7$S$_4$ hollow submicrospheres as the template, the photocatalytic activity of which was higher than those of Ag/Ag$_2$S, Cu$_2$O, Cu$_7$S$_4$ and P25 for the photodegradation of MO under visible light irradiation. Superoxide radicals and holes were confirmed to be the main reactive species for MO degradation through radical scavenger experiments. **Figure 6** shows photogagradation plot for MO under visible light of the as-obtained Ag/Ag$_2$S/Ag$_3$CuS$_2$ hollow microspheres.

4. Application
4.1. Sewage Treatment

Organic pollutants durably presented in sewage often have reducibility. Consequently, inorganic hollow sphere photocatalysts can randomly oxidize organic pollutants by photo-generated holes and hydroxyl to carbon dioxide (CO$_2$), water and other nontoxic inorganic substances to purify water, as well as avoiding secondary water pollution. At present, photocatalytic technology can effectively degrade halogenated aliphatic hydrocarbon, dye, nitroaromatic compounds, polycyclic aromatic hydrocarbon, heterocyclic compounds, hydrocarbon, phenols, surfactants, pesticides and so on. For example, nitrogen doped TiO$_2$ hollow spheres can decompose bisphenol A [83], carbon, nitrogen co-doped TiO$_2$ hollow spheres can remove dye X-3B [84], copper ion doped TiO$_2$ hollow spheres can degrade chlorotetracycline [85], zirconium oxide doped TiO$_2$ hollow spheres can decompose Rh B [58], bismuth molybdate hollow spheres can decompose phenol [86] and Zn$_2$GeO$_4$ hollow spheres can degrade antibiotic metronidazole [87]. While inorganic hollow spheres as water purification material have achieved great progress, it also exists limitations such as unable to efficiently resolve high concentration wastewater and difficult to recycle for powder catalyst.

4.2. Air Purification

Due to the vast application of organic materials and continuous off-gas emissions from automobile, there is a large amount of poisonous organic gas and nitrogen oxides suspending in our breathing air. Air purification by photocatalysts can operate at normal temperature and pressure, do no harm to environmental and human body, costs little, completely decompose organics. Dong *et al.* [88] prepared hierarchical (BiO)$_2$CO$_3$ hollow microspheres which can be reused as NO removal material in indoor air. The indium vanadate hollow spheres prepared by Ai *et al.* [89] can oxidize NO to nitric acid under visible light and maintain their photocatalytic ratio after reused. Ikeda *et al.* [90] encapsulates TiO$_2$ into a hollow SiO$_2$ shell to attain a type of composite material, which can decompose gas-phase acetone and isopropanol into CO$_2$. Nevertheless, absence of applied research on

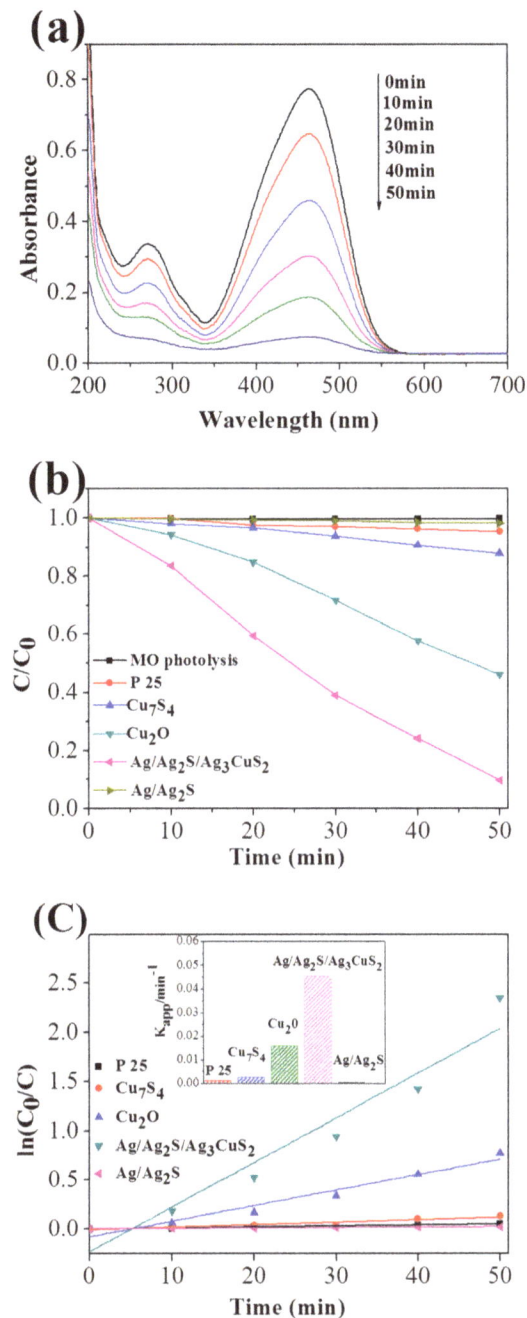

Figure 6. (a) UV-visible absorption spectra of degradation of MO by $Ag/Ag_2S/Ag_3CuS_2$ under visible light ($\lambda > 420$ nm); (b) photocatalytic degradation of MO over different photocatalysts; (c) Linear transformation $\ln C_0/C = Kappt$ of the kinetic curves of MO degradation over different photocatalysts. The inset shows the $Kapp$ for MO degradation over different photocatalysts [82].

air-handling equipment gives rise to the unrealizable industrialization of air purification by nanosize photocatalysts.

4.3. Hydrogen Production

Hydrogen gas is a kind of clean energy sources with a high combustion value, superior efficiency and environmental friendly. The hydrogen production chiefly depends on coal and natural gas so far, which could aggravate consumption of nonrenewable resources and environmental pollution. The most ideal way to settle down the problem is to transfer solar energy into hydrogen energy with photocatalyst using renewable substance as raw materials such as water and biomass. The hierarchical $Sn_2Nb_2O_7$ hollow spheres synthesized by Zhou et al. shows a superior visible-light-driven photocatalytic H_2 production activity($3 \mu mol \cdot h^{-1}$), which is about 4 times higher than that of the bulk $Sn_2Nb_2O_7$ sample prepared by a conventional high temperature solid state reaction method [91]. Zhang [92] and co-workers prepared composite CdS-TiO_2 hollow sphere by coupling TiO_2 to CdS which had a conduction band potential more negative than that of H^+/H_2 via a hydrothermal process, second step impregnation method and sol-gel method in sequence. The hydrogen production rate by the composite hollow spheres from water under ultraviolet light irradiation in one hour is 0.81 ml/g. After that, they prepared NiO-CdS hollow spheres with p-n junction, the hydrogen production rate by which under ultraviolet light irradiation in one hour (1.81 ml/g) precede the common composite CdS-TiO_2 hollow spheres [93]. The reason is that interior electric field of p-n junction boosting transport rate of photo-production electronics. While great progresses have been achieved in hydrogen production by photocatalysts until now, it fails to meet practical applications for the low efficiency of directly water splitting.

4.4. CO_2 Reduction

Greenhouse gas CO_2 is a major cause to global warming. It is also a kind of potential carbon resources. Hence, effective control and utilization of CO_2 becomes research focus. Present emission reduction technology requires numerous energy and exists potential safety problems in application. In the process of constant exploring emission reduction technology, it has been found that photogenerated electronics can change CO_2 to organic compounds with high application value such as methane and methanol. Tu et al. [94] prepared hollow spheres consisting of alternating $Ti_{0.91}O_2$ nanosheets and graphene nanosheets with polymer beads as sacrificial templates and via a microwave irradiation technique. The ultrathin $Ti_{0.91}O_2$ nanosheets allow charge carriers to move rapidly onto the surface to take part in the photoreduction reaction and the alternating compact stacking structure allow the photogenerated electron to transfer fast from $Ti_{0.91}O_2$ nanosheets to graphene nanosheets to extend lifetime of the charge carriers. The photocatalytic activity of the hollow spheres is 9 times higher than P25. It can reduce CO_2 to carbonic oxide and methane. Photocatalytic reduction for CO_2 operates at normal temperature and pressure, directly ultilizes solar light with no energy consumption, does no harm to environment, is the most promising conversion method.

4.5. Other Application

Photogenerated holes and electronics can directly react with compounds which deactive cytoderm, cytomembrane and cell leading to the death of bacterium. So, photocatalysts can be used as a new type of antibacterial materials. Feng et al. [95] prepared Ag-doped TiO_2 hollow spheres by hydrothermal precipitation method using carbon spheres as the templates. TiO_2 hollow sphere doping 9.4 mol% Ag has excellent antibacterial activities against scherichia coli, staphylococcus aurels and candida albicans at room temperature. Furthermore, hollow spheres with photocatalytic properties can be photoelectric detector [96].

5. Outlook

Inorganic hollow spheres show their excellent performance in photocatalytic area, which belongs to their large specific surface area, controllable structure and their own special optical, electrical, magnetic properties. They can be widely applied in sewage treatment, hydrogen production, air purification and other field. Metal oxide, metal sulfide and composite hollow spheres formed by metal oxide or sulfide have successfully prepared and primarily improve their photocatalytic property by controlling surface morphology and doping various other semiconductors. However, owing to the limitations of synthetic method, there exist some drawbacks including poor structure controllability, uneven shell thickness and broad particle size distribution, which are adverse to photocatalytic activity. Therefore, inorganic hollow spheres with well structure and property controllability remain to be one of the principal development trends in photocatalytic field. On the other hand, various types of inorganic hollow spheres own photocatalytic property, but the photodegradation efficiency is low. Rare earth elements possess unique f electronic configuration and abundant storage. So, hollow spheres composed of rare earth elements will be one of the research focuses in photocatalytic area.

Acknowledgements

The support from National Natural Science Foundation of China (No. 21106084), Shanghai Science and Research Innovation Foundation (No. 14zz164) and Shanghai Rising-Star Program (14QA143300) was appreciated.

References

[1]   Hu, J., Chen, M., Fang, X. and Wu, L. (2011) Fabrication and Application of Inorganic Hollow Spheres. *Chemical Society Reviews*, **40**, 5472-5491. http://dx.doi.org/10.1039/c1cs15103g

[2]   Gao, R., Chen, M., Li, W., Zhou, S. and Wu, L. (2013) Facile Fabrication and Some Specific Properties of Polymeric/ Inorganic Bilayer Hybrid Hollow Spheres. *Journal of Materials Chemistry A*, **1**, 2183-2191. http://dx.doi.org/10.1039/c2ta00837h

[3]   Zhu, B.T., Wang, Z., Ding, S., Chen, J.S. and Lou, X.W. (2011) Hierarchical Nickel Sulfide Hollow Spheres for High Performance Supercapacitors. *RSC Advances*, **1**, 397-400. http://dx.doi.org/10.1039/c1ra00240f

[4]   Guan, J., Mou, F., Sun, Z. and Shi, W. (2010) Preparation of Hollow Spheres with Controllable Interior Structures by Heterogeneous Contraction. *Chemical Communications*, **46**, 6605-6607. http://dx.doi.org/10.1039/c0cc01044h

[5]   Li, P., Fan, H., Cai, Y. and Xu, M. (2014) Zn-Doped In_2O_3 Hollow Spheres: Mild Solution Reaction Synthesis and Enhanced Cl_2 Sensing Performance. *CrystEngComm*, **16**, 2715-2722. http://dx.doi.org/10.1039/c3ce42325e

[6]   Xu, S., Hessel, C.M., Ren, H., Yu, R.B., Jin, Q., Yang, M., *et al.* (2014) α-Fe_2O_3 Multi-Shelled Hollow Microspheres for Lithium Ion Battery Anodes with Superior Capacity and Charge Retention. *Energy & Environmental Science*, **7**, 632-637. http://dx.doi.org/10.1039/c3ee43319f

[7]   Tian, W., Zhang, C., Zhai, T.Y., Li, S.L., Wang, X., Liao, M.Y., *et al.* (2013) Flexible SnO_2 Hollow Nanosphere Film Based High-Performance Ultraviolet Photodetector. *Chemical Communications*, **49**, 3739-3741. http://dx.doi.org/10.1039/c3cc39273b

[8]   Cong, H., Wang, Y., Yu, B., Wang, J. and Jiao, M. (2014) Synthesis of Anisotropic TiO_2 Hollow Microspheres Using Cave Particles as Templates and Application in Water Treatment. *New Journal of Chemistry*, **38**. 2564-2568. http://dx.doi.org/10.1039/c3nj01302b

[9]   Xue, C., Wang, T., Yang, G., Yang, B. and Ding, S. (2014) A Facile Strategy for the Synthesis of Hierarchical TiO_2/ CdS Hollow Sphere Heterostructures with Excellent Visible Light Activity. *Journal of Materials Chemistry A*, **2**, 7674-7679. http://dx.doi.org/10.1039/c4ta01190b

[10] Wu, L., Feng, H., Liu, M., Zhang, K. and Li, J. (2013) Graphene-Based Hollow Spheres as Efficient Electrocatalysts for Oxygen Reduction. *Nanoscale*, **5**, 10839-10843. http://dx.doi.org/10.1039/c3nr03794k

[11] Li, R., Li, L., Han, Y., Gai, S., He, F. and Yang, P. (2014) Core-Shell Structured Gd_2O_3:Ln@$m$$SiO_2$ Hollow Nanospheres: Synthesis, Photoluminescence and Drug Release Properties. *Journal of Materials Chemistry B*, **2**, 2127-2135. http://dx.doi.org/10.1039/c3tb21718c

[12] Sun, Z., Liao, T., Kim, J.G., Liu, K., Jiang, L., Kim, J.H., *et al.* (2013) Architecture Designed ZnO Hollow Microspheres with Wide-Range Visible-Light Photoresponses. *Journal of Materials Chemistry C*, **1**, 6924-6929. http://dx.doi.org/10.1039/c3tc31649a

[13] Ye, M., Chen, Z., Wang, W., Shen, J. and Ma, J. (2010) Hydrothermal Synthesis of TiO_2 Hollow Microspheres for the Photocatalytic Degradation of 4-Chloronitrobenzene. *Journal of Hazardous Materials*, **184**, 612-619. http://dx.doi.org/10.1016/j.jhazmat.2010.08.080

[14] Li, G., Zhang, H., Lan, J., Li, J., Chen, Q., Liu, J., *et al.* (2013) Hierarchical Hollow TiO_2 Spheres: Facile Synthesis and Improved Visible-Light Photocatalytic Activity. *Dalton Transactions*, **42**, 8541-8544. http://dx.doi.org/10.1039/c3dt50503k

[15] Gong, X.Q. and Selloni, A. (2005) Reactivity of Anatase TiO_2 Nanoparticles: The Role of the Minority (001) Surface. *Journal of Physical Chemistry B*, **109**, 19560-19562. http://dx.doi.org/10.1021/jp055311g

[16] Nakamura, R., Ohashi, N., Imanishi, A., Osawa, T., Matsumoto, Y., Koinuma, H., *et al.* (2005) Crystal-Face Dependences of Surface Band Edges and Hole Reactivity, Revealed by Preparation of Essentially Atomically Smooth and Stable (110) and (100) n-TiO_2 (Rutile) Surfaces. *Journal of Physical Chemistry B*, **109**, 1648-1651. http://dx.doi.org/10.1021/jp044710t

[17] Ong, W.J., Tan, L.L., Chai, S.P., Yong, S.T. and Mohamed, A.R. (2014) Highly Reactive {001} Facets of TiO_2-Based Composites: Synthesis, Formation Mechanism and Characterization. *Nanoscale*, **6**, 1946-2008. http://dx.doi.org/10.1039/c3nr04655a

[18] Wang, X., He, H., Chen, Y., Zhao, J. and Zhang, X. (2012) Anatase TiO_2 Hollow Microspheres with Exposed {001}

Facets: Facile Synthesis and Enhanced Photocatalysis. *Applied Surface Science*, **258**, 5863-5868. http://dx.doi.org/10.1016/j.apsusc.2012.02.117

[19] Jiao, Y., Peng, C., Guo, F., Bao, Z., Yang, J., Schmidt-Mende, L., *et al.* (2011) Facile Synthesis and Photocatalysis of Size-Distributed TiO_2 Hollow Spheres Consisting of {116} Plane-Oriented Nanocrystallites. *The Journal of Physical Chemistry C*, **115**, 6405-6409. http://dx.doi.org/10.1021/jp200491u

[20] Li, S., Chen, J., Zheng, F., Li, Y. and Huang, F. (2013) Synthesis of the Double-Shell Anatase-Rutile TiO_2 Hollow Spheres with Enhanced Photocatalytic Activity. *Nanoscale*, **5**, 12150-12155. http://dx.doi.org/10.1039/c3nr04043g

[21] Zeng, Y., Wang, X., Wang, H., Dong, Y., Ma, Y. and Yao, J. (2010) Multi-Shelled Titania Hollow Spheres Fabricated by a Hard Template Strategy: Enhanced Photocatalytic Activity. *Chemical Communications*, **46**, 4312-4314. http://dx.doi.org/10.1039/c0cc00706d

[22] Tao, Y.G., Xu, Y., Pan, J., Gu, H., Qin, C. and Zhou, P. (2012) Glycine Assisted Synthesis of Flower-Like TiO_2 Hierarchical Spheres and Its Application in Photocatalysis. *Materials Science and Engineering: B*, **177**, 1664-1671. http://dx.doi.org/10.1016/j.mseb.2012.08.010

[23] Liu, J., Chang, L., Wang, J., Zhu, M. and Zhang, W. (2010) A Facile One-Step Approach to Visible-Light-Sensitive Vanadium-Doped TiO_2 Hollow Microspheres. *Materials Science and Engineering: B*, **172**, 142-145. http://dx.doi.org/10.1016/j.mseb.2010.04.037

[24] Feng, X., Yang, O., Yang, L., Guang, Q. and Liu, Y.L. (2010) Preparation of Pt-Doped TiO_2 Hollow Spheres by Hydrothermal Precipitation and Their Photocatalytic Activities in Rh B Degradation. *Journal of Jinan University*, **31**, 495-499.

[25] Wang, C., Ao, Y., Wang, P., Hou, J., Qian, J. and Zhang, S. (2010) Preparation, Characterization, Photocatalytic Properties of Titania Hollow Sphere Doped with Cerium. *Journal of Hazardous Materials*, **178**, 517-521. http://dx.doi.org/10.1016/j.jhazmat.2010.01.111

[26] Wang, C., Ao, Y., Wang, P., Hou, J. and Qian, J. (2010) Preparation, Characterization and Photocatalytic Activity of the Neodymium-Doped TiO_2 Hollow Spheres. *Applied Surface Science*, **257**, 227-231. http://dx.doi.org/10.1016/j.apsusc.2010.06.071

[27] Xu, J., Ao, Y. and Chen, M. (2009) Preparation of B-Doped Titania Hollow Sphere and Its Photocatalytic Activity under Visible Light. *Materials Letters*, **63**, 2442-2444. http://dx.doi.org/10.1016/j.matlet.2009.08.031

[28] Ao, Y., Xu, J., Fu, D. and Yuan, C. (2009) A Simple Method to Prepare N-Doped Titania Hollow Spheres with High Photocatalytic Activity under Visible Light. *Journal of Hazardous Materials*, **167**, 413-417. http://dx.doi.org/10.1016/j.jhazmat.2008.12.139

[29] Ao, Y., Xu, J., Zhang, S. and Fu, D. (2010) A One-Pot Method to Prepare N-Doped Titania Hollow Spheres with High Photocatalytic Activity under Visible Light. *Applied Surface Science*, **256**, 2754-2758. http://dx.doi.org/10.1016/j.apsusc.2009.11.023

[30] Zhuang, J., Tian, Q., Zhou, H., Liu, Q., Liu, P. and Zhong, H. (2012) Hierarchical Porous TiO_2@C Hollow Microspheres: One-Pot Synthesis and Enhanced Visible-Light Photocatalysis. *Journal of Materials Chemistry*, **22**, 7036-7042. http://dx.doi.org/10.1039/c2jm16924j

[31] Shi, J.W., Chen, J.W., Cui, H.J., Fu, M.L., Luo, H.Y., Xu, B., *et al.* (2012) One Template Approach to Synthesize C-Doped Titania Hollow Spheres with High Visible-Light Photocatalytic Activity. *Chemical Engineering Journal*, **195-196**, 226-232. http://dx.doi.org/10.1016/j.cej.2012.04.095

[32] Zhou, J.K., Lv, L., Yu, J., Li, H.L., Guo, P.Z., Sun, H., *et al.* (2008) Synthesis of Self-Organized Polycrystalline F-Doped TiO_2 Hollow Microspheres and Their Photocatalytic Activity under Visible Light. *The Journal of Physical Chemistry C*, **112**, 5316-5321. http://dx.doi.org/10.1021/jp709615x

[33] Yu, J. and Shi, L. (2010) One-Pot Hydrothermal Synthesis and Enhanced Photocatalytic Activity of Trifluoroacetic Acid Modified TiO_2 Hollow Microspheres. *Journal of Molecular Catalysis A: Chemical*, **326**, 8-14. http://dx.doi.org/10.1016/j.molcata.2010.04.016

[34] Wang, C., Ao, Y., Wang, P., Hou, J. and Qian, J. (2011) Preparation of Cerium and Nitrogen Co-Doped Titania Hollow Spheres with Enhanced Visible Light Photocatalytic Performance. *Powder Technology*, **210**, 203-207. http://dx.doi.org/10.1016/j.powtec.2011.03.015

[35] Lu, J., Su, F., Huang, Z., Zhang, C., Liu, Y., Ma, X., *et al.* (2013) N-Doped Ag/TiO_2 Hollow Spheres for Highly Efficient Photocatalysis under Visible-Light Irradiation. *RSC Advances*, **3**, 720-724. http://dx.doi.org/10.1039/c2ra22713d

[36] Wang, S., Qian, H., Hu, Y., Dai, W., Zhong, Y., Chen, J., *et al.* (2013) Facile One-Pot Synthesis of Uniform TiO_2-Ag Hybrid Hollow Spheres with Enhanced Photocatalytic Activity. *Dalton Transactions*, **42**, 1122-1128. http://dx.doi.org/10.1039/c2dt32040a

[37] Xiang, Q., Yu, J., Cheng, B. and Ong, H.C. (2010) Microwave-Hydrothermal Preparation and Visible-Light Photoactivity of Plasmonic Photocatalyst Ag-TiO_2 Nanocomposite Hollow Spheres. *Chemistry: An Asian Journal*, **5**, 1466-1474.

http://dx.doi.org/10.1002/asia.200900695

[38] Jiang, Z., Zhu, J., Liu, D., Wei, W., Xie, J. and Chen, M. (2014) *In Situ* Synthesis of Bimetallic Ag/Pt Loaded Single-Crystalline Anatase TiO_2 Hollow Nano-Hemispheres and Their Improved Photocatalytic Properties. *CrystEngComm*, **16**, 2384-2394. http://dx.doi.org/10.1039/c3ce41949e

[39] Fei, H.L., Zhou, H.J., Wang, J.G., Sun, P.C., Ding, D.T. and Chen, T.H. (2008) Synthesis of Hollow V_2O_5 Microspheres and Application to Photocatalysis. *Solid State Sciences*, **10**, 1276-1284. http://dx.doi.org/10.1016/j.solidstatesciences.2007.12.026

[40] Duan, Z.Y. and Chen, Y.C. (2010) Hydrothermal Preparation for Self-Assembled γ-MnO_2 Hollow Spheres and Their Photocatalytic Activity. *Journal of Anqing Teachers College* (*Natural Science*), **16**, 88-91. http://dx.doi.org/10.3969/j.issn.1007-4260.2010.02.026

[41] Song, X.F. and Gao, L. (2008) Facile Synthesis and Hierarchical Assembly of Hollow Nickel Oxide Architectures Bearing Enhanced Photocatalytic Properties. *Journal of Physical Chemistry C*, **112**, 15299-15305. http://dx.doi.org/10.1021/jp804921g

[42] Guo, G.L. and Huang, J.H. (2011) Ta_2O_5 Hollow Sphere: Fabrication and Photocatalytic Activity. *Chinese Journal of Inorganic Chemistry*, **27**, 214-218. http://d.wanfangdata.com.cn/Periodical_wjhxxb201102002.aspx

[43] Xu, J.S. and Zhu, Y.J. (2011) α-Fe_2O_3 Hierarchically Hollow Microspheres Self-Assembled with Nanosheets: Surfactant-Free Solvothermal Synthesis, Magnetic and Photocatalytic Properties. *CrystEngComm*, **13**, 5162-5169. http://dx.doi.org/10.1039/c1ce05252g

[44] Yu, Y., Zhang, L., Wang, J., Yang, Z., Long, M., Hu, N., *et al.* (2012) Preparation of Hollow Porous Cu_2O Microspheres and Photocatalytic Activity under Visible Light Irradiation. *Nanoscale Research Letters*, **7**, 347-353.

[45] Huang, J., Xu, X., Gu, C., Fu, G., Wang, W. and Liu, J. (2012) Flower-Like and Hollow Sphere-Like WO_3 Porous Nanostructures: Selective Synthesis and Their Photocatalysis Property. *Materials Research Bulletin*, **47**, 3224-3232. http://dx.doi.org/10.1016/j.materresbull.2012.08.009

[46] Yu, J., Qi, L., Cheng, B. and Zhao, X. (2008) Effect of Calcination Temperatures on Microstructures and Photocatalytic Activity of Tungsten Trioxide Hollow Microspheres. *Journal of Hazardous Materials*, **160**, 621-628. http://dx.doi.org/10.1016/j.jhazmat.2008.03.047

[47] Yu, J.G. and Yu, X.X. (2008) Hydrothermal Synthesis and Photocatalytic Activity of Zinc Oxide Hollow Spheres. *Environmental Science &Technology*, **42**, 4902-4907. http://dx.doi.org/10.1021/es800036n

[48] Khoa, N.T., Kim, S.W., Thuan, D.V., Yoo, D.H., Kim, E.J. and Hahn, S.H. (2014) Hydrothermally Controlled ZnO Nanosheet Self-Assembled Hollow Spheres/Hierarchical Aggregates and Their Photocatalytic Activities. *CrystEngComm*, **16**, 1344-1350. http://dx.doi.org/10.1039/c3ce41763h

[49] Patrinoiu, G., Tudose, M., Calderón-Moreno, J.M., Birjega, R., Budrugeac, P., Ene, R., *et al.* (2012) A Green Chemical Approach to the Synthesis of Photoluminescent ZnO Hollow Spheres with Enhanced Photocatalytic Properties. *Journal of Solid State Chemistry*, **186**, 17-22. http://dx.doi.org/10.1016/j.jssc.2011.11.024

[50] Cao, S.W. and Zhu, Y.J. (2008) Hierarchically Nanostructured α-Fe_2O_3 Hollow Spheres: Preparation, Growth Mechanism, Photocatalytic Property and Application in Water Treatment. *Journal of Physical Chemistry C*, **112**, 6253-6257. http://dx.doi.org/10.1021/jp8000465

[51] Li, L., Deng, J., Yu, R., Chen, J., Wang, Z. and Xing, X. (2013) Niobium Pentoxide Hollow Nanospheres with Enhanced Visible Light Photocatalytic Activity. *Journal of Materials Chemistry A*, **1**, 11894-11900. http://dx.doi.org/10.1039/c3ta12599h

[52] Manjula, P., Boppella, R. and Manorama, S.V. (2012) A Facile and Green Approach for the Controlled Synthesis of Porous SnO_2 Nanospheres: Application as an Efficient Photocatalyst and an Excellent Gas Sensing Material. *ACS Applied Materials & Interfaces*, **4**, 6252-6260. http://dx.doi.org/10.1021/am301840s

[53] Li, B.X., Xie, Y., Jing, M., Rong, G., Tang, Y. and Zhang, G. (2006) In_2O_3 Hollow Microspheres: Sythesis from Designed In(OH)$_3$ Precursors and Applications in Gas Sensors and Photocatalysis. *Langmuir*, **22**, 9380-9385. http://dx.doi.org/10.1021/la061844k

[54] Ma, B., Guo, J., Dai, W.L. and Fan, K. (2012) Ag-AgCl/WO_3 Hollow Sphere with Flower-Like Structure and Superior Visible Photocatalytic Activity. *Applied Catalysis B: Environmental*, **123-124**, 193-199. http://dx.doi.org/10.1016/j.apcatb.2012.04.029

[55] Rahimi, R., Tadjarodi, A., Rabbani, M., Kerdari, H. and Imani, M. (2012) Preparation, Characterization and Photocatalytic Properties of Ba-Cd-Sr-Ti Doped Fe_3O_4 Nanohollow Spheres on Removal of Congo Red under Visible-Light Irradiation. *Journal of Superconductivity and Novel Magnetism*, **26**, 219-228. http://dx.doi.org/10.1007/s10948-012-1716-9

[56] Agrawal, M., Gupta, S., Pich, A., Zafeiropoulos, N.E. and Stamm, M. (2009) A Facile Approach to Fabrication of ZnO-TiO_2 Hollow Spheres. *Chemistry of Materials*, **21**, 5343-5348. http://dx.doi.org/10.1021/cm9028098

[57] Xuan, S.H., Jiang, W.Q., Gong, X.L., Hu, Y. and Chen, Z.Y. (2009) Magnetically Separable Fe_3O_4/TiO_2Hollow Spheres: Fabrication and Photocatalytic Activity. *Journal of Physical Chemistry C*, **113**, 553-558. http://dx.doi.org/10.1021/jp8073859

[58] Sun, C., Liu, L., Qi, L., Li, H., Zhang, H., Li, C., *et al.* (2011) Efficient Fabrication of ZrO_2-Doped TiO_2 Hollow Nanospheres with Enhanced Photocatalytic Activity of Rhodamine B Degradation. *Journal of Colloid and Interface Science*, **364**, 288-297. http://dx.doi.org/10.1016/j.jcis.2011.07.055

[59] Li, W.B., Bu, Y.Y. and Yu, J.Q. (2012) Preparation of ZnO/In_2O_3 Composite Hollow Spheres and Their Photoelectrocatalytic Properties to Glucose Degradation. *Acta Physical-Chimica Sinica*, **28**, 2676-2682. http://dx.doi.org/10.3866/PKU.WHXB201207101

[60] Zhang, C., Yin, L., Zhang, L., Qi, Y. and Lun, N. (2012) Preparation and Photocatalytic Activity of Hollow ZnO and ZnO-CuO Composite Spheres. *Materials Letters*, **67**, 303-307. http://dx.doi.org/10.1016/j.matlet.2011.09.073

[61] Wang, W.W., Zhu, Y.J. and Yang, L.X. (2007) $ZnO-SnO_2$ Hollow Spheres and Hierarchical Nanosheets: Hydrothermal Preparation, Formation Mechanism, and Photocatalytic Properties. *Advanced Functional Materials*, **17**, 59-64. http://dx.doi.org/10.1002/adfm.200600431

[62] Liang, J., Xu, J., Long, J., Zhang, Z. and Wang, X. (2013) Self-Assembled Micro/Nano-Structured Zn_2GeO_4 Hollow Spheres: Direct Synthesis and Enhanced Photocatalytic Activity. *Journal of Materials Chemistry A*, **1**, 10622-10625. http://dx.doi.org/10.1039/c3ta12183f

[63] Chen, A.X. (2007) Preparation of $ZrTiO_4$ Composite Hollow Spheres and Their Photocatalytic Properties (Jinan). M.E. Thesis, Shangdong University, Jinan.

[64] Dai, X.J., Luo, Y.S., Zhang, W.D. and Fu, S.Y. (2010) Facile Hydrothermal Synthesis and Photocatalytic Activity of Bismuth Tungstate Hierarchical Hollow Spheres with an Ultrahigh Surface Area. *Dalton Transactions*, **39**, 3426-3432. http://dx.doi.org/10.1039/b923443h

[65] Lu, Y., Luo, Y., Xiao, H.M. and Fu, S.Y. (2014) Novel Core-Shell Structured $BiVO_4$ Hollow Spheres with an Ultra-High Surface Area as Visible-Light-Driven Catalyst. *CrystEngComm*, **16**, 5059-6065. http://dx.doi.org/10.1039/c4ce00379a

[66] Chen, X., Liu, J., Wang, H., Ding, Y., Sun, Y. and Yan, H. (2013) One-Step Approach to Novel $Bi_4V_2O_{11}$ Hierarchical Hollow Microspheres with High Visible-Light-Driven Photocatalytic Activities. *Journal of Materials Chemistry A*, **1**, 877-883. http://dx.doi.org/10.1039/c2ta00312k

[67] Tian, G., Chen, Y., Zhou, W., Pan, K., Dong, Y., Tian, C., *et al.* (2011) Facile Solvothermal Synthesis of Hierarchical Flower-Like Bi_2MoO_6 Hollow Spheres as High Performance Visible-Light Driven Photocatalysts. *Journal of Materials Chemistry*, **21**, 887-892. http://dx.doi.org/10.1039/c0jm03040f

[68] Li, X., Huang, R., Hu, Y., Chen, Y., Liu, W., Yuan, R., *et al.* (2012) A Templated Method to Bi_2WO_6 Hollow Microspheres and Their Conversion to Double-Shell Bi_2O_3/Bi_2WO_6 Hollow Microspheres with Improved Photocatalytic Performance. *Inorganic Chemistry*, **51**, 6245-6250. http://dx.doi.org/10.1021/ic300454q

[69] Zhang, N., Ouyang, S., Kako, T. and Ye, J. (2012) Synthesis of Hierarchical Ag_2ZnGeO_4 Hollow Spheres for Enhanced Photocatalytic Property. *Chemical Communications*, **48**, 9894-9896. http://dx.doi.org/10.1039/c2cc34738e

[70] Li, G., Liu, F. and Zhang, Z. (2010) Enhanced Photocatalytic Activity of Silica-Embedded TiO_2 Hollow Microspheres Prepared by One-Pot Approach. *Journal of Alloys and Compounds*, **493**, L1-L7. http://dx.doi.org/10.1016/j.jallcom.2009.12.046

[71] Zhao, W., Feng, L., Yang, R., Zheng, J. and Li, X. (2011) Synthesis, Characterization, and Photocatalytic Properties of Ag Modified Hollow SiO_2/TiO_2 Hybrid Microspheres. *Applied Catalysis B: Environmental*, **103**, 181-189. http://dx.doi.org/10.1016/j.apcatb.2011.01.025

[72] Zhang, L., Li, X., Chang, Z. and Li, D. (2011) Preparation, Characterization and Photoactivity of Hollow N, Co Co-Doped TiO_2/SiO_2 Microspheres. *Materials Science in Semiconductor Processing*, **14**, 52-57. http://dx.doi.org/10.1016/j.mssp.2011.01.004

[73] Li, X., Gao, Y., Yu, L. and Zheng, L. (2010) Template-Free Synthesis of CdS Hollow Nanospheres Based on an Ionic Liquid Assisted Hydrothermal Process and Their Application in Photocatalysis. *Journal of Solid State Chemistry*, **183**, 1423-1432. http://dx.doi.org/10.1016/j.jssc.2010.04.001

[74] Luo, M., Liu, Y., Hu, J., Liu, H. and Li, J. (2012) One-Pot Synthesis of CdS and Ni-Doped CdS Hollow Spheres with Enhanced Photocatalytic Activity and Durability. *ACS Applied Materials & Interfaces*, **4**, 1813-1821. http://dx.doi.org/10.1021/am3000903

[75] Yu, X., Yu, J., Cheng, B. and Huang, B. (2009) One-Pot Template-Free Synthesis of Monodisperse Zinc Sulfide Hollow Spheres and Their Photocatalytic Properties. *Chemistry: A European Journal*, **15**, 6731-6739. http://dx.doi.org/10.1002/chem.200900204

[76] Zhang, J., Liu, S., Yu, J. and Jaroniec, M. (2011) A Simple Cation Exchange Approach to Bi-Doped ZnS Hollow

Spheres with Enhanced UV and Visible-Light Photocatalytic H_2-Production Activity. *Journal of Materials Chemistry*, **21**, 14655-14662. http://dx.doi.org/10.1039/c1jm12596f

[77] Rengaraj, S., Venkataraj, S., Tai, C., Kim, Y., Repo, E. and Sillanpää, M. (2011) Self-Assembled Mesoporous Hierarchical-Like In_2S_3 Hollow Microspheres Composed of Nanofibers and Nanosheets and Their Photocatalytic Activity. *Langmuir*, **27**, 5534-5541. http://dx.doi.org/10.1021/la104780d

[78] Meng, X., Tian, G., Chen, Y., Zhai, R., Zhou, J., Shi, Y., *et al.* (2013) Hierarchical CuS Hollow Nanospheres and Their Structure-Enhanced Visible Light Photocatalytic Properties. *CrystEngComm*, **15**, 5144-5149. http://dx.doi.org/10.1039/c3ce40195b

[79] Zhu, H., Lian, C. and Wu, D. (2011) Room-Temperature Synthesis of (Ag, Cu)$_2$S Hollow Spheres by Cation Exchange and Their Optical Properties. *Materials Chemistry and Physics*, **127**, 24-27. http://dx.doi.org/10.1016/j.matchemphys.2011.02.023

[80] Yu, J.G., Zhang, J. and Liu, S.W. (2010) Ion-Exchange Synthesis and Enhanced Visible-Light Photoactivity of CuS/ZnS Nanocomposite Hollow Spheres. *Journal of Physical Chemistry C*, **114**, 13642-13649. http://dx.doi.org/10.1021/jp101816c

[81] Bhirud, A., Chaudhari, N., Nikam, L., Sonawane, R., Patil, K., Baeg, J.O., *et al.* (2011) Surfactant Tunable Hierarchical Nanostructures of $CdIn_2S_4$ and Their Photohydrogen Production under Solar Light. *International Journal of Hydrogen Energy*, **36**, 11628-11639. http://dx.doi.org/10.1016/j.ijhydene.2011.06.061

[82] Xing, C., Zhang, Y., Wu, Z., Jiang, D. and Chen, M. (2014) Ion-Exchange Synthesis of Ag/Ag$_2$S/Ag$_3$CuS$_2$ Ternary Hollow Microspheres with Efficient Visible-Light Photocatalytic Activity. *Dalton Transactions*, **43**, 2772-2780. http://dx.doi.org/10.1039/c3dt52875h

[83] Subagio, D.P., Srinivasan, M., Lim, M. and Lim, T.T. (2010) Photocatalytic Degradation of Bisphenol-A by Nitrogen-Doped TiO_2 Hollow Sphere in a Vis-LED Photoreactor. *Applied Catalysis B: Environmental*, **95**, 414-422. http://dx.doi.org/10.1016/j.apcatb.2010.01.021

[84] Ao, Y., Xu, J., Fu, D. and Yuan, C. (2009) Visible-Light Responsive C,N-Codoped Titania Hollow Spheres for X-3B Dye Photodegradation. *Microporous and Mesoporous Materials*, **118**, 382-386. http://dx.doi.org/10.1016/j.micromeso.2008.09.010

[85] Bu, D. and Zhuang, H. (2013) Biotemplated Synthesis of High Specific Surface Area Copper-Doped Hollow Spherical Titania and Its Photocatalytic Research for Degradating Chlorotetracycline. *Applied Surface Science*, **265**, 677-685. http://dx.doi.org/10.1016/j.apsusc.2012.11.080

[86] Yin, W., Wang, W. and Sun, S. (2010) Photocatalytic Degradation of Phenol over Cage-Like Bi_2MoO_6 Hollow Spheres under Visible-Light Irradiation. *Catalysis Communications*, **11**, 647-650. http://dx.doi.org/10.1016/j.catcom.2010.01.014

[87] Liu, J., Zhang, G., Yu, J.C. and Guo, Y. (2013) *In Situ* Synthesis of Zn_2GeO_4 Hollow Spheres and Their Enhanced Photocatalytic Activity for the Degradation of Antibiotic Metronidazole. *Dalton Transactions*, **42**, 5092-5099. http://dx.doi.org/10.1039/c2dt32623j

[88] Dong, F., Lee, S.C., Wu, Z., Huang, Y., Fu, M., Ho, W.K., *et al.* (2011) Rose-Like Monodisperse Bismuth Subcarbonate Hierarchical Hollow Microspheres: One-Pot Template-Free Fabrication and Excellent Visible Light Photocatalytic Activity and Photochemical Stability for NO Removal in Indoor Air. *Journal of Hazardous Materials*, **195**, 346-354. http://dx.doi.org/10.1016/j.jhazmat.2011.08.050

[89] Ai, Z.H., Zhang, L.Z. and Lee, S.C. (2010) Efficient Visible Light Photocatalytic Oxidation of NO on Aerosol Flow-Synthesized Nano-Crystalline $InVO_4$ Hollow Microspheres. *Journal of Physical Chemistry C*, **114**, 18594-18600. http://dx.doi.org/10.1021/jp106906s

[90] Ikeda, S., Kobayashi, H., Ikoma, Y., Harada, T., Yamazaki, S. and Matsumura, M. (2009) Structural Effects of Titanium(IV) Oxide Encapsulated in a Hollow Silica Shell on Photocatalytic Activity for Gas-Phase Decomposition of Organics. *Applied Catalysis A: General*, **369**, 113-118. http://dx.doi.org/10.1016/j.apcata.2009.09.008

[91] Zhou, C., Zhao, Y., Bian, T., Shang, L., Yu, H., Wu, L.Z., *et al.* (2013) Bubble Template Synthesis of $Sn_2Nb_2O_7$ Hollow Spheres for Enhanced Visible-Light-Driven Photocatalytic Hydrogen Production. *Chemical Communications*, **49**, 9872-9874. http://dx.doi.org/10.1039/c3cc45683h

[92] Zhang, Y.J., Li, X.J., Min, C., Wang, Y.C. and Li, S. (2010) Preparation of NiO-CdS Composite Hollow Spheres and Their Application for H_2-Production to Water Photodegradation. China Patent No. CN101623644.

[93] Zhang, Y.J., Li, X.J., Min, C., Wang, Y.C. and Li, S. (2010) Preparation of Composite Hollow Spheres with p-n Junction and Their Application for H_2-Production to Water Photodegradation. China Patent No. CN101623645.

[94] Tu, W.G., Zhou, Y., Liu, Q., Tian, Z., Gao, J., Chen, X., *et al.* (2012) Robust Hollow Spheres Consisting of Alternating Titania Nanosheets and Graphene Nanosheets with High Photocatalytic Activity for CO_2 Conversion into Renewable Fuels. *Advanced Functional Materials*, **22**, 1215-1221. http://dx.doi.org/10.1002/adfm.201102566

[95] Feng, X., Yang, L., Guan, Q. and Liu, Y.L. (2010) Preparation of Ag-Doped TiO$_2$ Hollow Spheres and Their Antibacterial Activity. *Ecological Science*, **29**, 251-255.

[96] Hu, L., Chen, M., Shan, W., Zhan, T., Liao, M., Fang, X., *et al.* (2012) Stacking-Order-Dependent Optoelectronic Properties of Bilayer Nanofilm Photodetectors Made from Hollow ZnS and ZnO Microspheres. *Advanced Materials*, **24**, 5872-5877. http://dx.doi.org/10.1002/adma.201202749

11

Synthesis and Characterization of Some Selenium Nanometric Compounds: Spectroscopic, Biological and Antioxidant Assessments

Aly H. Atta[1,2*], Ahmed I. El-Shenawy[1,3], Fathy A. Koura[1,4], Moamen S. Refat[5,6]

[1]Department of Chemistry, Faculty of Education, University of Dammam, Dammam, KSA
[2]Department of Chemistry, Faculty of Science, Suez University, Suez, Egypt
[3]Department of Chemistry, Faculty of Science, Benha University, Benha, Egypt
[4]Department of Chemistry, Faculty of Science, Al-Azhar University, Cairo, Egypt
[5]Department of Chemistry, Faculty of Science, Taif University, Al-Hawiah, KSA
[6]Department of Chemistry, Faculty of Science, Port Said University, Port Said, Egypt
Email: *aly_atta@yahoo.com

Abstract

Selenium (IV) vitamin A complex as antioxidant drug design was prepared and characterized by microanalysis, conductance, infrared spectra, Raman laser spectra, ^{1}HNMR spectra, scanning electron microscopy (SEM), X-ray powder diffraction (XRD) and thermogravimetric (TG/DTG and DTA) tool of analyses. Vitamin A chelate was coordinated as a mono-dentate ligand through the oxygen atom of –OH hydroxyl group. Thermal degradation analyses discussed the removal of terminal methyl molecules in the first and second decomposition stage while the organic ligand moieties existed in the third and subsequence steps. The Se (IV) complex in comparable with free vitamin A ligand has been assessed against some kinds of bacteria and fungi which gave a significant inhibition. The surface morphology and nano scale size of selenium metal and its vitamin complex were proved. The activation energy and other thermodynamic parameters (ΔH^*, ΔS^* and ΔG^*) of Se (IV) complex were calculated using Coats-Redfern and Horowitz-Metzger equations.

Keywords

Vitamin A, Selenium, Nano-Scale, Spectroscopic, Thermal Analysis, Antimicrobial Activity

1. Introduction

Selenium element was mentioned important due to its interesting role in the life construction of plants and biological systems, which absorb organoselenium compounds accumulated in the soils [1] [2]. It is an essential nutrient at trace level but toxic in excess [3]. Selenium is a trace mineral that is essential to good health but required only in small amounts [4]. Selenium is incorporated into proteins to make selenoproteins, which are important as antioxidant enzymes that help to prevent cellular damage from free radicals and to the development of chronic diseases such as cancer and heart disease [5]. Other selenoproteins help to regulate thyroid function and play a role in the immune system [6]. Because of its antioxidant role, selenium has been studied for its potential to protect the body from many degenerative diseases, including Parkinson's and cancer [7]. Vitamin A (**vit-OH**; **Figure 1**) is an essential nutrient for human body. The vitamin A has biological activities [8]. Vitamin A deficiency was a major nutritional deficiency disorder in many developing countries, especially affects young children, in whom it can cause xerophthalmia and lead to blindness, and can also limit growth, weaken innate and acquired host defenses, exacerbate infection and increase the risk of death [9]. Generally, metal ions were required for many critical functions in humans [10]-[12]. The ability to recognize to the molecular level and to treat diseases caused by inadequate metal-ion function constitutes an important aspect of medicinal bioinorganic chemistry [10] [12]. Metal complexes have been playing a key role in the development of modern chemotherapy [13], for example, complexation of non-steroidal anti-inflammatory drugs to copper overcomes some of the gastric side effects of these drugs [14]. The release of cytotoxins such as nitrogen mustards from redox-active metals such as cobalt in the hypoxic regions of solid tumors has the potential to improve drug activity and reduce toxicity [15]. The metal based drugs were also being used for the treatment of a variety of ailments viz. diabetes, rheumatoid arthritis, inflammatory and cardiovascular diseases as well as diagnostic [16]. A number of drugs and potential pharmaceutical agents also contain metal-binding and potentially influence their bioactivities [17]-[19]. Metal-organic frameworks were a burgeoning field in the last two decades, not only stems from their tremendous potential applications in areas such as catalysis, molecular adsorption, magnetism, nonlinear optics, and molecular sensing, but also from their novel topologies and intriguing structural diversities [20] [21]. On the other hand, many organic drugs, which possess modified pharmacological and toxicological properties administered in the form of metallic complexes [22], have the potential to act as ligands and the resulting metal-drug complexes are particularly important both in coordination chemistry and biochemistry [23] [24], however, the study of metal-drug complexes is still in its early stages, thus representing a great challenge in current synthetic chemistry and coordination chemistry. Recently, the trend of metal drug complexes has significant interest in order to achieve an enhanced therapeutic effect in combination with decreased toxicity. To the best of our knowledge, little attentions have been made to discuss the interaction between vitamin A and selenium (IV) metal ions and the literature is still poor in such spectroscopic characterizations. The interpretation is based on the ability of the cited drug to form complex associate with vital antioxidant metal ions like Se (IV). The spectral characteristic and the stability of the formed complex were also discussed.

2. Experimental

All chemicals used throughout this study were Analar or extra pure grade and received from Aldrich chemical company.

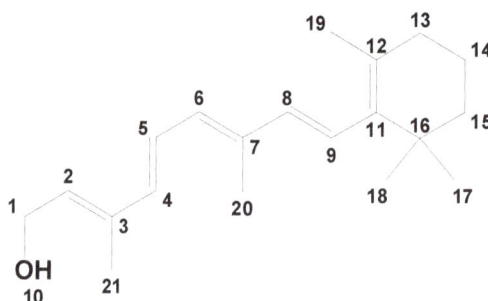

Figure 1. 3, 7-Dimethyl-9-(2, 6, 6-trimethyl-cyclohex-1-enyl)-nona-2, 4, 6, 8-tetraen-1-ol (Vitamin A; vit-OH).

2.1. Preparation of Selenium Nano Particles

The Se metal nano particles was prepared previously [25] by mixing equal volumes of aqueous solutions of 0.01 M of SeO_2 with 0.1 M of acetamide as following equation.

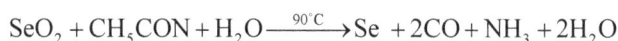

$$SeO_2 + CH_5CON + H_2O \xrightarrow{90^\circ C} Se + 2CO + NH_3 + 2H_2O$$

The mixture was heated on a water bath to approximately 90°C for about ~24 hrs. The precipitate filtered off, washed several times with bidistilled water and dried in *vacuo* over $CaCl_2$. The selenium metal product was received in powder solid form. The elemental analysis for the obtained product shows the absence of carbon and nitrogen elements. The yield of selenium was ~70%.

2.2. Preparation of Solid Selenium (IV) vit-OH Complex

A mixture of solid powder of selenium metal (1 mmole) and the **vit-OH** (4 mmole) in toluene solvent (50 mL) was refluxed for 12 hr at 60°C. The un-reacted of selenium metal powder was removed by filtration, and the resultant brownish red solution were reduced to *ca.* 1/3 of its volume and cooled to room temperature. The solid brown complex obtained was then collected by filtration, washed with little amount of toluene and dried in *vacuo* over anhydrous calcium (II) chloride. Elemental analysis, Calculated: %C = 78.45, %H = 9.88, %Se = 6.45; Found: %C = 78.32, %H = 9.56, %Se = 6.31.

2.3. Instrumental Analyses

The elemental analyses of carbon, and hydrogen contents were performed using a Perkin Elmer CHN 2400 (USA). The molar conductivities of freshly prepared 1.0×10^{-3} mol/cm^3 dimethylsulfoxide (DMSO) solutions were measured for the dissolved compounds using Jenway 4010 conductivity meter. The electronic absorption spectrum of Se (IV) complex was recorded in DMSO solvent within 600 - 200 nm range using a UV2 Unicam UV/Vis Spectrophotometer fitted with a quartz cell of 1.0 cm path length. The infrared spectra with KBr discs were recorded on a Bruker FT-IR Spectrophotometer (4000 - 400 cm^{-1}). Raman laser of samples were measured on the Bruker FT Raman with laser 50 mW. The ^{1}H-NMR spectra were recorded on Varian Mercury VX-300 NMR spectrometer. ^{1}H spectra were run at 300 MHz spectra in deuterated dimethylsulphoxide (DMSO-d$_6$). Chemical shifts are quoted in δ and were related to that of the solvents. The thermal studies TG/DTG-50H and DTA-50 H were carried out on a Shimadzu thermo-gravimetric analyzer under nitrogen till 800°C. Scanning electron microscopy (SEM) images were taken in Quanta FEG 250 equipment. The X-ray diffraction patterns for the complexes were recorded on X 'Pert PRO PANanalytical X-ray powder diffraction, target copper with secondary monochromate.

2.4. Antimicrobial Test

According to Gupta *et al.*, 1995 [26], the hole well method was applied. The investigated isolates of bacteria and fungi seeded in tubes with nutrient broth (NB) and Dox's broth (DB), respectively. The seeded (NB) for bacteria and (DB) for fungi (1 ml) were homogenized in the tubes with 9 ml of melted (45°C) nutrient agar (NA) for bacteria and (DA) for fungi. The homogenous suspensions poured into Petri dishes. The holes (diameter 0.5 cm) done in the cool medium. After cooling in these holes, about 100 µl of the investigated compounds applied using a micropipette. After incubation for 24 h in an incubator at 37°C and 28°C for bacteria and fungi, respectively, the inhibition zone diameter were measured and expressed in cm. The antimicrobial activities of the investigated compounds were tested against some kinds of bacteria as *Escherichia coli* (Gram − ve) and *Staph albus* (Gram + ve) as well as some kinds of fungi as *Aspergillus flavus* and *Aspergillus niger*. In the same time with the antimicrobial investigations of the compounds, the pure solvent also tested. The concentration of each solution was 1.0×10^{-3} mol/L. Commercial DMSO was employed to dissolve the tested samples.

3. Results and Discussion

3.1. Characterization of Selenium Nano Scale

The formation of brown colored selenium metal in nano scale size was prepared upon the heating of an aqueous mixture of SeO_2 with acetamide at 90°C. The isolated of selenium metal powder occurs during the decomposition of acetamide according to the important role of metal ion on the decomposition process via aqueous media. The infrared spectrum of the SeO_2 with acetamide at 90°C product is shown in **Figure 2**. The infrared spectrum of the obtained selenium metal product shows no bands due to characteristic groups of acetamide (carbonyl and amide groups). The XRD patterns of Se metal are shown in **Figure 3**. All definite peaks of Se metal are indexed as Se metal structure, which are matched and compared with the standard data. The SEM micrograph of the prepared Se metal is shown in **Figure 4**. The average grain size of selenium metal were calculated and found in between 0.65 μm. From **Figure 4** it is so difficult to observe inhomogeneitiy within the same micrograph due to that the solution route synthesis with acetamide precursor is very fine and the particle size must be in nano-range. The grain size for nano compounds were calculated according to Scherrer's formula [27].

$$B = 0.87\, \lambda/D\cos\theta$$

where D is the crystalline grain size in nm, θ, half of the diffraction angle in degree, λ is the wavelength of X-ray source (Cu-K_α) in nm, and B, degree of widening of diffraction peak which is equal to the difference of full width at half maximum (FWHM) of the peak at the same diffraction angle between the measured sample and standard one.

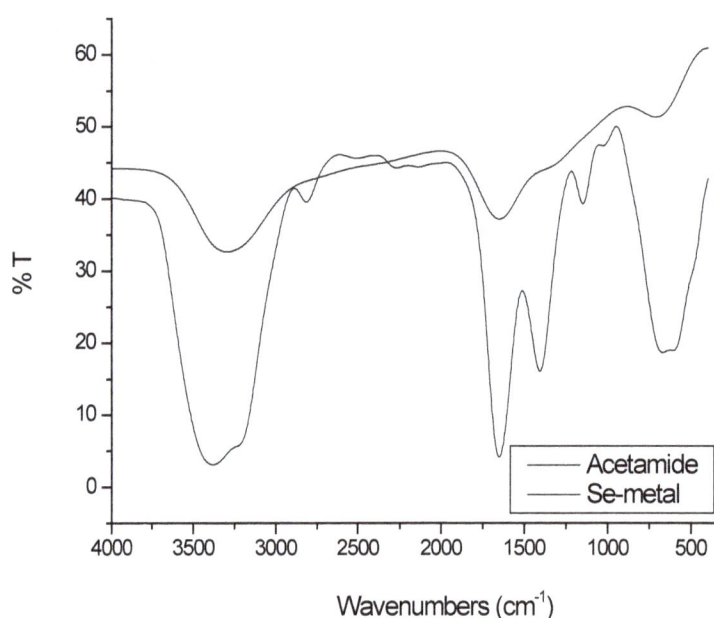

Figure 2. Infrared spectra of acetamide as a free ligand and Se metal.

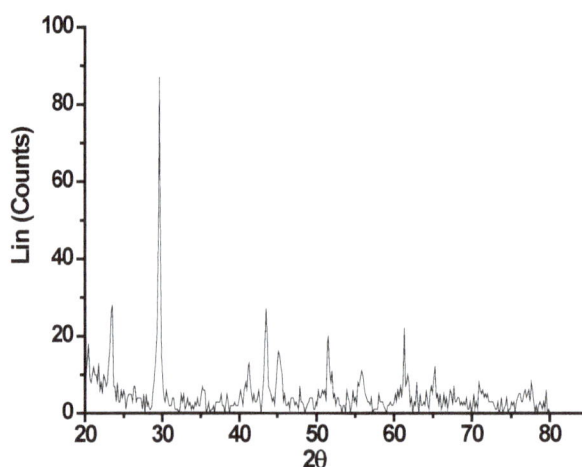

Figure 3. XRD spectrum of the resulted Se metal.

Figure 4. SEM micrographs of Se metal.

The Raman laser spectrum of Se metal is reported in **Figure 5**. For the pure selenium sample the spectrum is characterized by a strong band at 237 cm^{-1} which is attributed to the vibrational mode of –Se-Se-Se-chains [28] [29]. Additional weaker features can be revealed at 141 cm^{-1} which are assigned to the presence of Se$_8$ rings and to the bending modes of Se units.

3.2. Characterization of Selenium (IV) Vitamin A Complex

The reaction of **vit-OH** with selenium (0) metal in toluene solvent gave a brown colored solid complex. The found and calculated percentage of elemental analysis is in a well agreement with each other and proves the suggested molecular formula of the resulted selenium complex with molar ratio 1:4 (Se: **vit-OH**). The complexes have high melting points ~220°C. The molar conductivity of selenium complex in DMSO has 15 $\Omega^{-1}\cdot$cm$^2\cdot$mol^{-1}, so that, it is non-electrolytes nature [30] [31]. Vitamin A ligand behaves as monodentate ligand and coordinates to selenium metal through the oxygen of hydroxyl group.

3.2.1. Electronic Absorption Spectra

The formation of the Se-**vit-OH** complex was also confirmed by UV-Vis spectroscopic technique. **Figure 6**, show the electronic absorption spectrum of the complex in DMSO in the 200 - 600 nm range. It can see that the free vitamin A has one distinct absorption band at 350 nm which may be attributed to n → π* intra-ligand transition. In the spectra of the selenium complex this band is blue shifted clearly to 310 nm, where the complex show two bands at 239 and 310 nm assigned to π → π* and n → π* respectively, suggesting the ligand is binding to selenium metal through the lone pair of electrons on oxygen atom of –OH group. The electronic absorption spectrum of the selenium complex in DMSO solution has significant band at 370 nm due to the charge transfer transition from metal-to-ligand.

3.2.2. Infrared Spectra

The infrared spectra of **vit-OH** free ligand and its selenium complex were recorded in **Figure 7**. The spectra are similar but there are some differences which could give indication on the type of coordination. The IR spectrum of **vit-OH** shows a very strong broad band at 3406 cm^{-1} which assigned to ν(O-H) stretching vibration of alcoholic OH group, ionization of the alcoholic OH group with subsequent ligation through oxygen atom seems a plausible explanation [32]. To ascertain the involvement of ν(OH) of alcoholic group of **vit-OH** in the coordination process to be followed the stretching vibration bands of ν(C-O) in selenium **vit-OH complex**, the examination of this complex show that the ν(C-O) is shifted to lower wavenumber from 1073 and 1176 cm^{-1} in case of free ligand to 1084 and 1143 cm^{-1} in the complex. This result indicates that the alcoholic group is participated in the complexation [33] and the **vit-OH** acts as monodentate ligand. The **vit-OH** complex show two weak bands observed at 2955 and 2929 cm^{-1} which assigned to the stretching vibration of ν(C-H) of CH$_3$ and CH$_2$ groups. Involvement of the oxygen atoms of hydroxyl group are also confirmed by the presence of new bands in the selenium complex at 520 and 488 cm^{-1} due to the ν(M-O) stretching vibrations respectively [34].

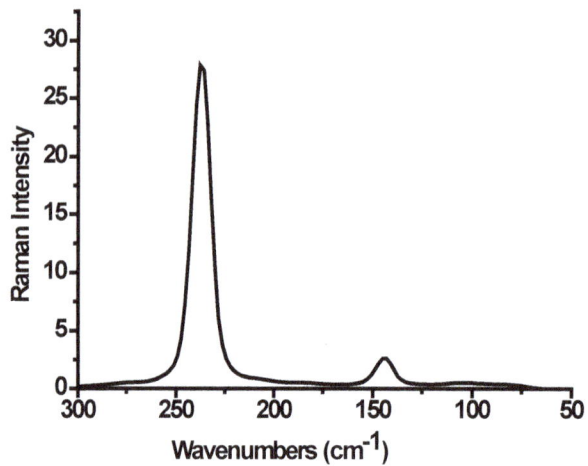

Figure 5. Raman laser spectrum of Se metal.

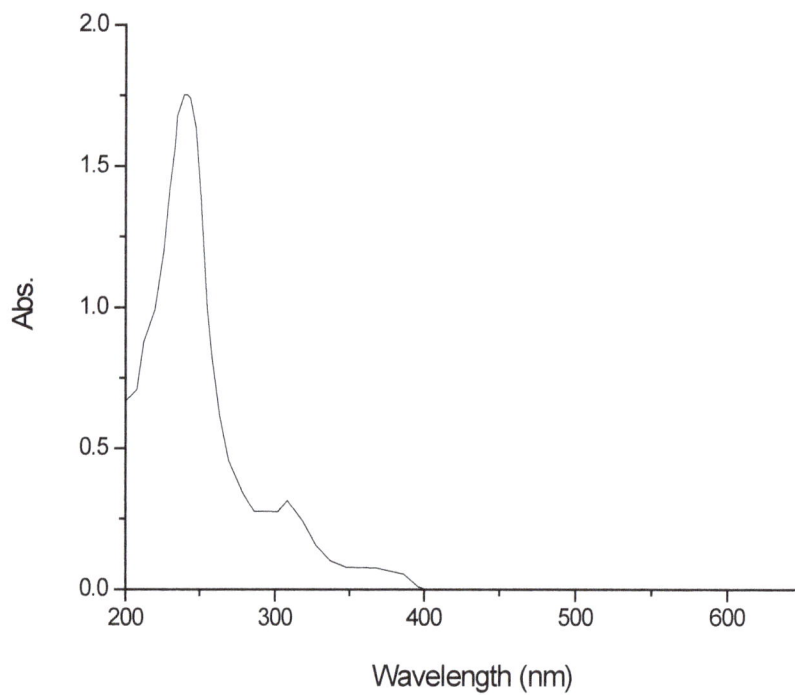

Figure 6. UV-vis electronic spectrum of Se-**vit-OH** complex.

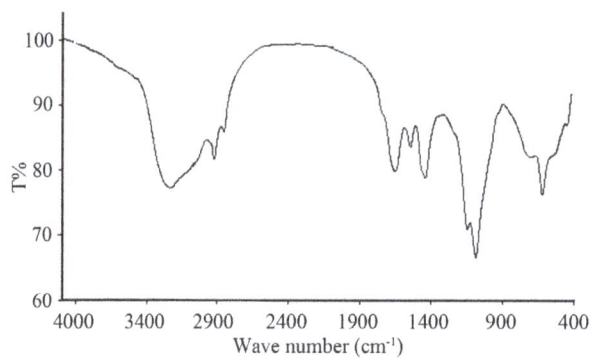

Figure 7. Infrared spectrum of Se-**vit-OH** complex.

3.2.3. ¹H-NMR Spectra

The ¹H-NMR data of **vit-OH** and its Se (0) complex is shown in **Figure 8**. ¹H-NMR spectrum of **vit-OH** show the signal at $\delta = \sim 9$ ppm, which assigned to the proton of alcoholic OH group that is disappear in the Se-**vit-OH** complex. The disappearance of the signal of the proton of the hydroxyl group in the ¹H-NMR spectrum of the complex confirms that the hydroxyl group contribute in the complexation between **vit-OH** and Se (0) metal [35]. The proton NMR spectrum for Se (0) complex shows a singlet peaks at 1.30, 3.20 and 4.50 ppm. These peaks are assigned to protons of -CH_3, -CH_2, and =CH groups, supporting the complex formula as suggested in **Figure 9**.

3.2.4. Thermal Analysis

The obtained Se-**vit-OH** complex was studied by thermogravimetric TG/DTG and DTA differential thermal analysis from ambient temperature to 800°C under N_2 atmospheres. The TG curves were redrawn as mg mass loss versus temperature and DTG curve was redrawn as rate of loss of mass versus temperature (**Figure 10** and **Figure 11**). The thermal decomposition of [Se(**vit-OH**)$_4$] complex occurs at four decomposition steps. The first degradation step takes place within a temperature range of 50°C - 120°C at $DTG_{max} = 80$°C and it correspond to the loss of three terminal methyl groups with an observed weight loss 3.30% (calcd. = 3.67%). The second decomposition step occur within temperature range 120°C - 225°C at $DTG_{max} = 185$°C and DTA = 195°C (exothermic) that assigned to the loss of another 14CH_3 molecules with a weight loss (obs =17.50%, calcd. = 17.15%). The third step occur within temperature range 225°C - 330°C at $DTG_{max} = 285$°C and DTA= 300°C (exothermic) due to the loss of $C_{22}H_{22}O_4$ organic molecule with mass loss (obs. = 28.65%, calcd. = 28.58%). The fourth step occur within a temperature range 330°C - 435°C at $DTG_{max} = 365$°C and DTA = 370°C (exo) due to the loss of $C_{23}H_{47}$ (organic moiety) with a mass loss (obs. = 26.60%, calcd. = 26.37%). The Se-metal contaminated with polluted carbon atoms remains as final residual till 800°C.

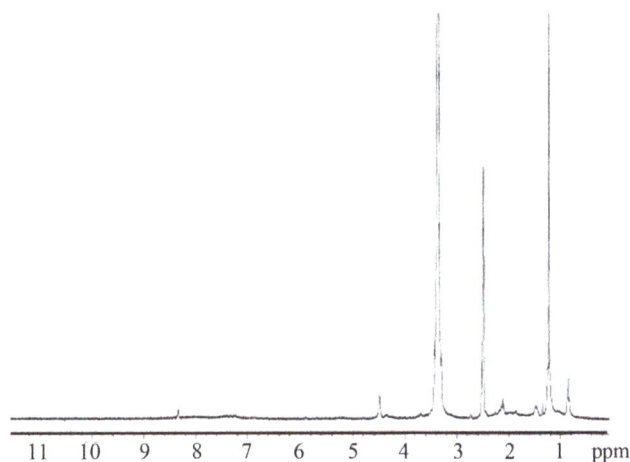

Figure 8. ¹HNMR spectrum of Se-**vit-OH** complex.

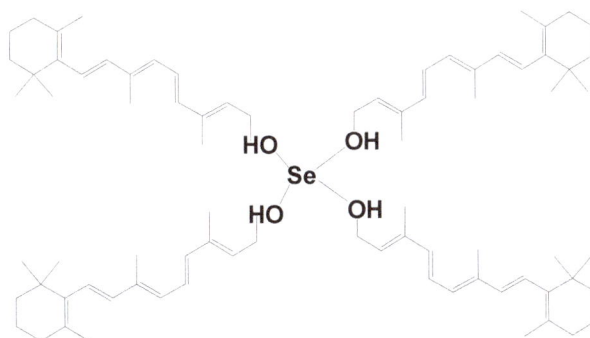

Figure 9. Suggested structure of Se-**vit-OH** complex.

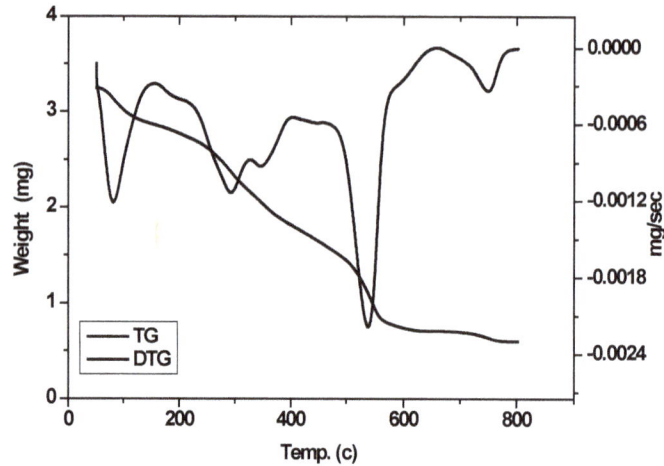

Figure 10. TG/DTG curves of Se-**vit-OH** complex.

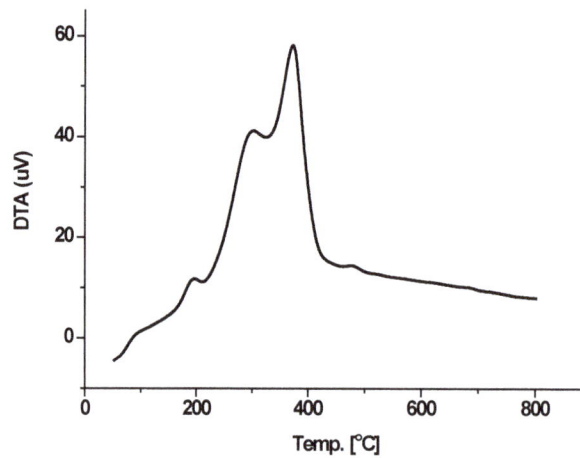

Figure 11. DTA curve of Se-**vit-OH** complex.

3.2.5. Kinetic Studies

Most commonly used methods are the differential method of Freeman and Carroll [36] integral method of Coat and Redfern [37] and the approximation method of Horowitz and Metzger [38]. In the present investigation, the general thermal behaviors of the vitamin A comple in terms of stability ranges, peak temperatures and values of kinetic parameters are discussed. The kinetic parameters have been evaluated using the Coats-Redfern equation:

$$\int_0^\alpha \frac{d\alpha}{(1-\alpha)^n} = \frac{A}{\varphi} \int_{T_1}^{T_2} \exp\left(-\frac{E^*}{RT}\right) dt \tag{1}$$

This equation on integration gives:

$$\ln\left[-\frac{\ln(1-\alpha)}{T^2}\right] = -\frac{E^*}{RT} + \ln\left(\frac{AR}{\varphi E^*}\right) \tag{2}$$

where φ is the linear heating rate, R is the gas constant, T is the DTG temperature peak, α, is the fraction of the sample decomposed at time t, A is the pre-exponential factor. A plot of left-hand side against $1/T$ was drawn. E^* is the energy of activation in J·mol^{-1} and calculated from the slop and A in (s^{-1}) from the intercept value. The entropy of activation ΔS^* in (JK^{-1}·mol^{-1}) was calculated by using the equation:

$$\Delta S^* = R \cdot \ln\left(Ah/k_B T_s\right) \tag{3}$$

where k_B is the Boltzmann constant, h is the Plank's constant and T_s is the DTG peak temperature [39].

The Horowitz-Metzger equation is an illustrative of the approximation methods.

$$\log\left[\left\{1-(1-\alpha)^{1-n}\right\}\big/(1-n)\right] = E^*\theta/2.303RT_s^2 \quad \text{for } n \neq 1 \tag{4}$$

When $n = 1$, the LHS of equation 4 would be $\log[-\log(1 - \alpha)]$. For a first-order kinetic process the Horowitz-Metzger equation may be written in the form:

$$\log\left(\log\left(w_\alpha/w_\gamma\right)\right) = E^*\theta/2.303RT_s^2 - \log 2.303$$

where $\theta = T - T_s$, $w_\gamma = w_\alpha - w$, w_α = mass loss at the completion of the reaction. The plot of $\log[\log(w_\alpha/w_\gamma)]$ vs θ was drawn and found to be linear from the slope of which E^* was calculated. The pre-exponential factor, A, was calculated from the equation:

$$E^*/RT_s^2 = A\big/\left[\varphi\exp\left(-E^*/RT_s\right)\right]$$

The entropy of activation, ΔS^*, was calculated from Equation (3). The enthalpy activation, ΔH^*, and Gibbs free energy, ΔG^*, were calculated from; $\Delta H^* = E^* - RT$ and $\Delta G^* = \Delta H^* - T\Delta S^*$, respectively. The kinetic parameters such as activation energy (ΔE^*), enthalpy (ΔH^*), entropy (ΔS^*) and free energy change of the decomposition (ΔG^*) have been evaluated graphically as shown in **Figure 12** and **Figure 13** by employing the Coats-Redfern

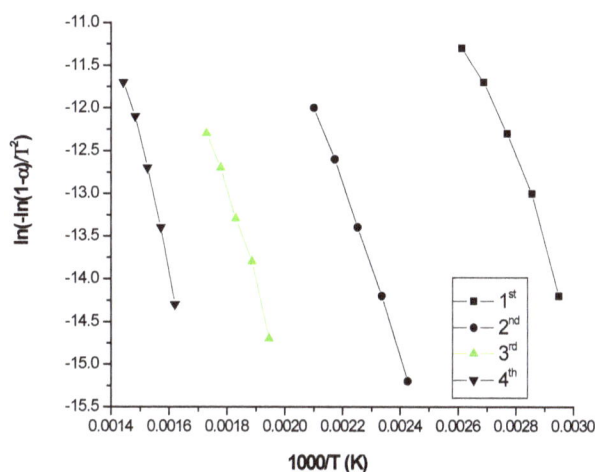

Figure 12. Coats-Redfern plots of the first, second, third and fourth thermal decomposition steps of Se-**vit-OH** complex.

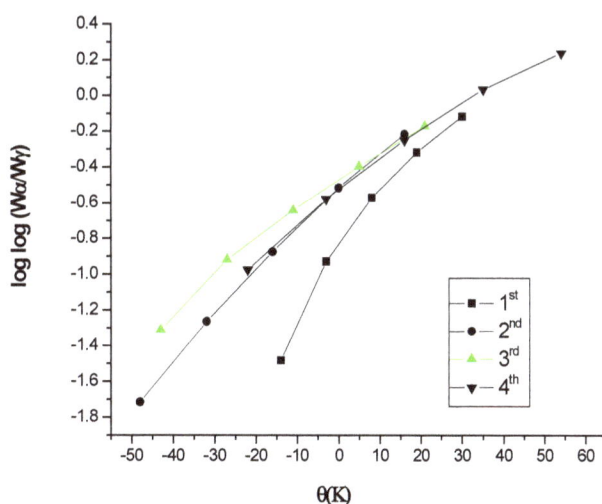

Figure 13. Horowitz-Metzger plots of the first, second, third and fourth thermal decomposition steps of Se-**vit-OH** complex.

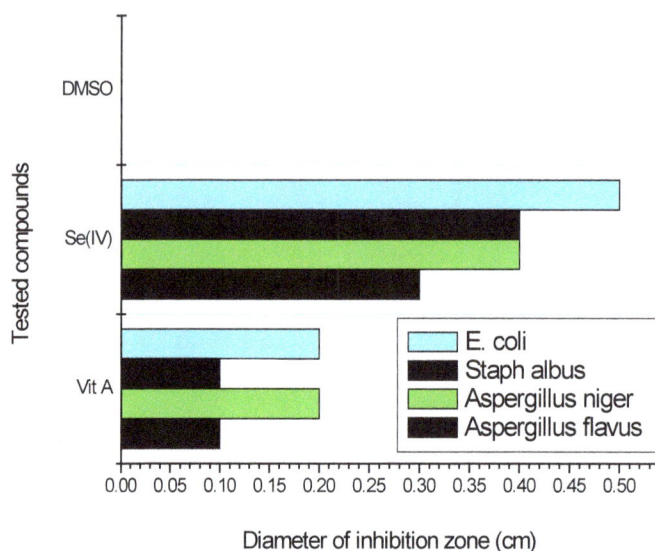

Figure 14. Biological evaluation of **vit-OH**, **Se-vit-OH** complex and DMSO solvent.

and Horwitz-Mitzger relations. The most significant result is the considerable thermal stability of the Se-**vit-OH** complex reflected from the high values of the activation energy of the decomposition. The entropy change ΔS^* of the formation activated complex from the starting reactants is in most cases of negative values. The negative sign of the ΔS^* suggests that the thermodynamic behavior is non-spontaneous (more ordered) reactions and the degree of structural "complexity" (arrangement, "organization") of the activated complex was lower than that of the starting reactants, also the thermodynamic behavior of selenium complex is endothermic reactions ($\Delta H > 0$) and endergonic ($\Delta G > 0$), during the reactions. The thermodynamic data obtained with the two methods are in harmony with each other. The correlation coefficients of the Arrhenius plots of the thermal decomposition steps were found to lie in the range 0.97 to 0.99, showing a good fit with linear function.

3.2.6. Antimicrobial Assessment of Se-vit-OH Complex

Peng *et al*. [40] reported that 36 nm of Se metal has lower toxicity than selenite or selenomethionine with higher particle sizes, but all of these forms of (Se) possess similar ability to increase selenoenzyme levels. They deduced also that the size of nanoparticles plays an important role in their biological activity: as expected, from range 5 - 200 nm for Se which can be directly scavenge free radicals *in vitro* in a size-dependent fashion. Accordingly the present techniques of synthesizing selenium metal within the range of (15 - 35 nm) is a unique and pioneer to increase the validity for selenium in biological reaction and leading new trends to decrease toxicity of selenium of higher particle size as reported in literature [40]. Antibacterial and antifungal activities of Se-**vit-OH** complex are carried out against some kinds of bacteria as *Escherichia coli* (Gram − ve) and *Staphalbus* (Gram + ve) as well as some kinds of fungi as *Aspergillus niger* and *Aspergillus flavus*. The antimicrobial activity was estimated based on the size of inhibition zone around dishes. The free vitamin A was found to have the lowest activity against four types of bacteria and fungi, while the selenium complex was found to be more potent than the original chelate in their inhibition properties (**Figure 14**). This has been explained in terms of the greater lipid solubility and cellular penetration of the complexes [41] [42].

Acknowledgements

This work was funded By Deanship of Scientific Research at the University of Dammam, Kingdom Saudi Arabia under Project Grants No. 2013173.

References

[1]   Odom, J.D., Dawson, W.H. and Ellis, P.D. (1979) Selenium-77 Relaxation Time Studies on Compounds of Biological Importance: Dialkyl Selenides, Dialkyl Diselenides, Selenols, Selenonium Compounds, and Seleno Oxyacids. *Journal*

of the American Chemical Society, **101**, 5815-5822. http://dx.doi.org/10.1021/ja00513a058

[2]   Robberecht, H.J. and Deelstra, H.A. (1984) Selenium in Human Urine Determination, Speciation and Concentration Levels. *Talanta*, **31**, 497-508. http://dx.doi.org/10.1016/0039-9140(84)80129-0

[3]   Andrews, R.W. and Johnson, D.C. (1975) Voltammetric Deposition and Stripping of Selenium(IV) at a Rotating Gold-Disk Electrode in 0.1M Perchloric Acid. *Analytical Chemistry*, **47**, 294-299. http://dx.doi.org/10.1021/ac60352a005

[4]   Thomson, C.D. (2004) Assessment of Requirements for Selenium and Adequacy of Selenium Status: A Review. *European Journal of Clinical Nutrition*, **58**, 391-402. http://dx.doi.org/10.1038/sj.ejcn.1601800

[5]   Combs, G.F. and Gray, W.P. (1998) Chemopreventive Agents: Selenium. *Pharmacology & Therapeutics*, **79**, 179-192. http://dx.doi.org/10.1016/S0163-7258(98)00014-X

[6]   Levander, O.A. (1997) Nutrition and Newly Emerging Viral Diseases: An Overview. *Journal of Nutrition*, **127**, 948S-950S.

[7]   Corvilain, B., Contempre, B., Longombe, A.O., Goyens, P., Gervy-Decoster, C., Lamy, F., Vanderpas, J.B. and Dumont, J.E. (1993) Selenium and the Thyroid: How the Relationship Was Established. *The American Journal of Clinical Nutrition*, **57**, 244S-248S.

[8]   Sauberlich, H.E., Hodges, R.E., Wallace, D.L., Kolder, H., Canham, J.E., Hood, J., Raica Jr., N. and Lowry, L.K. (1974) A Metabolism and Requirements in the Human Studied with the Use of Labeled Retinol. *Vitamins and Hormones: Advances on Research and Applications*, **32**, 251.

[9]   Sommer, A. and West, K.P. (1996) Vitamin A Deficiency: Health, Survival, and Vision. Oxford University Press, New York, 100-116.

[10] Drevensek, P., Zupancic, T., Pihlar, B., Jerala, R., Kolitsch, U., Plaper, A. and Turel, I. (2005) Mixed-Valence Cu(II)/Cu(I) Complex of Quinolone Ciprofloxacin Isolated by a Hydrothermal Reaction in the Presence of l-Histidine: Comparison of Biological Activities of Various Copper-Ciprofloxacin Compounds. *Journal of Inorganic Biochemistry*, **99**, 432-442. http://dx.doi.org/10.1016/j.jinorgbio.2004.10.018

[11] He, J.H., Xiao, D.R., Chen, H.Y., Yan, S.W., Sun, D.Z., Wang, X., Yang, J., Yuan, R. and Wang, E.B. (2012) Two Novel Entangled Metal-Quinolone Complexes with Self-Threading and Polythreaded Characters. *Inorganica Chimica Acta*, **385**, 170-177. http://dx.doi.org/10.1016/j.ica.2012.01.056

[12] Kathawate, L., Sproules, S., Pawar, O., Markad, G., Haram, S., Puranik, V. and Salunke-Gawali, S. (2013) Synthesis and Molecular Structure of a Zinc Complex of the Vitamin K_3 Analogue Phthiocol. *Journal of Molecular Structure*, **1048**, 223-229. http://dx.doi.org/10.1016/j.molstruc.2013.05.057

[13] Gielen, M. and Tiekink, E.R.T. (2005) Metallotherapeutic Drugs and Metal-Based Diagnostic Agents, the Use of Metals in Medicine. Wiley, Chichester. http://dx.doi.org/10.1002/0470864052

[14] Weder, J.E., Dillon, C.T., Hambley, T.W., Kennedy, B.J., Lay, P.A., Biffin, J.R., Regtop, H.L. and Daview, N.M. (2002) Copper Complexes of Non-Steroidal Anti-Inflammatory Drugs: An Opportunity Yet to Be Realized. *Coordination Chemistry Reviews*, **232**, 95-126. http://dx.doi.org/10.1016/S0010-8545(02)00086-3

[15] Ware, D.C., Brothers, P.J. and Clark, G.R. (2000) Synthesis, Structures and Hypoxia-Selective Cytotoxicity of Cobalt(III) Complexes Containing Tridentate Amine and Nitrogen Mustard Ligands. *Journal of the Chemical Society, Dalton Transactions*, **6**, 925-932. http://dx.doi.org/10.1039/a909447d

[16] Nakai, M., Sekiguchi, F., Obata, M., Ohtsuki, C., Adachi, Y., Sakurai, H., Orvig, C., Rehder, D. and Yano, S. (2005) Synthesis and Insulin-Mimetic Activities of Metal Complexes with 3-Hydroxypyridine-2-Carboxylic Acid. *Journal of Inorganic Biochemistry*, **99**, 1275-1282. http://dx.doi.org/10.1016/j.jinorgbio.2005.02.026

[17] Muller, J.G. and Burrows, C.J. (1998) Metallodrug Complexes That Mediate DNA and Lipid Damage via Sulfite Autoxidation: Copper(II) Famotidine and Iron(III) Bis(Salicyglycine). *Inorganica Chimica Acta*, **275-276**, 314-319. http://dx.doi.org/10.1016/S0020-1693(97)06179-3

[18] Duda, A.M., Kowalik-Jankowska, T., Kozlowski, H. and Kupka, T. (1995) Histamine H_2 Antagonists: Powerful Ligands for Copper(II). Reinterpretation of the Famotidine-Copper(II) System. *Journal of the Chemical Society, Dalton Transactions*, No. 17, 2909-2913. http://dx.doi.org/10.1039/dt9950002909

[19] Kubiak, M., Duda, A.M., Ganadu, M.L. and Kozlowski, H. (1996) Crystal Structure of a Copper(II)-Famotidine Complex and Solution Studies of the Cu^{2+}-Famotidine-Histidine Ternary System. *Journal of the Chemical Society, Dalton Transactions*, No. 9, 1905-1908. http://dx.doi.org/10.1039/dt9960001905

[20] Wu, C.-D., Lu, C.-Z., Zhuang, H.-H. and Huang, J.-S. (2002) Hydrothermal Assembly of a Novel Three-Dimensional Framework Formed by $[GdMo_{12}O_{42}]^{9-}$ Anions and Nine Coordinated Gd^{III} Cations. *Journal of the American Chemical Society*, **124**, 3836-3837. http://dx.doi.org/10.1021/ja017782w

[21] Dybtsev, D.N., Chun, H. and Kim, K. (2004) Rigid and Flexible: A Highly Porous Metal-Organic Framework with Unusual Guest-Dependent Dynamic Behavior. *Angewandte Chemie International Edition*, **43**, 5033-5036. http://dx.doi.org/10.1002/anie.200460712

[22] López-Gresa, M.P., Ortiz, R., Perelló, L., Latorre, J., Liu-González, M., García-Grand, S., Pérez-Priede, M. and Cantón, E. (2002) Interactions of Metal Ions with Two Quinolone Antimicrobial Agents (Cinoxacin and Ciprofloxacin): Spectroscopic and X-Ray Structural Characterization. Antibacterial Studies. *Journal of Inorganic Biochemistry*, **92**, 65-74. http://dx.doi.org/10.1016/S0162-0134(02)00487-7

[23] Xiao, D.-R., Wang, E.-B., An, H.-Y., Li, Y.-G. and Xu, L. (2007) Syntheses and Structures of Three Unprecedented Metal-Ciprofloxacin Complexes with Helical Character. *Crystal Growth & Design*, **7**, 506-512. http://dx.doi.org/10.1021/cg060492c

[24] Xiao, D.R., He, J.H., Sun, D.Z., Chen, H.Y., Yan, S.W., Wang, X., Yang, J., Yuan, R. and Wang, E.B. (2012) Three 3D Metal-Quinolone Complexes Based on Trimetallic or Rod-Shaped Secondary Building Units. *European Journal of Inorganic Chemistry*, **2012**, 1783-1789. http://dx.doi.org/10.1002/ejic.201101229

[25] Refat, M.S. and El-Sabawy, K.M. (2011) Infrared Spectra, Raman Laser, XRD, DSC/TGA and SEM Investigations on the Preparations of Selenium Metal, (Sb_2O_3, Ga_2O_3, SnO and HgO) Oxides and Lead Carbonate with Pure Grade Using Acetamide Precursors. *Bulletin of Materials Science*, **34**, 873-881. http://dx.doi.org/10.1007/s12034-011-0208-z

[26] Gupta, R., Saxena, R.K., Chatarvedi, P. and Virdi, J.S. (1995) Chitinase Production by *Streptomyces viridificans*: Its Potential in Fungal Cell Wall Lysis. *Journal of Applied Bacteriology*, **78**, 378-383. http://dx.doi.org/10.1111/j.1365-2672.1995.tb03421.x

[27] Scherrer, P. (1918) Bestimmung der Größe und der inneren Struktur von Kolloidteilchen mittels Röntgenstrahlen. *Nachrichten von der Gesellschaft der Wissenschaften zu Göttingen*, **2**, 98-100.

[28] Iovu, M.S., Kamitsos, E.I., Varsamis, C.P.E., Boolchand, P. and Popescu, M. (2005) Raman Spectra of $AsxSe_{100-x}$ and $As_{40}Se_{60}$ Glassesdoped with Metals. *Chalcogenide Letters*, **2**, 21-25.

[29] Kovanda, V., Vlček, M. and Jain, H. (2003) Structure of As-Se and As-P-Se Glasses Studied by Raman Spectroscopy. *Journal of Non-Crystalline Solids*, **326-327**, 88-92. http://dx.doi.org/10.1016/S0022-3093(03)00383-1

[30] Refat, M.S. (2007) Complexes of Uranyl(II), Vanadyl(II) and Zirconyl(II) with Orotic Acid "Vitamin B13": Synthesis, Spectroscopic, Thermal Studies and Antibacterial Activity. *Journal of Molecular Structure*, **842**, 24-37. http://dx.doi.org/10.1016/j.molstruc.2006.12.006

[31] Vogel, T. (1989) Vogel Textbook of Practical Organic Chemistry. 4th Edition, Wiley & Sons, Inc., New York, 133-325.

[32] Nakamoto, K. (1970) Infrared Spectra of Inorganic and Coordination Compounds. Wiley InterScience, New York.

[33] Mohamed, G.G., Zayed, M.A., Nour El-Dien, F.A. and El-Nahas, R.G. (2004) IR, UV-Vis, Magnetic and Thermal Characterization of Chelates of Some Catecholamines and 4-Aminoantipyrine with Fe(III) and Cu(II). *Spectrochimica Acta Part A: Molecular and Biomolecular Spectroscopy*, **60**, 1775-1781. http://dx.doi.org/10.1016/j.saa.2003.08.027

[34] Santi, E., Torre, M.H., Kremer, E., Etcheverry, S.B. and Baran, E. (1993) Vibrational Spectra of the Copper(II) and Nickel(II) Complexes of Piroxicam. *Vibrational Spectroscopy*, **5**, 285-293. http://dx.doi.org/10.1016/0924-2031(93)87004-D

[35] Arumuganathan, T., Srinivasarao, A., Kumar, T.V. and Das, S.K. (2008) Two Different Zinc(II)-Aqua Complexes Held Up by a Metal-Oxide Based Support: Synthesis, Crystal Structure and Catalytic Activity of [HMTAH]$_2$[{Zn(H$_2$O)$_5$}{Zn(H$_2$O)$_4$}{Mo$_7$O$_{24}$}]·2H$_2$O (HMTAH = Protonated Hexamethylenetetramine). *Journal of Chemical Sciences*, **120**, 95-103. http://dx.doi.org/10.1007/s12039-008-0012-5

[36] Freeman, E.S. and Carroll, B. (1958) The Application of Thermoanalytical Techniques to Reaction Kinetics: The Thermogravimetric Evaluation of the Kinetics of the Decomposition of Calcium Oxalate Monohydrate. *The Journal of Physical Chemistry*, **62**, 394-397. http://dx.doi.org/10.1021/j150562a003

[37] Coats, A.W. and Redfern, J.P. (1964) Kinetic Parameters from Thermogravimetric Data. *Nature*, **201**, 68-69. http://dx.doi.org/10.1038/201068a0

[38] Horowitz, H.W. and Metzger, G. (1963) A New Analysis of Thermogravimetric Traces. *Analytical Chemistry*, **35**, 1464-1468. http://dx.doi.org/10.1021/ac60203a013

[39] Flynn, J.H.F. and Wall, L.A. (1966) General Treatment of the Thermogravimetry of Polymers. *Journal of Research of the National Bureau of Standards*, **70A**, 487-523. http://dx.doi.org/10.6028/jres.070A.043

[40] Peng, D.G., Zhang, J.S., Liu, Q.L. and Taylor, E.W. (2007) Size Effect of Elemental Selenium Nanoparticles (Nano-Se) at Supranutritional Levels on Selenium Accumulation and Glutathione S-Transferase Activity. *Journal of Inorganic Biochemistry*, **101**, 1457-1463. http://dx.doi.org/10.1016/j.jinorgbio.2007.06.021

[41] Jam, P. and Chaturvedi, K.K. (1976) Potentiometric Study of the Complexes of Sulphamethazine Salicylaldimine with Cu(II), Ni(II) and Co(II). *Journal of Inorganic and Nuclear Chemistry*, **38**, 799-800. http://dx.doi.org/10.1016/0022-1902(76)80359-4

[42] Jam, P. and Chaturvedi, K.K. (1977) Complexes of Cu(II), Ni(II) and Co(II) with Sulphamerazine. *Journal of Inorganic and Nuclear Chemistry*, **39**, 901-903. http://dx.doi.org/10.1016/0022-1902(77)80182-6

12

Surface Effect of Silica Nano-Particles with Different Size on Thermotropic Liquid Crystalline Polyester Composites

Hiroki Fukatsu[1,2], Masaaki Kuno[3], Yasuhiro Matsuda[3], Shigeru Tasaka[1,3]

[1]Graduate School of Science and Engineering, Shizuoka University, Hamamatsu, Japan
[2]Technical Solution Center, Polyplastics Co., Ltd., Fuji, Japan
[3]Graduate School of Engineering, Shizuoka University, Hamamatsu, Japan
Email: tsctasa@ipc.shizuoka.ac.jp

Abstract

Interface properties of nano-silica/thermotropic liquid crystalline polyesters (TLCP) composites were investigated by X-ray diffraction analysis and differential scanning calorimetory. The crystallinity of TLCP in the composites drastically decreased with an increase of nano-silica content, depending on the surface area of the silica particles. Little size effects (40 - 400 nm) in the particles and strong interaction between silica surface and the C=O moieties of TLCP were observed by IR analysis. The glass transition temperature of TLCP (<20 nm) in silica surface increased about 20°C higher than that in bulk.

Keywords

Component, Nano-Silica, Thermotropic Crystalline Polyester, Interface, Differential Scanning Calorimetry

1. Introduction

Thermotropic main-chain liquid crystalline polymers (TLCPs) usually have rigid main chains and good mechanical properties, high chemical and thermal stability, which make TLCPs high performance engineering plastics widely used for electronics, automotive and aerospace applications [1]. The main chains of molten TLCPs align spontaneously for their rigidity [2], and this orientated structure affects the excellent physical and chemical properties of TLCP. In general, such as glass fibers (GFs), or mineral fillers are added to TLCP to improve dimensional stability, thermal properties and processability of neat TLCP. In order to achieve above properties, understanding the mutual interaction at the interface between the inorganic filler and TLCP is important.

There have been many studies on the influence of various kinds of fillers; GFs or carbon fibers (CFs) to reduce the anisotropy of TLCPs [3]-[8]. Bhama and Stupp [7] synthesized a random terpolymer of p-acetoxybenzoic acid, diacetoxyhydroquine, and pimelic acid, and added CFs to investigate the interfacial structure of this TLCP. The orientation time of the TLCP determined by NMR was decreased by adding only 1 wt% of CFs, and polarizing optical micrographs (POM) of the composites of the TLCP and CFs indicated that the orientation rate of the TLCP was enhanced at the interfacial zones.

Lee *et al.* [8] prepared composites of a copolymer composed of 70 mol% of 4-hydroxybenzoic acid (HBA) and 30 mol% of 6-hydroxy-2-naphthoic acid (HNA) and GFs. To investigate the influence of chemical interaction between TLCP and GFs, they used three types of GFs; GFs treated with γ-methacrylate propyltrimethoxysilane (γ-MPS), which reacts with the TLCP, GFs treated with phenyltriethoxysilane (PTS) which does not react with the TLCP, and GFs without any treatments. A thin layer of TLCP with banded optical texture parallel to GFs was identified in the composite of TLCP and untreated GFs based on the experimental results of POM, scanning electron micrography, and X-ray diffractometry (XRD). This thin layer was induced by the guiding, anchoring, and packing effects of GFs. The same thin layer was also observed in the composite of TLCP and GFs treated with γ-MPS, but there was only thinner layer where TLCP randomly arranged in the composites of TLCP and GFs treated with PTS.

The conventional thermal analysis (DSC), X-ray diffraction (XRD) can obtain even enough interface information to increase the surface area by adding an excess of heterogeneous material in the polymer in our laboratory. Then unique phenomenon was observed in TLCP and SNPs with different sizes [9]-[11]. Our final purpose is to provide a novel TLCP composite with various fillers which obtained interfacial information by the conventional method. Comparing the researches on the influence of adding fiber on the properties of TLCP, there have been much fewer reports on the effect of adding nano-particles on the properties of TLCPs.

It is expected to be difficult for the rigid chains of TLCPs to align at the curved surfaces of nano-particles, while it is easy for the rigid chains of TLCPs to align parallel to the fibers. The rigid chains of TLCPs in composites with nano-particles may be forced to form structure which is different from the structure of TLCPs without fillers. This suggests that the influence of nano-particles on the properties of TLCPs can be different from that of fibers.

In this study, XRD, IR, differential scanning calorimetry (DSC) and dynamic mechanical analysis (DMA) were carried out for composites of a copolymer of HBA and HNA, and SNPs with different sizes to investigate the size effect of SNPs with on the interfacial structural of this TLCP.

This template, created in MS Word 2007, provides authors with most of the formatting specifications needed for preparing electronic versions of their papers. All standard paper components have been specified for three reasons: 1) ease of use when formatting individual papers, 2) automatic compliance to electronic requirements that facilitate the concurrent or later production of electronic products, and 3) conformity of style throughout a journal paper. Margins, column widths, line spacing, and type styles are built-in; examples of the type styles are provided throughout this document and are identified in italic type, within parentheses, following the example. Some components, such as multi-leveled equations, graphics, and tables are not prescribed, although the various table text styles are provided. The formatter will need to create these components, incorporating the applicable criteria that follow.

2. Experimental

The TLCP used in this study is VECTRA A950 supplied by Polyplastics Co. Ltd, which is composed of 73 mol% of HBA and 27 mol% of HNA. The chemical structure of this TLCP is shown in **Figure 1**. The weight average molecular weight and molecular weight distribution of VECTRA A950 were reported to be 30,000 [12], and 2.45 [13], respectively. The random sequence distribution of this sample was also confirmed [14]-[19]. SNPs with average diameter of 51 nm were purchased from Kanto Chemical, and those with average diameter of 240 and 485 nm were the products of Amatecs. SNPs were dried for an hour in a vacuum oven heated at 300°C to remove adsorbed water. The densities of the TLCP and SNPs are 1.4 g·cm^{-3}, and 2.2 g·cm^{-3}, respectively.

The composites of the TLCP and SNPs were prepared by melt-compounding with a twin screw melt mixer (Labo-plastmill 4C150-01, Toyo Seiki Seisaku-sho) at 300°C for 8 min. The screw rotation speed was 100 rpm,

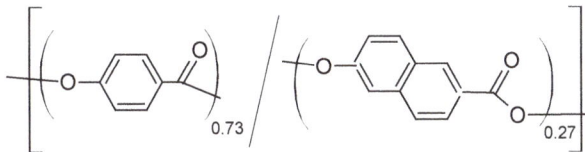

Figure 1. Chemical structure of TLCP used in this study.

and the final rotating torque of mixing was saturated under this condition. The contents of SNPs in the composites were confirmed with a thermogravimetric analyzer (TGA Q-500, TA Instruments). The films of the composites for the measurements were prepared by hot-pressing at 300°C, and then cooled slowly at room temperature.

The thermal properties of the composites were measured with a Q1000 calorimeter manufactured by TA Instruments calibrated with analytical grade indium. The samples were heated and cooled at the rate of 10°C/min under nitrogen atmosphere. XRD was carried out at room temperature with a RINT2500 manufactured by Rigaku using Cu Kα radiation as an X-ray source. Fourier transfer Infrared (FTIR) spectra were obtained using a FTIR-8900 of Shimadzu device of resolution 4 cm^{-1}. Dynamic mechanical measurement (DMA) was performed by a rheometric solids analyzer RSA III (Rheometric Scientific). The measurements were carried out in tensile mode, both the frequency (Fr) and the strain amplitude (γ) are kept constant in each test interval; the only variable parameter is temperature. The samples were heated with a ramp rate of 5°C/min at constant γ = 0.1% and Fr = 1 Hz under a nitrogen environment.

3. Results and Discussion

Figure 2 shows XRD patterns for the composites of TLCP and SNPs with different diameters. The fraction of SNPs was fixed to be 50 wt% (39 vol%). The sample designated as "Unfilled" is TLCP without SNPs. The intensity of the peak at 19.5°, which is caused by the psudeohexagonal structure of VECTRA A950 [20] [21], was decreased by adding SNPs. SNPs with smaller diameter decreased the peak intensity more significantly, and the same tendency was also measured for the composites with different weight fraction of SNPs. This result suggests smaller SNPs prevent the formation of crystal structure of TLCP. The heating and cooling curves of DSC for the composites of TLCP and SNPs with different diameters are shown in **Figure 3**. The endothermic peak at 280°C in the heating curve and exothermic peak at 230°C in the cooling curve are caused by the phase transition from crystal to nematic and from nematic to crystal, respectively. Both the endothermic peak and exothermic peak became weaker by adding SNPs with smaller diameters. In addition, the abrupt change of the specific heat around 90°C is caused by the glass transition of the amorphous regions of TLCP. The change of the specific heat around the glass transition temperature (Tg) became more significant by adding SNPs with smaller diameters. Moreover, Tg decreases with an increase of amorphous contents, and increases more than 20°C with further adding SNPs of 51 nm. These DSC results suggest that smaller SNPs prevent the formation of crystal structure of the TLCP and increase the amorphous region, which was supported by the results of XRD. **Figure 4** and **Figure 5** summarize the results of DSC for the composites of the TLCP and SNPs. ΔHm and ΔCp are the heat of fusion at 280°C and the change of the specific heat at 90°C, respectively. These clearly show a decrease of ΔHm and an increase of ΔCp by adding SNPs, and smaller SNPs was influenced on the changes of ΔHm and ΔCp. The results of XRD and DSC showed that the addition of SNPs to the composite of the TLCP and SNPs prevents the crystallization and induces the formation of amorphous regions of the TLCP, and that these tendencies became more significant when SNPs with smaller sizes were used for the sample preparation. This type of interfacial structures has been seen in polyamide/nano-silica composites [22]. **Figure 6** shows IR changes of C=O stretching band sift for the composites of TLCP/SNPs. The doublet bands (~1740 cm^{-1}) in TLCP and shift to lower and higher wave numbers with broadening, and changed in strength by mixing with SNPs. This behavior has no size dependences of SNPs in range of this experiment. The C=O groups get two states in amorphous. One is a constrained state and the other is a relatively free state. This means that a part of the C=O groups in TLCP chain is strongly bound in the surface of SNPs. **Figure 7** shows storage modulus (a) and tanδ (b) for the composites of the TLCP and SNPs with different size as a function of temperature. The storage modulus of smaller SNPs showed much lower and especially drastic decrease over 150°C. This result showed that stiffness decreased at high temperature due to increase of the amorphous region and decrease of the crystalline region as

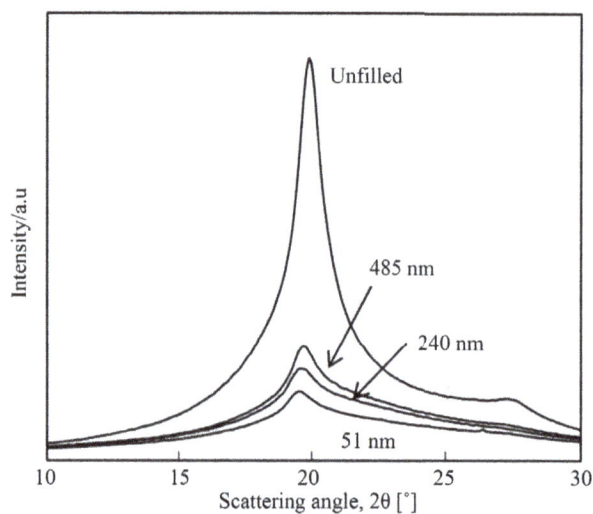

Figure 2. XRD patterns of the composites of TLCP and SNPs with different diameters indicated in the figure. The weight fraction of SNPs was fixed to be 50%. The sample designated as "Unfilled" indicates TLCP without SNPs.

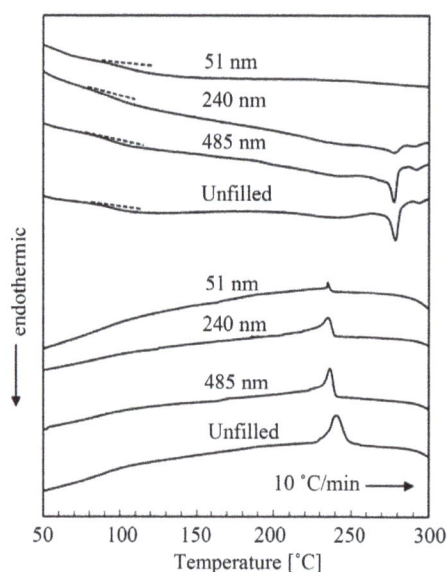

Figure 3. DSC curves of the composites of TLCP and SNPs with different diameters indicated in the figure. The weight fraction of SNPs was fixed to be 50%. The sample designated as "Unfilled" indicates TLCP without SNPs. The 4 upper curves and 4 lower curves were obtained by heating and cooling measurements, respectively.

suggested from the results DSC, the XRD. Because around 100°C peak is corresponded with cooperative motions and glass transition, the results suggest that the presence of 51 nm SNP slow down the chain dynamics of TLCP thus increasing the glass transition. These results agree with those obtained from DSC. It is considered that the electrostatic interaction between negative charged carbonyl group and positive charged metal ions make TLCP chains tie up compared to the bulk state. This has been discussed about acrylate adhesives by IR analysis. [23].

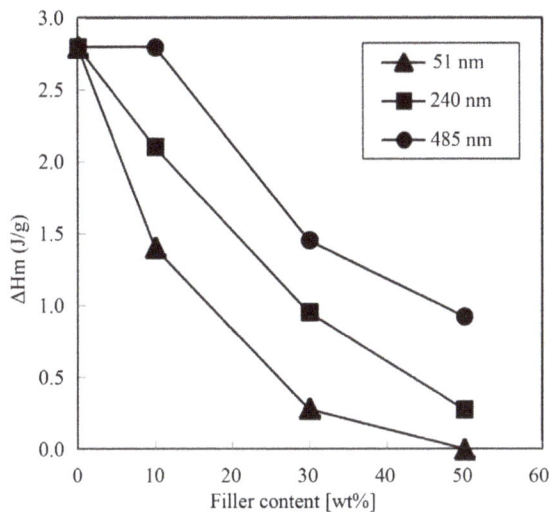

Figure 4. The heat of fusion detected for the composites at 230°C.

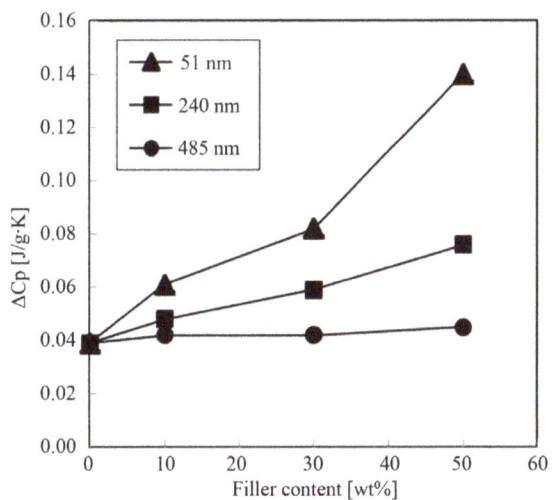

Figure 5. The change of the specific heat of Tg.

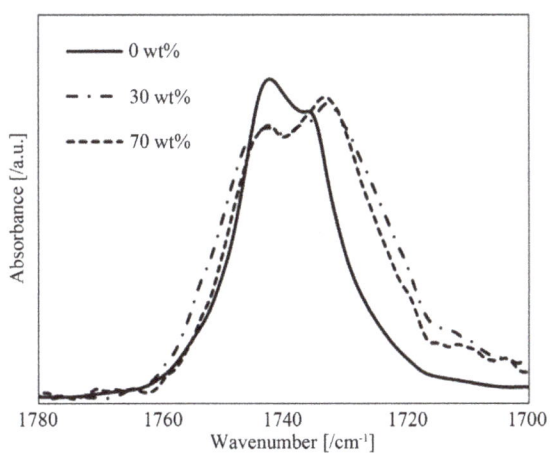

Figure 6. IR spectral changes in TLCP/SNPs (51 nm) composites.

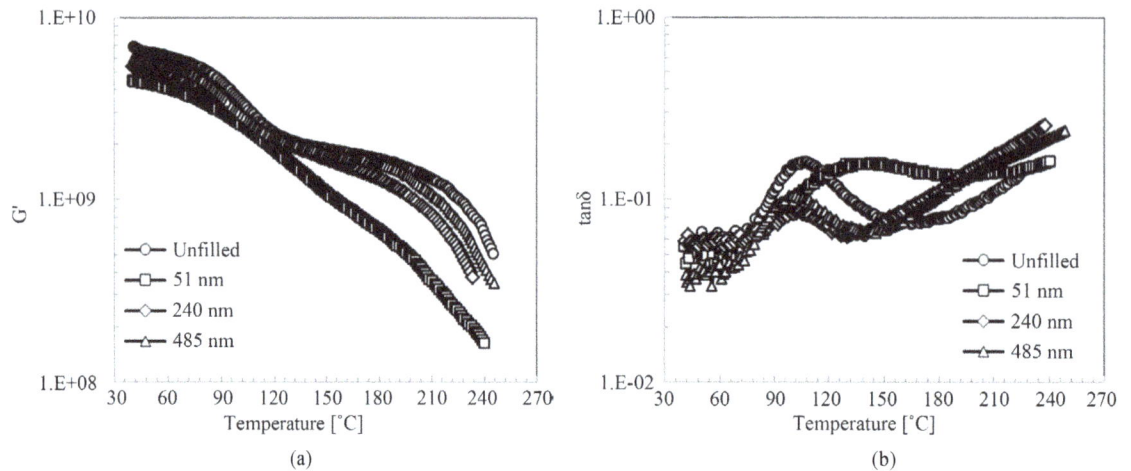

Figure 7. Temperature dependence of (a) storage modulus (G'); (b) tanδ of the composites of TLCP (0.7) and SNPs (0.3) with different diameters.

When the TLCP chains align parallel to the curved interface of SNPs, the TLCP chains are forced to bend along the curved interface of SNPs. This deformation makes it difficult for the TLCP chains to form crystal structure because the chains of the TLCP without SNPs have high rigidity. We have no definite evidence that the TLCP chains align parallel to the curved interface of SNPs at this time, but there have been many reports that polymer chains of other TLCPs in composites with GFs or CFs aligned parallel to the interface with fibers [8] [9]. The stronger effect for SNPs with smaller sizes to depress the crystallization can be explained by the increase of the surface area. The surface area per weight of SNPs with diameter of 51 nm is 4.7 times as large as that of SNPs with diameter of 240 nm, and 9.5 times as large as that of SNPs with diameter of 485 nm. The decrease of ΔHm for the composites of the TLCP and SNPs with diameters of 240 nm and 485 nm shown in **Figure 4** can be explained by assuming that SNPs were completely dispersed in the composites, and that there are interfacial phase with thickness of over 100 nm where the TLCP cannot be crystallized. It is generally accepted that the thickness of interfacial phase between polymers and fillers is comparable to the dimension of a single polymer chain, and that the molecular mobility and structure of the polymers at this interfacial phase are different from those of the polymers at the bulk phase. The thick interfacial phase of the TLCP estimated in this study may be caused by the rigidity of the TLCP because rigid polymers usually have larger dimension than flexible polymers. The composite of SNPs with 51 nm has a small ΔHm value, which may be caused by the partial aggregation giving a decrease of the effective surface area of SNPs.

4. Conclusion

In this study, X-ray diffractometry (XRD) and differential scanning calorimetry (DSC) were carried out for composites of TLCP and SNPs with different diameters. Results show that the addition of SNPs to the composite of the TLCP and SNPs prevents the crystallization and induces the formation of amorphous regions of the TLCP, and that these tendencies became more significant when SNPs with smaller diameters were used for the sample preparation. This seems to make the strongly bended structure of TLCP molecules with distortion at the interface. The formation of these interfaces is important not only for physical interests, but also for the application of TLCPs as engineering plastics because the liquid crystalline structure dominates the mechanical properties of TLCPs.

References

[1]   Sawyer, L.C., Linstid, H.C. and Romer, M. (1998) Emerging Applications for Neat LCPs. *Plastics Engineering* (*N.Y.*), **54**, 37-41.

[2]   Park, S.K., Kim, S.H. and Hwang, J.T. (2009) Effect of Fumed Silica Nanoparticles on Glass Fiber Filled Thermotropic Liquid Crystalline Polymer Composites. *Polymer Composites*, **30**, 309-317. http://dx.doi.org/10.1002/pc.20557

[3]   Voss, H. and Friedrich, K. (1986) Influence of Short-Fibre Reinforcement on the Fracture Behaviour of a Bulk Liquid

Crystal Polymer. *Journal of Materials Science*, **21**, 2889-2900. http://dx.doi.org/10.1007/BF00551508

[4] Chivers, R.A. and Moore, D.R. (1991) Influence of Fibre Reinforcement on the Mechanical Anisotropy of Liquid Crystal Polymers. *Polymer*, **32**, 2190-2198. http://dx.doi.org/10.1016/0032-3861(91)90045-K

[5] Plummer, C.J.G., Zülle, B., Demarmels, A. and Kausch, H.-H. (1993) The Structure of Filled and Unfilled Thermotropic Liquid Crystalline Polymer Injection Moldings. *Journal of Applied Polymer Science*, **48**, 751-766. http://dx.doi.org/10.1002/app.1993.070480501

[6] Scaffaro, R., Pedretti, U. and La Mantia, F.P. (1996) Effects of Filler Type and Mixing Method on the Physical Properties of a Reinforced Semirigid Liquid Crystal Polymer. *European Polymer Journal*, **32**, 869-875. http://dx.doi.org/10.1016/0014-3057(96)00009-2

[7] Bhama, S. and Stupp, S.I. (1990) Liquid Crystal Polymer-Carbon Fiber Composites. Molecular Orientation. *Polymer Engineering & Science*, **30**, 228-234. http://dx.doi.org/10.1002/pen.760300406

[8] Lee, W.-G., Hsu, T.-C.J. and Su, A.C. (1994) Interphase Morphology of Liquid Crystalline Polymer/Glass Fiber Composites: Effect of Fiber Surface Treatment. *Macromolecules*, **27**, 6551-6558. http://dx.doi.org/10.1021/ma00100a046

[9] Zhang, X., Tasaka, S. and Inagakim N. (2000) Surface Mechanical Properties of Low-Molecular-Weight Polystyrene below Its Glass-Transition Temperatures. *Journal of Polymer Science Part B*: *Polymer Physics*, **28**, 654-658. http://dx.doi.org/10.1002/(SICI)1099-0488(20000301)38:5%3C654::AID-POLB2%3E3.0.CO;2-Z

[10] Zhang, X., Tasaka, S. and Inagaki, N. (2000) Studies on Surface Molecular Motion of Oligomeric Polystyrene by Differential Scanning Calorimetry. *Polymers for Advanced Technologies*, **11**, 40-47. http://dx.doi.org/10.1002/(SICI)1099-1581(200001)11:1%3C40::AID-PAT935%3E3.0.CO;2-%23

[11] Zhang, X., Tasaka, S. and Inagaki, N. (2003) Adhesion Behavior of Blends of Polybutadiene and Tackifiers. *Journal of Adhesion Science and Technology*, **17**, 423-434. http://dx.doi.org/10.1163/156856103762864714

[12] Romo-Uribe, A. and Windle, A.H. (1996) "Log-Rolling" Alignment in Main-Chain Thermotropic Liquid Crystalline Polymer Melts under Shear: An *In-Situ* WAXS Study. *Macromolecules*, **29**, 6246-6255. http://dx.doi.org/10.1021/ma960211h

[13] Kromer, H., Khun, R., Pielartzik, H., Siebke, W., Eckhardt, V. and Schmidt, M. (1991) Persistence Length and Molecular Mass Distribution of a Thermotropic Main-Chain Liquid-Crystal Polymer. *Macromolecules*, **24**, 1950-1954. http://dx.doi.org/10.1021/ma00008a036

[14] Chivers, R.A., Blackwell, J. and Gutierrez, G.A. (1984) The Structure of Copoly(4-hydroxybenzoic acid/2-hydroxy-6-naphthoic Acid): 2. An Atomic Model for the Copolyester Chain. *Polymer*, **25**, 435-440. http://dx.doi.org/10.1016/0032-3861(84)90198-8

[15] Blackwell, J., Gutierrez, G.A. and Chivers, R.A. (1984) Diffraction by Aperiodic Polymer Chains: The Structure of Liquid Crystalline Copolyesters. *Macromolecules*, **17**, 1219-1224. http://dx.doi.org/10.1021/ma00136a019

[16] Blackwell, J., Biswas, A. and Bonart, R.C. (1985) X-Ray Studies of the Structure of Liquid-Crystalline Copolyesters: Treatment of an Atomic Model as a One-Dimensional Paracrystal. *Macromolecules*, **18**, 2126-2130. http://dx.doi.org/10.1021/ma00153a010

[17] Chivers, R.A. and Blackwell, J. (1985) Three-Dimensional Structure of Copolymers of *p*-Hydroxybenzoic Acid and 2-Hydroxy-6-Naphthoic Acid: A Model for Diffraction from a Nematic Structure. *Polymer*, **26**, 997-1002. http://dx.doi.org/10.1016/0032-3861(85)90219-8

[18] Biswas, A. and Blackwell, J. (1987) X-Ray Diffraction from Liquid-Crystalline Copolyesters: Matrix Methods for Intensity Calculations Using a One-Dimensional Paracrystalline Model. *Macromolecules*, **20**, 2997-3002. http://dx.doi.org/10.1021/ma00178a008

[19] Mitchell, G.R. and Windle, A.H. (1985) Diffraction from Thermotropic Copolyester Molecules. *Colloid and Polymer Science*, **263**, 230-244. http://dx.doi.org/10.1007/BF01415509

[20] Gutierrez, G.A., Chivers, R.A., Blackwell, J., Stamatoff, J.B. and Yoon, H. (1983) The Structure of Liquid Crystalline Aromatic Copolyesters Prepared from 4-Hydroxybenzoic Acid and 2-Hydroxy-6-Naphthoic Acid. *Polymer*, **24**, 937-942. http://dx.doi.org/10.1016/0032-3861(83)90141-6

[21] Mitchell, G.R. and Windle, A.H. (1983) Measurement of Molecular Orientation in Thermotropic Liquid Crystalline Polymers. *Polymer*, **24**, 1513-1520. http://dx.doi.org/10.1016/0032-3861(83)90164-7

[22] Achiha, O., Kyogoku, Y., Matsuda, Y. and Tasaka, S. (2012) Interfacial Structure of Composites of Poly(*m*-xylylen adipamide) and Silica Nano-Particles. *Japanese Journal of Applied Physics*, **51**, 100204-1-3. http://dx.doi.org/10.1143/JJAP.51.100204

[23] Soga, I. and Granick, S. (2000) Segmental Orientations of Trains versus Loops and Tails: The Adsorbed Polymethylmethacrylate System When the Surface Coverage Is Incomplete. *Colloids and Surface A*, **170**, 113-117. http://dx.doi.org/10.1016/S0927-7757(00)00409-X

13

Vibration of Gold Nano-Beam with Variable Young's Modulus Due to Thermal Shock

Eman A. N. Al-Lehaibi[1], Hamdy M. Youssef[2]

[1]Mathematics Department, College of Science and Arts—Sharoura, Najran University, Najran, KSA
[2]Mechanics Department, Faculty of Engineering, Umm Al-Qura University, Makkah, KSA
Email: sa_1993@hotmail.com, youssefanne2005@gmail.com

Abstract

In this paper, we will study the most important effects in the nano-scale resonator: the coupling effect of temperature and strain rate, and the non-Fourier effect in heat conduction. A solution for the generalized thermoelastic vibration of nano-resonator induced by thermal loading has been developed. The Young's modulus is taken as a linear function of the reference temperature. The effects of the thermal loading and the reference temperature in all the studied fields have been studied and represented in graphs with some comparisons. The Young's modulus makes significant effects on all the studied fields where the values of the temperature, the vibration of the deflection, stress, displacement, strain, stress-strain energy increase when the Young's modulus has taken to be variable.

Keywords

Thermoelasticity, Euler-Bernoulli Equation, Goldnano-Beam, Young's Modulus

1. Introduction

Diao *et al.* [1] were the first who discussed the effects of the free surfaces on the structure and the elastic properties of the gold nanowires by atomistic simulations. Although the atomistic simulation is a good method to calculate the elastic parameters of the nano-structured materials, it is only used to homogeneous nano-structured materials (e.g., nano-plates, nano-wires, nano-beams, … , etc.) with a finite number of atoms.

Recently, nano mechanical resonators have attracted considerable attention due to their many applications on technology. The analysis of various effects on the characteristics of resonators, such as the resonant frequencies and the quality factorsis crucial for designing high-performance components. Many authors have studied the vibration and the heat transfer process of nano-beams [2]-[8]. Kidawa [2] studied a problem of transverse vibrations of a beam induced by a mobile heat source. The analytical solution of the problem was obtained by using the Green's functions method. While, Kidawa did not consider the thermoelastic coupling effect between the governing equations. Boley [3] studied the vibrations of a simply supported rectangular nano-beam affected by a thermal shock distributed along its span. Manolis and Beskos [4] discussed the thermally induced vibration of structures consisting of nano-beams, exposed to rapid surface heating. They have also studied the effects of the damping and the axial loads on the structural response. Al-Huniti *et al.* [5] investigated the thermally induced displacements and stresses of a rod using the Laplace transforms technique. Ai Kah Soh *et al.* studied the vibration of micro/nano-scale beam resonators induced by ultra-short-pulsed laser by considering the thermoelastic coupling term in [6] and [7]. The propagation characteristics of the longitudinal wave in nano-plates with small-scale effects are studied by Wang *et al.* [8].

2. Variable Young's Modulus

The temperature dependence of the Young's modulus for some materials was measured in the range of 293K and 973 K, using the impulse excitation method and compared with literature data reported. The data could be fitted with [9]

$$E = E_0 - BTe^{(-T_0/T)}. \tag{1}$$

The values of parameters E_0 and T_0 are related to the temperature and the parameter B to the harmonic character of the medium.

Farraro and Rex found that no departure from linearity was detected when they studied the dependency of the Young's modulus on the temperature, and the get the linear relation [10]

$$E = E_0 - E_1 T. \tag{2}$$

where E_0 is the Young's modulus in the standard case and E_1 is constant, and they measured it for pure Nickel, Platinum, and Molybdenum.

Now, we will consider the Young's modulus depends on the temperature by the following function

$$E(T) \approx E_0(1 - \gamma T_0) = E_0 E^*, \tag{3}$$

where γ is constant and

$$E^* = (1 - \gamma T_0). \tag{4}$$

In this paper, the non-Fourier effect on heat conduction, and the coupling effect between temperature and strain rate in the nano-scale beam will be studied when Young's modulus is variable as a function of temperature. A general solution for the generalized thermoelastic vibration of gold nano-beam resonator induced by thermal shock will be developed. Laplace transforms and direct method will be used to get the lateral vibration, the temperature, the displacement, the stress-strain energy of the beam. The effects of Young's modulus will be studied and represented graphically.

3. Problem Formulation

Since nano-beams with rectangular cross-sections are easier to fabricate, such cross-sections are commonly adopted in the design of NEMS resonators. Consider small flexural deflections of a thin elastic beam of length $\ell(0 \le x \le \ell)$, width $b\left(-\dfrac{b}{2} \le y \le \dfrac{b}{2}\right)$ and thickness $h\left(-\dfrac{h}{2} \le z \le \dfrac{h}{2}\right)$, for which the x, y and z-axes are defined along the longitudinal, width and thickness directions of the beam, respectively (**Figure 1**). In equilibrium, the beam is unstrained, unstressed, without damping mechanism, and the temperature is T_0 everywhere [6].

In the present work, the Euler-Bernoulli equation is considered, and then, any plane cross-section, initially perpendicular to the axis of the beam remains plane and perpendicular to the neutral surface during bending. Thus, the displacements are given by [6] [7]:

$$u = -z\frac{\partial w(x,t)}{\partial x}, \quad v = 0, \quad w(x,y,z,t) = w(x,t). \tag{5}$$

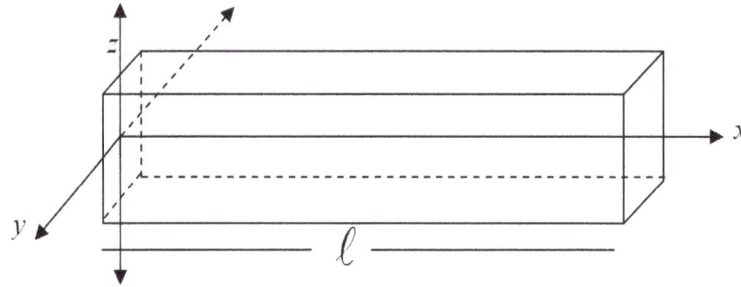

Figure 1. Cross-sections in the design of NEMS resonators.

Thus, the differential equation of thermally induced lateral vibration of the beam may be expressed in the form [6] [7] [11]-[13]:

$$\frac{\partial^4 w}{\partial x^4} + \frac{\rho A}{E_0 E^* I}\frac{\partial^2 w}{\partial t^2} + \alpha_T \frac{\partial^2 M_T}{\partial x^2} = 0 , \tag{6}$$

where ρ the density of the beam, E is Young's modulus, I [$= bh^3/12$] the inertial moment about x-axis, α_T the coefficient of linear thermal expansion, $w(x,t)$ the lateral deflection, x the distance along the length of the beam, $A = hb$ is the area of the cross section and t the time and M_T is the thermal moment as follows [6] [7] [11]-[13]:

$$M_T = \frac{12}{h^3}\int_{-h/2}^{h/2} \theta z \mathrm{d}z , \tag{7}$$

where $\theta = (T - T_0)$ is the dynamical temperature increment of the resonator, $T(x, z, t)$ is the temperature distribution, and T_0 the room temperature.

According to Lord-Shulman model (L-S), the non-Fourier heat conduction equation has the following form [6] [7] [11]-[14]:

$$\frac{\partial^2 \theta}{\partial x^2} + \frac{\partial^2 \theta}{\partial z^2} = \left(\frac{\partial}{\partial t} + \tau_o \frac{\partial^2}{\partial t^2}\right)\left(\frac{\rho C_v}{K}\theta + \frac{\beta T_0}{K}e\right), \tag{8}$$

Where $e = \dfrac{\partial u}{\partial x} + \dfrac{\partial v}{\partial y} + \dfrac{\partial w}{\partial z}$ is the volumetric strain, C_v is the specific heat at constant volume, τ_0 the thermal relaxation time, K the thermal conductivity, $\beta = \dfrac{E_0 E^* \alpha_T}{1 - 2v}$ and v is Poisson's ratio. Where there is no heat flow across the upper and lower surfaces of the beam, so that $\dfrac{\partial \theta}{\partial z} = 0$ at $z = \pm h/2$ For a very thin nano-beam and assuming the temperature varies in terms of a $\sin(pz)$ function along the thickness direction [6] [7] [11]-[13], where $p = \pi/h$, gives

$$\theta(x,z,t) = \theta_1(x,t)\sin(pz) . \tag{9}$$

Hence, Equation (6) gives

$$\frac{\partial^4 w}{\partial x^4} + \frac{\rho A}{E_0 E^* I}\frac{\partial^2 w}{\partial t^2} + \frac{12\alpha_T}{h^3}\frac{\partial^2 \theta_1}{\partial x^2}\int_{-h/2}^{h/2} z \sin(pz)\mathrm{d}z = 0 \tag{10}$$

Moreover, Equation (8) gives

$$\frac{\partial^2 \theta_1}{\partial x^2}\sin(pz) - p^2 \theta_1 \sin(pz) = \left(\frac{\partial}{\partial t} + \tau_o \frac{\partial^2}{\partial t^2}\right)\left(\frac{\rho C_v}{K}\theta_1 \sin(pz) - \frac{\beta T_0}{K}z\frac{\partial^2 w}{\partial x^2}\right) \tag{11}$$

After doing the integrations, Equation (10) takes the form

$$\frac{\partial^4 w}{\partial x^4} + \frac{\rho A}{E_0 E^* I}\frac{\partial^2 w}{\partial t^2} + \frac{24\alpha_T}{h\pi^2}\frac{\partial^2 \theta_1}{\partial x^2} = 0 \,. \tag{12}$$

In Equation (11), we multiply the both sides by z and integrating with respect to z from $-\dfrac{h}{2}$ to $\dfrac{h}{2}$, and then we obtain

$$\left(\frac{\partial^2 \theta_1}{\partial x^2} - p^2\theta_1\right) = \left(\frac{\partial}{\partial t} + \tau_o\frac{\partial^2}{\partial t^2}\right)\left(\varepsilon\theta_1 - \frac{\beta T_0\pi^2 h}{24K}\frac{\partial^2 w}{\partial x^2}\right), \tag{13}$$

where $\varepsilon = \dfrac{\rho C_v}{K}$.

For simplicity, we will use the following dimensionless variables [15]:

$$\left(x', w', h'\right) = \varepsilon c_o\left(x, w, h\right), \; \left(t', \tau_o'\right) = \varepsilon c_o^2\left(t, \tau_o\right), \;\; \sigma' = \frac{\sigma}{E_0}, \;\; \theta_1' = \frac{\theta_1}{T_o}, \; c_o^2 = \frac{E_0}{\rho}. \tag{14}$$

Then, we have

$$\frac{\partial^4 w}{\partial x^4} + A_1\frac{\partial^2 w}{\partial t^2} + A_2\frac{\partial^2 \theta_1}{\partial x^2} = 0 \,, \tag{15}$$

and

$$\frac{\partial^2 \theta_1}{\partial x^2} - A_3\theta_1 = \left(\frac{\partial}{\partial t} + \tau_o\frac{\partial^2}{\partial t^2}\right)\left(\theta_1 - A_4\frac{\partial^2 w}{\partial x^2}\right), \tag{16}$$

where

$$A_1 = \frac{12}{E^* h^2}, \;\; A_2 = \frac{24\alpha_t T_o}{\pi^2 h}, \;\; A_3 = p^2, \;\; A_4 = \frac{\pi^2\beta h}{24K\varepsilon}\,. $$

For convenience, we dropped the prime.

4. Formulation the Problem in the Laplace Transform Domain

Applying the Laplace transform for Equations (14) and (15), this is defined by the following formula

$$\bar{f}(s) = L\left[f(t)\right] = \int_0^\infty f(t)\mathrm{e}^{-st}\mathrm{d}t \,. $$

Hence, we obtain the following system

$$\frac{\mathrm{d}^4\bar{w}}{\mathrm{d}x^4} + A_1 s^2\bar{w} + A_2\frac{\mathrm{d}^2\bar{\theta}_1}{\mathrm{d}x^2} = 0 \,, \tag{17}$$

and

$$\frac{\mathrm{d}^2\bar{\theta}_1}{\mathrm{d}x^2} - A_3\bar{\theta}_1 = \left(s + \tau_o s^2\right)\left(\bar{\theta}_1 - A_4\frac{\mathrm{d}^2\bar{w}}{\mathrm{d}x^2}\right). \tag{18}$$

We will consider the function $\bar{\eta}$ as follows:

$$\frac{\mathrm{d}^2\bar{w}}{\mathrm{d}x^2} = \bar{\eta} \,, \tag{19}$$

Then, we have

$$\frac{\mathrm{d}^2\bar{\theta}_1}{\mathrm{d}x^2} = \alpha_1\bar{\theta}_1 - \alpha_2\bar{\eta} \,, \tag{20}$$

and

$$\frac{d^2\bar{\eta}}{dx^2} = -\alpha_3\bar{w} - \alpha_4\bar{\theta}_1 + \alpha_5\bar{\eta},$$ (21)

where $\alpha_1 = \left(A_3 + s + \tau_o s^2\right)$, $\alpha_2 = A_4\left(s + \tau_o s^2\right)$, $\alpha_3 = A_1 s^2$, $\alpha_4 = A_2\left(A_3 + s + \tau_o s^2\right)$, $\alpha_5 = A_2 A_4\left(s + \tau_o s^2\right)$.

Consider the first end of the beam $x = 0$ is clamped and loaded thermally, which gives [6] [7]:

$$w(0,t) = \eta(0,t) = 0,$$ (22)

and

$$\theta_1(0,t) = \theta_0 f(t),$$ (23)

where θ_0 is constant.

By using Laplace transform, the conditions will take the forms

$$\bar{w}(0,s) = \bar{\eta}(0,s) = 0,$$ (24)

and

$$\bar{\theta}_1(0,s) = \theta_0\bar{f}(s).$$ (25)

Consider the other end of the beam $x = \ell$ is clamped and remains at zero increments of temperature as follows:

$$w(\ell,t) = \theta_1(\ell,t) = \eta(\ell,t) = 0.$$ (26)

After using Laplace transform, we have

$$\bar{w}(\ell,s) = \bar{\theta}_1(\ell,s) = \bar{\eta}(\ell,s) = 0.$$ (27)

After some simplifications by using *MAPLE* programme, we get the final solutions in the Laplace transform domain as follows:

The lateral deflection

$$\bar{w}(x,s) = \frac{\Delta\sinh\left(k_1(\ell-x)\right)}{\left(k_1^2 - k_2^2\right)\left(k_1^2 - k_3^2\right)\sinh\left(k_1\ell\right)} + \frac{\Delta\sinh\left(k_2(\ell-x)\right)}{\left(k_2^2 - k_1^2\right)\left(k_2^2 - k_3^2\right)\sinh\left(k_2\ell\right)} + \frac{\Delta\sinh\left(k_3(\ell-x)\right)}{\left(k_3^2 - k_1^2\right)\left(k_3^2 - k_2^2\right)\sinh\left(k_3\ell\right)}.$$ (28)

The temperature

$$\bar{\theta}(z,x,s) = -\frac{\alpha_2 k_1^2\Delta\sin(pz)\sinh\left(k_1(\ell-x)\right)}{\left(k_1^2 - \alpha_1\right)\left(k_1^2 - k_2^2\right)\left(k_1^2 - k_3^2\right)\sinh\left(k_1\ell\right)} - \frac{\alpha_2 k_2^2\Delta\sin(pz)\sinh\left(k_2(\ell-x)\right)}{\left(k_2^2 - \alpha_1\right)\left(k_2^2 - k_1^2\right)\left(k_2^2 - k_1^2\right)\sinh\left(k_2\ell\right)} - \frac{\alpha_3 k_3^2\Delta\sin(pz)\sinh\left(k_3(\ell-x)\right)}{\left(k_3^2 - \alpha_1\right)\left(k_3^2 - k_1^2\right)\left(k_3^2 - k_2^2\right)\sinh\left(k_3\ell\right)}.$$ (29)

The displacement

$$\bar{u}(z,x,s) = -\frac{z\Delta k_1\cosh\left(k_1(\ell-x)\right)}{\left(k_1^2 - k_2^2\right)\left(k_1^2 - k_3^2\right)\sinh\left(k_1\ell\right)} - \frac{z\Delta k_2\cosh\left(k_2(\ell-x)\right)}{\left(k_2^2 - k_1^2\right)\left(k_2^2 - k_1^2\right)\sinh\left(k_2\ell\right)} - \frac{z\Delta k_3\cosh\left(k_3(\ell-x)\right)}{\left(k_3^2 - k_1^2\right)\left(k_3^2 - k_2^2\right)\sinh\left(k_3\ell\right)}.$$ (30)

The Strain

$$\bar{e}(z,x,s) = \frac{z\Delta k_1^2 \sinh\left(k_1\left(\ell-x\right)\right)}{\left(k_1^2-k_2^2\right)\left(k_1^2-k_3^2\right)\sinh\left(k_1\ell\right)} + \frac{z\Delta k_2^2 \sinh\left(k_2\left(\ell-x\right)\right)}{\left(k_2^2-k_3^2\right)\left(k_2^2-k_1^2\right)\sinh\left(k_2\ell\right)}$$

$$+ \frac{z\Delta k_3^2 \sinh\left(k_3\left(\ell-x\right)\right)}{\left(k_3^2-k_1^2\right)\left(k_3^2-k_2^2\right)\sinh\left(k_3\ell\right)}. \tag{31}$$

where

$$\Delta = \frac{\theta_0 \bar{f}(s)}{\alpha_1\alpha_2}\left(\alpha_1-k_1^2\right)\left(\alpha_1-k_2^2\right)\left(\alpha_1-k_3^2\right) \quad \text{and} \quad \pm k_1, \pm k_2, \pm k_3 \quad \text{are the roots of the equation}$$

$$k^6 - lk^4 + mk^2 - n = 0, \tag{32}$$

and

$$l = \alpha_1 + \alpha_5, \quad m = \alpha_1\alpha_5 - \alpha_2\alpha_4 + \alpha_3 \quad \text{and} \quad n = \alpha_1\alpha_3.$$

5. The Stress-Strain Energy

The stress on the x-axis, according to Hooke's law is:

$$\sigma_{xx}(x,z,t) = E_0 E^* \left(e - \alpha_T \theta\right). \tag{33}$$

By using the non-dimensional variables in (13), we obtain the stress in the form

$$\sigma_{xx}(x,z,t) = E^* \left(e - \alpha_T T_0 \theta\right). \tag{34}$$

By using Laplace transform, the above equation takes the form:

$$\bar{\sigma}_{xx}(x,z,s) = E^* \left(\bar{e} - \alpha_T T_0 \bar{\theta}\right). \tag{35}$$

The stress-strain energy, which is generated by the beam, is given by

$$W(x,z,t) = \sum_{i,j=1}^{3} \frac{1}{2}\sigma_{ij}e_{ij} = \frac{1}{2}\sigma_{xx}e_{xx} = -\frac{1}{2}z\sigma_{xx}\eta, \tag{36}$$

We can re-write Equation (36) to be in the form

$$W(x,z,t) = -\frac{1}{2}z\left[L^{-1}\left(\bar{\sigma}_{xx}\right)\right]\left[L^{-1}\left(\bar{\eta}\right)\right], \tag{37}$$

$L^{-1}[\bullet]$ is the inversion of Laplace transform.

To complete the solution in the Laplace transform domain, we have to determine the type of heating which we have used to load the boundary of the medium thermally.

We have applied harmonic thermal loading as follows [16]:

$$f(t) = \sin(\omega t), \tag{38}$$

after using Laplace transform, we obtain

$$\bar{f}(s) = \frac{\omega}{s^2+\omega^2}, \tag{39}$$

ω is the angular frequency of thermal vibration.

6. Numerical Inversion of the Laplace Transform

To determine the solutions in the time domain, the Riemann-sum approximation method is used to obtain the numerical results. In this method, any function in Laplace domain can be inverted to the time domain as

$$f(t) = \frac{e^{\kappa t}}{t}\left[\frac{1}{2}\bar{f}(\kappa) + Re\sum_{n=1}^{N}(-1)^n \bar{f}\left(\kappa + \frac{in\pi}{t}\right)\right] \tag{40}$$

where Re is the real part and i is imaginary number unit. For faster convergence, numerous numerical experiments have shown that the value of κ satisfies the relation $\kappa t \approx 4.7$ Tzou [17].

7. Numerical Results and Discussion

Now, we will consider a numerical example for which computational results are given. For this purpose, Gold (Au) is taken as the thermoelastic material for which we take the following values of the different physical constants [18]:

$k = 318\,\text{W}/(\text{m}\cdot\text{K})$, $\alpha_T = 14.2 \times (10)^{-6}\,\text{K}^{-1}$, $\rho = 1930\,\text{kg}/\text{m}^3$, $T_0 = 293\,\text{K}$, $C_\upsilon = 130\,\text{J}/(\text{kg}\cdot\text{K})$, $E = 180\,\text{GPa}$, $\upsilon = 0.44$.

The aspect ratios of the beam are fixed as $\ell/h = 10$ and $b/h = 1/2$ when h is varied, ℓ and b change accordingly with h.

For the nano-scale beam, we will take the range of the beam length ℓ $(1 - 100) \times 10^{-9}\,\text{m}$. The original time t will be considered in the picoseconds $(1 - 100) \times 10^{-12}$ sec and the relaxation time τ_0 in the range $(1 - 100) \times 10^{-14}$ sec.

The figures (**Figure 2-7**) were prepared by using the non-dimensional variables which are defined in (9) for beam length $\ell = 1.0$, $\theta_0 = 1.0$ $z = h/6$ and $t = 0.1$.

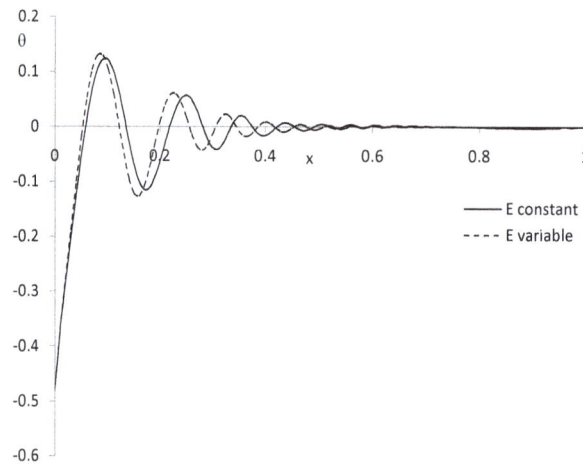

Figure 2. The temperature distribution with different cases of Young's modulus.

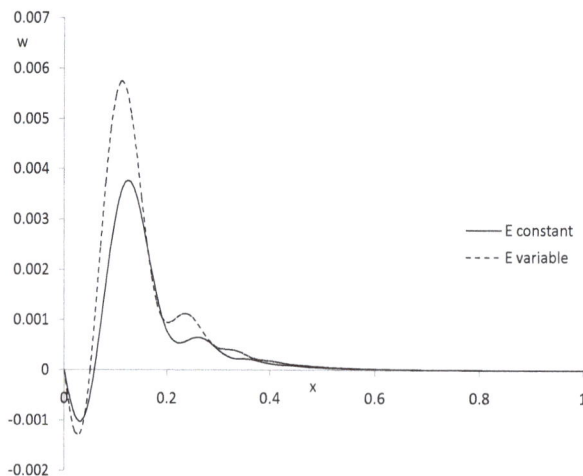

Figure 3. The lateral vibration distribution with different cases of Young's modulus.

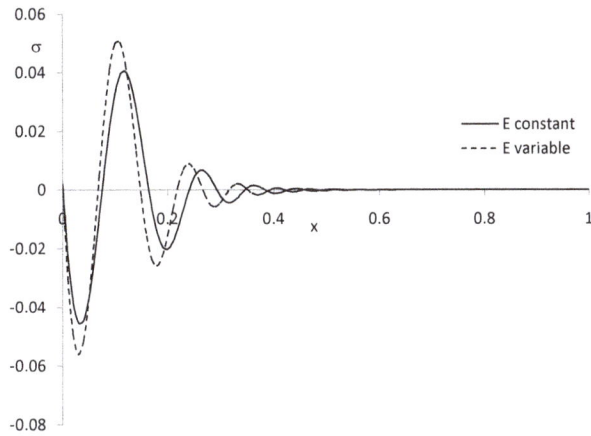

Figure 4. The stress distribution with different cases of Young's modulus.

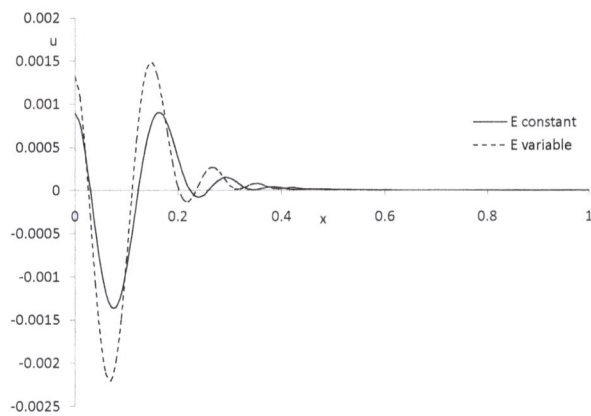

Figure 5. The displacement distribution with different cases of Young's modulus.

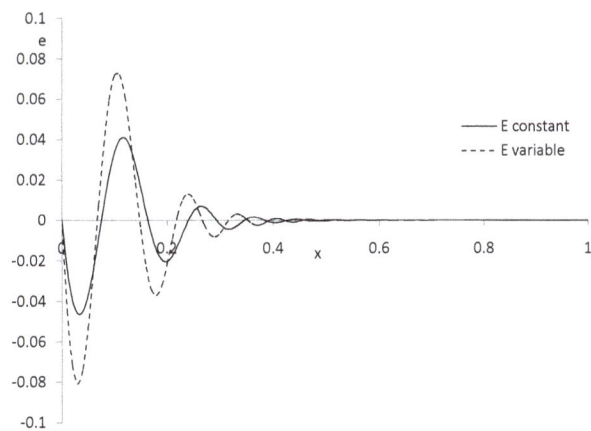

Figure 6. The strain distribution with different cases of Young's modulus.

8. Conclusion

The Young's modulus has significant effects on all the studied fields. The values of the temperature, the vibra-

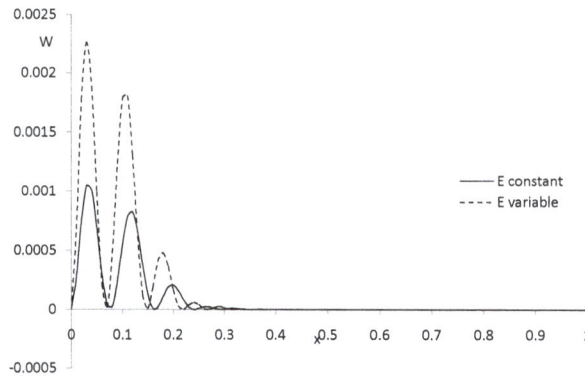

Figure 7. The stress-strainenergy distribution with different cases of Young's modulus.

tion of the deflection, stress, displacement, strain, stress-strain energy increase when the Young's modulus is variable. The peak points of all the distributions increase when the Young's modulus is variable with large differences in the case of Young's modulus is constant.

References

[1]   Diao, J.K., Gall, K. and Dunn, M.L. (2004) Atomistic Simulation of the Structure and Elastic Properties of Gold Nanowires. *Journal of the Mechanics and Physics of Solids*, **52**, 1935-1962. http://dx.doi.org/10.1016/j.jmps.2004.03.009

[2]   Kidawa-Kukla, J. (2003) Application of the Green Functions to the Problem of the Thermally Induced Vibration of a Beam. *Journal of Sound and Vibration*, **262**, 865-876. http://dx.doi.org/10.1016/S0022-460X(02)01133-1

[3]   Boley, B.A. (1972) Approximate Analyzes of Thermally Induced Vibrations of Beams and Plates. *Journal of Applied Mechanics*, **39**, 212-216. http://dx.doi.org/10.1115/1.3422615

[4]   Manolis, G.D. and Beskos, D.E. (1980) Thermally Induced Vibrations of Beam Structures. *Computer Methods in Applied Mechanics and Engineering*, **21**, 337-355. http://dx.doi.org/10.1016/0045-7825(80)90101-2

[5]   Al-Huniti, N.S., Al-Nimr, M.A. and Naij, M. (2001) Dynamic Response of a Rod Due to a Moving Heat Source under the Hyperbolic Heat Conduction Model. *Journal of Sound and Vibration*, **242**, 629-640. http://dx.doi.org/10.1006/jsvi.2000.3383

[6]   Soh, A.K., Sun, Y.X. and Fang, D.N. (2008) Vibration of Microscale Beam Induced by Laser Pulse. *Journal of Sound and Vibration*, **311**, 243-253. http://dx.doi.org/10.1016/j.jsv.2007.09.002

[7]   Sun, Y.X., Fang, D.N., Saka, M. and Soh, A.K. (2008) Laser-Induced Vibrations of Micro-Beams under Different Boundary Conditions. *International Journal of Solids and Structures*, **45**, 1993-2013. http://dx.doi.org/10.1016/j.ijsolstr.2007.11.006

[8]   Wang, Y.Z., Li, F.M. and Kishimoto, K. (2010) Scale Effects on the Longitudinal Wave Propagation in Nanoplates. *Physica E*, **42**, 1356-1360. http://dx.doi.org/10.1016/j.physe.2009.11.036

[9]   Bruls, R.J., Hintzen, H.T., De With, G. and Metselaar, R. (2001) The Temperature Dependence of the Young's Modulus of MgSiN2, AlN and Si3N4. *Journal of the European Ceramic Society*, **21**, 263-268. http://dx.doi.org/10.1016/S0955-2219(00)00210-7

[10] Farraro, R. and McLellan, R.B. (1977) Temperature Dependence of the Young's Modulus and Shear Modulus of Pure Nickel, Platinum, and Molybdenum. *Metallurgical Transactions A*, **8**, 1565-1563. http://dx.doi.org/10.1007/BF02644859

[11] Sun, Y.X., Fang, D.N. and Soh, A.K. (2006) Thermoelastic Damping in Micro-Beam Resonators. *International Journal of Solids and Structures*, **43**, 3213-3229. http://dx.doi.org/10.1016/j.ijsolstr.2005.08.011

[12] Fang, D.N., Sun, Y.X. and Soh, A.K. (2006) Analysis of Frequency Spectrum of Laser-Induced Vibration of Micro-beam Resonators. *Chinese Physics Letters*, **23**, 1554-1557. http://dx.doi.org/10.1088/0256-307X/23/6/055

[13] Duwel, A., Gorman, J., Weinstein, M., Borenstein, J. and Ward, P. (2003) Experimental Study of Thermoelastic Damping in MEMS Gyros. *Sensors and Actuators A*, **103**, 70-75. http://dx.doi.org/10.1016/S0924-4247(02)00318-7

[14] Lord, H. and Shulman, Y. (1967) A Generalized Dynamical Theory of Thermoelasticity. *Journal of the Mechanics and Physics of Solids*, **15**, 299-309. http://dx.doi.org/10.1016/0022-5096(67)90024-5

[15] Youssef, H.M. (2013) Vibration of Gold Nano-Beam with Variable Thermal Conductivity: State-Space Approach. *Ap-*

plied Nanoscience, **3**, 397-407.

[16] Youssef, H.M., Elsibai, K.A. and El-Bary, A.A. (2014) Vibration of Cylindrical Gold Nano-Beam with Fractional Order Thermoelastic Waves. *Jökull Journal*, **64**, 416-427.

[17] Tzou, D. (1996) Macro-to-Micro Heat Transfer. Taylor & Francis, Washington DC.

[18] Youssef, H.M., Elsibai, K.A. and El-Bary, A.A. (2014) Vibration of Gold NanoBeam in Context of Two-Temperature Generalized Thermoelasticity Subjected to Laser Pulse. *Latin American Journal of Solids and Structures*, **11**, 2460-2482. http://dx.doi.org/10.1590/S1679-78252014001300008

14

Titania-Silica Composites: A Review on the Photocatalytic Activity and Synthesis Methods

Yuri Hendrix, Alberto Lazaro, Qingliang Yu, Jos Brouwers

Department of the Built Environment, Eindhoven University of Technology, Eindhoven, The Netherlands
Email: Q.Yu@bwk.tue.nl

Abstract

The photocatalyic activity of titania is a very promising mechanism that has many possible applications like purification of air and water [1]-[4]. To make it even more attractive, titania can be combined with silica to increase the photocatalytic efficiency and durability of the photocatalytic material, while lowering the production costs [1]. In this article, relevant literature is reviewed to obtain an overview about the chemistry and physics behind some of the different parameters that lead to cost-effective photocatalytic titania-silica composites. The first part of this review deals with the mechanisms involved in the photocatalytic activity, then the chemistry behind certain methods for the synthesis of the titania-silica composites is discussed, and in the last and third part of this review, the influence of silica supports on titania is discussed. These three sections represent three different fields of research that are combined in this review to obtain better insights on the photocatalytic titania-silica composites. While many research subjects in these fields have been well known for some time now, some subjects are only more recently resolved and some subjects are still under discussion (e.g. the cause for the increased hydrophilic surface of titania after illumination). This article aims to review the most important literature to give an overview of the current situation of the fundamentals of photocatalysis and synthesis of the cost-effective photocatalyic composites. It is found that the most cost-effective photocatalytic titania-silica composites are the ones that have a thin anatase layer coated on silica with a large specific surface area, and are prepared with the precipitation or sol-gel methods.

Keywords

Photocatalysis, Photocatalytic TiO$_2$, Titania-Silica Composites, Low Cost Synthesis

1. Introduction

Composites made out of silica and titania can have the photocatalytic properties from titania, the high stability from silica and extra properties coming from chemical bonds between the two materials [1]. Titania is photocatalytic because it is able to absorb energy from light, and then use that energy to catalyze the degradation of organic molecules and the oxidation of some inorganic pollutants like nitrogen oxides (NO_x) [1]-[12]. As the photocatalytic activity takes place only on the exposed surface area of titania, the amount of titania needed for the same photocatalytic efficiency can be reduced enormously by coating a thin layer of titania on silica [5] [13]-[42]. As the production of silica can be cheaper than that of titania, the costs of the photocatalyic material can then be significantly reduced. In addition to lower production costs, the durability increases as silica has a higher mechanical and thermal stability than titania. So when the composites are used instead of pure titania, the photocatalytic material can be used for a longer time with high photocatalytic efficiencies. In addition, because of the enhanced thermal stability, the photocatalytic material can be used in applications that require higher preparation temperatures. In addition to the lower costs and increased durability, the photocatalytic efficiency of the material can be increased with the addition of silica because silica can have a large specific surface area and is able to adsorb some pollutants and intermediates for a longer time than pure titania.

One promising application of the photocatalytic materials is the degradation of pollutants. The main reasons why using photocatalytic materials for air purification is promising include: lower costs of materials and energy needed than the other current purification methods, the ability of many photocatalytic materials to oxidize pollutants even if they are present in low concentrations, and the fact that the pollutants do not have to be stored but are converted into less harmful side-products (e.g. CO_2 from organic molecules and NO_3^- from NO_x after a complete photocatalytic oxidation (PCO)). The photocatalytic titania has been, and is being used in many other applications as well, including: photoelectrolysis of water, medical applications (where titania works as a disinfectant by destroying bacteria and viruses), municipal and industrial wastewater treatment, self-cleaning glass with anti-fogging abilities, and even in textiles that are self-cleaning.

A good method to have air purification is by the incorporation of photocatalytic material in building materials (including: concrete, wallpaper, gypsum and paint [2]-[4] [8]-[10] [43]-[51]), due to the large illuminated surfaces areas that many building materials have. Investigations into the photocatalytic building materials showed that the concentration of pollutants close to photocatalytic building materials indeed significantly decreased. Since large areas of building materials are often illuminated anyway with sun-light or indoor light and because these building materials become self-cleaning, the maintenance costs of these materials can be very low. Because of the large illuminated area and low maintenance cost, the potential of the photocatalytic building material for air purification is very promising.

However, in most of the research field of the applications of photocatalytic materials, only pure and doped titania are mentioned and not the titania-silica composites despite the large benefits these composites can have (e.g. lower costs, higher durability). An important reason for this absence of composites can be the complexity of the research field of the titania-silica composites. For the synthesis of the titania-silica composites alone, there are many different methods, each with their own parameters that can be changed in multiple ways. As many studies on the titania-silica composites have been done with different goals in mind, many different kinds of titania-silica composites have been produced [1] [5] [13]-[42] [52]-[84], from which some are either not suitable for photocatalysis or have a very expensive production method. Since the photocatalytic activity of titania alone is already a complex system [3] [4] [6] [11] [85]-[89], it can be understandable that adding more complexity to the system (for example, with silica) is not desirable. This review is written in order to provide insight on the low cost synthesis of titania-silica composites, and how each different parameter can be tuned to produce highly efficient photocatalytic material to show how the composites can be an attractive alternative to the titania for photocatalytic applications.

2. Photocatalytic Titania

2.1. Mechanism of Photocatalysis

The process of photocatalysis in titania starts when a photon is absorbed by an electron in the valance band of titania [3] [4] [6] [11] [85]-[89]. This electron is then excited to the conduction band, and by doing so, leaves a hole behind in the valance band (reaction 1). The valance and conduction bands of titania have the right energy levels for many important redox reactions. After the excitation of electrons, holes in the valance band have a redox potential of +2.53 V, which is enough for the oxidation of hydroxyl ions into OH⁺ (see reaction 2) or the oxidation of adsorbed organic molecules groups. The largest source of hydroxyl ions comes from the dissociation of water (see reaction 3). The redox potential of electrons in the conduction band is −0.52 V, which is strong enough to reduce oxygen to superoxide (see reaction 4). It is also possible that the excited electrons and holes will react with different adsorbed species depending on the environment. For example, if there is a high amount of adsorbed water, it is possible that more radical hydroxyls will form through the reaction of hydrogen peroxide as shown in reaction 5 and 6.

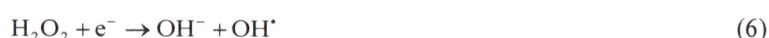

$$TiO_2 + light \rightarrow TiO_2 + e^- + h^+ \tag{1}$$

$$OH^- + h^+ \rightarrow OH^{\boldsymbol{\cdot}} \tag{2}$$

$$H_2O \rightleftarrows OH^- + H^+ \tag{3}$$

$$O_2 + e^- \rightarrow O_2^- \tag{4}$$

$$O_2 + 2H_2O + 2e^- \rightarrow 2H_2O_2 \tag{5}$$

$$H_2O_2 + e^- \rightarrow OH^- + OH^{\boldsymbol{\cdot}} \tag{6}$$

where e^- is an excited electron in the conduction band, h^+ is a hole in the valance band, $OH^{\boldsymbol{\cdot}}$ is a radical hydroxyl and O_2^- is a superoxide. Radical hydroxyls and superoxides are strong oxidants that can react with certain inorganic pollutants like NO_x and many organic molecules. In **Figure 1**, a schematic view is given for the photocatalytic mechanism.

An important property of titania, which influences the photocatalytic efficiency, is the amount of hydroxyl groups in its environment. In turn, the amount of hydroxyl groups is determined by the humidity in air, or the amount of water and its pH in liquids. This amount will determine how many hydroxyl groups will be chemically bonded to the surface of the titania. Bonded hydroxyl groups can either react with holes themselves and form radicals, or adsorb other hydroxyl groups and water molecules which can subsequently react with the holes and excited electrons [90] [91]. The photocatalytic activity in air can thus be higher with a higher humidity. However, it is also possible that a very high humidity will lower the photocatalytic activity by taking up more adsorption sides on the surface. For example, the photocatalytic oxidation of NO_x depends on the adsorption to titania and is thus lower with a very high humidity.

2.2. Oxygen Vacancies, Hydrophilicity and Self-Cleaning Surfaces

By reacting Ti^{4+} and O^{2-} into Ti^{3+} and O^-, excited electrons and holes can remain at the surface longer if there are no adsorbed species they can react with directly [3]. Because the difference in charge between titanium and oxygen atoms is then reduced, the oxygen atoms are much less stable and can, with relatively little energy, leave the crystal forming oxygen vacancies. These oxygen vacancies are important in titania for different mechanisms. For example, around an oxygen vacancy there is an excess of electrons, making titania a n-type semiconductor,

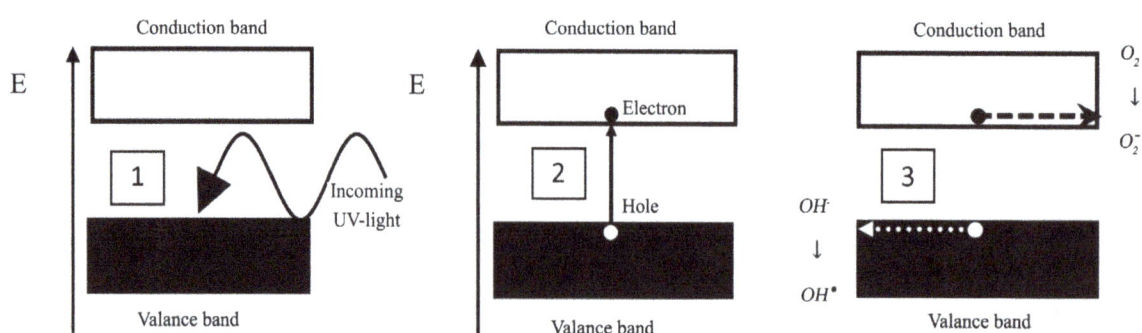

Figure 1. Schematic drawing of the photocatalytic activity of titania. 1: The absorption of a photon; 2: The excitation of an electron to the conduction band; 3: The transport of the electron and hole from the initial point to reach the surface of titania where the electron and hole can react with an adsorbed molecule.

which has a higher conductivity than when titania is an intrinsic semiconductor.

Another reason why oxygen vacancies are important is because the surface of titania becomes more hydrophilic when water molecules occupy these oxygen vacancies. After a water molecule occupies the vacancy, one hydrogen atom of the water molecule can react with a neighboring oxygen atom forming two hydroxyl groups [3]. The increase in hydroxyl groups can lead to an increase in photocatalytic efficiency and to an increase in hydrophilicity of the surface. This increase in hydrophilicity, was first reported by Wang *et al.* [92] in 1997 with a titania coating on glass. By illuminating the coated glass with UV-light, the glass became transparent since the water fog that was present on the glass, defogged as the contact angle between the water droplets and the glass decreased to zero. They also showed that after keeping the hydrophilic surface away from any light source for some days, the glass became more hydrophobic, which means that the formation of a hydrophilic surface is a reversible process. However, it has been reported that oxygen vacancies are not solely responsible for the hydrophilicity, as some studies showed that hydrophilicity was independent in some cases on the number of oxygen vacancies [93] [94]. While it is possible that the degradation of organic materials on the surface can also play a role on the hydrophilicity increase, it nevertheless has been shown not to be a determining factor [95].

The hydrophilicity and the degradation of organic materials on the surface of titania are two reasons why titania can be used for self-cleaning applications [3]. The degradation of organic materials through photocatalytic oxidation prevents organic substances to accumulate on the surface and can prevent the growth of bacteria and fungi. The hydrophilicity of the surface increases the water adsorption so that it can replace other adsorbed species and it lowers the energy required for water to slide over the surface so that contaminants can be washed off more easily.

2.3. Effect of Different Crystal Forms of Titania on the Photocatalytic Efficiency

Titania has several forms, but the two main crystal structures that most researchers focus on are rutile and anatase [3]. These two crystal forms are both tetragonal structures in which titanium atoms are 6-coordinated in an octahedral formation. The band gap of rutile is 3.0 eV and the band gap of anatase is 3.2 eV. For rutile and anatase to become photocatalytically active, they need to absorb electromagnetic radiation with wavelengths smaller than 413 nm and 387 nm respectively. While rutile is thus able to absorb more light in the visible range, anatase is more photocatalytically active. Luttrell *et al.* [96] showed this higher photocatalytic efficiency by studying the difference in PCO efficiencies of anatase and rutile thin films of different sizes for the PCO of methyl orange. They showed that for films thinner than 2.5 nm the difference was not significant between the two forms, but for thicker films, the anatase thin film had a higher efficiency. They measured that the maximum thickness, where the photocatalytic efficiency increases with increasing size, is 2.5 nm for rutile and 5 nm for anatase. Thus, from this study it can be concluded that excited electrons and holes in anatase can travel farther than in rutile so that more electrons and holes can reach the surface. The ability of exited electrons and holes to travel longer distances in anatase was contributed to a longer lifetime and higher conductivity [96]-[98].

An important reason why excited electrons and holes have a longer lifetime and higher conductivity in anatase is because of the differences between the oxygen vacancies that form in anatase and rutile. Oxygen vacancies cause extra energy levels within the band gap. Calculations done by Mattioli *et al.* [97] showed that oxygen vacancies in an anatase crystal can cause both shallow delocalized energy levels and deep localized energy levels in the band gap, while in rutile only deep localized levels can form. Since anatase has also shallow delocalized energy levels in its band gap, it has a higher conductivity and the excited electrons and holes have longer lifetimes than in rutile as they are less trapped in the deep localized energy levels where the chance of recombination is higher.

While anatase has a high photocatalytic efficiency, amorphous titania has the lowest efficiency [99]. The main reason for this lower efficiency is because, in amorphous titania, there are many spots where recombination of the electron-hole pair can happen. The recombination through defects is the most common way electrons and holes are lost. Thus, amorphous titania has a much higher recombination capacity. In addition, conductivity in amorphous materials is very low since energy levels in amorphous materials are much more localized. The high recombination rate and low conductivity means that only electrons which are excited directly at the surface play a part in the photocatalytic activity in amorphous titania.

Some researchers have measured higher PCO efficiencies in titania that contains both rutile and anatase than in titania with only anatase. Degussa P25 nanoparticles, which are commercial titania nanoparticles made out of around 80% anatase and 20% rutile, are well known for their high photocatalytic activity [100] and are often used as a reference material. The conduction band of rutile has been measured to start at a higher energy level than that of anatase even though its band gap is smaller, which is why titania with both crystal forms can have a higher photocatalytic efficiency [101]. Since electrons always go to a lower energy state if possible, excited electrons in rutile will go to the conduction band of anatase. As electron holes can be viewed as opposite electrons, electron holes in anatase will flow to the valance band of rutile because its top lies above the energy level of the valance band from anatase. Because holes move from anatase to rutile and excited electrons from rutile to anatase, the recombination chance is reduced and a difference in electron density is produced at the interface between the two forms, causing an increase in conductivity and lifetimes for the electrons and holes, resulting in a higher photocatalytic efficiency.

3. Titania-Silica Composite Synthesis

3.1. Silica Sources

Many different types of silica from different sources can be applied as a support for titania including: fumed silica, precipitation silica from alkali silicates, silica produced with the Stöber method, zeolites, clays, glass, silica from the dissolution of silica minerals and more [1] [5] [13]-[42] [52]-[84] [102]-[107]. Fumed silica is formed at high temperatures where silica compounds, like chlorosilanes, are transformed into silica [108] [109]. Silica from alkali silicates is formed by neutralization of the alkali solutions so that the silicate polymerizes to silica and precipitates from the solution. In both fumed silica and precipitated silica, amorphous aggregates are formed. These aggregates can have very large specific surface areas but also have complex undefined structures. Another silica which is often used because of the more defined structure, is silica made with a sol-gel method. The Stöber method [110] is the best known example of a sol-gel method for producing silica. During this method, Tetraethyl orthosilicate (TEOS) is slowly added to a solution of ethanol, water and ammonia. Depending on the composition of the solution, silica colloids of varies sizes and shapes can be formed. The advantage of this silica is that the resulting shape and size of the silica can be well controlled. However, the disadvantage is that this silica has a smaller specific surface area than fumed and precipitated silica. For both well-defined shapes and high specific surface areas, researchers have also used zeolites as support. However compared to the other mentioned supports, the zeolites are more expensive. For very low production costs and a high specific surface area, silica made during the dissolution of olivine has a great potential [111]-[114] but is still in its developing stage.

3.2. Titania-Silica Chemistry

The reaction of titania precursors with silica happens either directly with silanols or indirectly through hydrolysis into titania monomers ($Ti(OH)_4$) first and subsequently by condensation with silanols [1] [5] [13]-[42] [52]-[84] [109] [115]. Either way, the titania will form bonds with the silica through reaction 7.

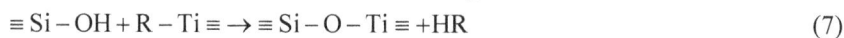

$$\equiv Si - OH + R - Ti \equiv \rightarrow \equiv Si - O - Ti \equiv + HR \tag{7}$$

where R is a side group of a titania precursor or a hydroxyl group of a titania monomer. The Si-O-Ti bond can be measured by using techniques like infra-red/Raman spectrometry and X-ray photoelectron spectroscopy [14]-[16] [32]-[34] [52] [56]-[60]. This condensation reaction between the titania precursor and the silica surface depends mostly on the hydroxyl groups of the silica [16] [56] [59] [63] [66] [67] since the rest of the silica is very inert. In turn, the amount of hydroxyl groups on the silica is dependent on the temperature during the pretreatment, the method used, and the amount of water and its pH used [67] [109]. For example, if the silica undergoes pre-heating temperatures higher than 800°C and no water is used during the synthesis, there will be only a low amount of hydroxyl groups on the silica surface left so that only a few titanium atoms can be found on the silica after the reaction [67]. On the other hand, if lower temperatures are used, the density of hydroxyl groups will be high enough on the silica surface that hydrogen bonds between silanols can form. Titania precursors react more with these hydrogen bonded silanols than isolated silanols [66] [67]. These silanols are close enough to each other that a titania precursor can react with multiple hydroxyl groups, making the reaction of hydrogen bonded silanols favorable over the reaction of isolated silanol.

When water is used during the synthesis method of the titania-silica composites, the titania precursor undergoes hydrolysis first. During the hydrolysis, the side groups of the precursor are replaced by hydroxyl groups [116]-[119]. After a full hydrolysis at a neutral pH, $Ti(OH)_4$ is the most common product, as titanium has four valance electrons. Below a pH of 4, the ions $Ti(OH)_3^+$ and $Ti(OH)_2^{2+}$ can also be formed [116] [117]. It is also possible that, a double bonded oxygen atom which stays bonded during the hydrolysis, forms during the reaction of the precursor so that only two hydroxyl groups can bond to the titanium. Titanium hydroxides are titania monomers that can form larger titania molecules by polymerization through condensation with other monomers if their concentration is high enough.

3.3. Different Synthesis Methods

There are many different methods to synthesize the titania-silica composites. An indirect way to prepare them is by adding premade titania nanoparticles to a silica support [103] [120] [121] at a pH of around 3 - 4. At that pH, the titania and silica have opposite charges so that the titania and silica will have electrical attraction. However, for more stability and a better homogenous coating, direct methods are often more favorable. The vapor-deposition methods (chemical vapor deposition (CVD) and physical vapor deposition (PVD)), for example, are such methods [5] [23]-[25] [66]-[70]. During the CVD method, the titania precursor is heated to the gas phase to react with dry silica in an inert environment and during the PVD method the titania is sputtered against a support surface for thin films. The impregnation [17]-[22] [52] [59]-[63] and the grafting [52] [64] [65] methods are also direct methods. During both these methods, the titania precursor is dissolved in an organic solvent like toluene or hexane. This solvent is then added to the silica support so that the precursor reacts with the silanols. During the grafting method, the solvent is removed through evaporation and during the impregnation method, the solvent is removed in some other way (e.g. filtration). During the vapor-deposition, the impregnation and the grafting methods, no water is used, which means that these methods do not have the option to form more than one layer of titania in one step, because no new hydroxyl groups can form on the coated titania during the reactions for further condensation. In addition, these methods are not optimal for low cost production since either very high temperatures or expensive organic solvents are required.

Methods that are more promising for low cost photocatalytic materials are the precipitation methods [13]-[16] [52]-[58] and the sol-gel methods [14] [32]-[42] [72]-[84]. These methods are capable of forming more than one monolayer titania on silica, and do not require expensive solvents. During the precipitation method, the titania precursor is dissolved in an aqueous solution with a low pH and low temperatures, where titania does not form. After mixing the aqueous solution containing the precursor with the silica, the solution is either neutralized with an alkaline solution and/or heated up to a specific temperature. This specific temperature depends on the pH and solvent used. By increasing the pH and/or temperature, titania slowly forms, which can be on a silica support for a coating if the hydrolysis is slow enough so that the concentration of titania monomers does not reach the critical supersaturation. Titanium chlorides ($TiCl_3$, $TiCl_4$) and titanium oxysulfate ($TiOSO_4$) are the precursors, which are often used in the precipitation methods. During the sol-gel methods, titanium alkoxides (e.g. titanium isopropoxide, titanium n-butoxide) are often used. To form a titania coating, the precursor is slowly added to a silica dispersion in an organic solvent (ethanol, n-propanol) which contains a low amount of water or to which a low amount of water is added after the precursor is added.

An important parameter in the methods that involves hydrolysis is the pH. Below pH 6, part of the $Ti(OH)_4$ is replaced by $Ti(OH)_3^+$, and also by $Ti(OH)_2^{2+}$ below pH 4. With decreasing pH, more $Ti(OH)_4$ is replaced by the ions which lead to a higher solubility [116]. A higher solubility means that the equilibrium between monomers and condensed titania is then more to the side of the monomers. Thus, when a low pH is used, more precursor is needed for the same amount of the condensated titania, as some of the titania monomers stay dissolved [116]. Since it is mostly the removal of OH^- groups that lead to the formation of ions, hydrated, amorphous and small sized titania particles are more dissolvable than crystalline titania and titania bonded to larger particles like the silica [122] [123]. The peptizing method, which is a different kind of sol-gel method, uses this constant equilibrium between titania monomers and condensated titania in an aqueous solution and the difference in dissolvability. During this method, hydrated precipitates are first formed in an aqueous solution and then slowly dissolved by reducing the pH to around 2 - 4. Using the Ostwald ripening process, crystalline titania nanoparticles are then formed [122] [123] or coated on a silica support [15].

Another way to use the sol-gel method for low cost photocatalytic material is by coating a support, like a glass plate, with a thin film using the dip-coating method [124]-[133]. During the dip-coating method, the support is dipped into a stable sol-gel mixture, and is then slowly pulled out of the mixture so that a thin layer of the mixture is adsorbed to the surface. During the drying, a thin titania film is then formed. Polymers (e.g. Poly (ethylene glycol)) can be used to obtain a higher porosity. By adding these large molecules in the sol-gel mixture, large pores are formed during the calcination step, when these molecules are removed.

Another method which is often used for the synthesis of titania-silica composites is the hydrothermal treatment [26]-[31] [71] [72]. The advantage of this method is that it can be used for both the coating step and the crystallization step (which will be discussed in 3.5). This method is done by adding a precursor, a solution containing some water and the silica (or silica source) to an autoclave. The solution is then heated up (e.g. to 200°C) for both the reaction and crystallization step.

3.4. Controlling the Hydrolysis Rate

For a homogenous coating with the seeded-growth process, the concentration of titanium monomers should not exceed the critical supersaturation, and thus the hydrolysis rate needs to be controlled. In aqueous solutions with a neutral pH, the hydrolysis of titania precursors happens so fast that the concentration of the monomers reaches the critical supersaturation point almost instantly, causing the titania to precipitate randomly in the solution instead of slowly forming on the silica surface.

The most important parameters on which the hydrolysis rate of titania precursors is dependent are: the pH, temperature, concentration of the precursor and of water, and type of precursor used. For example, in an aqueous solution with a pH below 1 and a temperature below 20°C, no titania will form [13]-[16] [52]-[58]. Having organic liquids (like n-propanol) [134] in the solvent increases the temperature and pH at which the titania is still soluble, because the dielectric constant of the solvent is then decreased. Having a low water content also prevents fast hydrolysis even when no acid is used [34] [81] [134]. However, as each hydrolysis-condensation reaction consumes a water molecule, enough water should be present, to add new hydroxyl groups on the surface of the forming silica-titania composites. Another way to slow down hydrolysis is by reacting the precursors first with molecules, like glycols, which are larger than the side-groups of the precursor. These molecules can replace the side-groups of the precursors [124] [135] if added in excess, so that new, less reactive titania precursors are formed. Depending on the exact method, another important variable is the speed at which a parameter is changed, for example, the change of pH during the neutralization method, the addition speed of a precursor during a sol-gel method and the speed at which the temperature increases during a hydrothermal treatment.

3.5. Transformation to Crystalline Titania

When the hydrolysis rate is very slow during the reaction, thermodynamics plays a more important role than kinetics. Since crystalline titania is more energetically favorable than amorphous titania, crystallization of the titania can then directly happen, especially at a low pH, where the solubility difference between amorphous titania and crystalline titania is larger [122] [134] [136]-[143]. However, the direct formation of crystalline titania is hard to control. If the hydrolysis is too slow, it can result in large rutile crystals with a low specific surface area, which is undesirable for the photocatalysis. In any other case, it is likely that most of the titania is amorphous titania after the reaction. Because amorphous titania has a much lower photocatalytic activity [96] [99], it can be beneficial to either use calcination or hydrothermal treatment to transform it into anatase.

During the calcination of pure titania, the transformation of amorphous titania to anatase happens at a temperature of about 400°C and at temperatures above 600°C the transformation to rutile occurs [143]. At these high temperatures, chemically bonded hydroxyl groups condensate with each other so that more bonds are formed between the titanium and oxygen atoms. Through rearrangements, the crystal structures are then slowly formed. Once a crystal is large enough to be stable, it will further increase in size by taking up more titania atoms, either through more rearrangements, or by merging with other crystals.

Besides the calcination in dry air, it is also possible to use hydrothermal treatment for the formation of crystalline titania [25]-[31] [71] [72] [145]-[149]. Since the formation of crystalline titania takes place in an aqueous environment during a hydrothermal treatment hydroxyl groups are incorporated into the formed structure which can be helpful for the photocatalytic activity. Hydrothermal treatment works at lower temperatures than calcination because the water increases the mobility of the atoms, reduces surface tension of the titania and catalyzes nucleation of crystals [145]-[149]. Wang and Ying [147] showed that using a hydrothermal treatment on amorphous titania, smaller and more stable titania nanoparticles were produced than with calcination.

The exact temperature at which the transformation to either anatase or rutile happens during both calcination and hydrothermal treatment depends on the size of the particles (according to Banfield *et al.* [144] below a size of 14 nm, anatase is more thermodynamically stable than rutile), the pH and other chemicals (e.g. adsorbed polymers, salts) that can influence the mobility of the atoms [117] [134] [136] [140] [142]. The formation of anatase or rutile from amorphous titania does not start at a single point where all amorphous material crystallizes into anatase or into rutile. By using higher temperatures, more amorphous titania will transform into anatase. However, higher temperatures will also transform anatase into rutile and increase the growth rate of the crystals, which leads to a smaller specific surface area [72] [134] [136]-[143].

When titania is chemically bonded to a substrate like silica, the substrate stabilizes the different structures of titania, and suppresses the transformation of amorphous titania to anatase and the transformation of anatase to rutile by decreasing the mobility of the titania atoms like an anchor [32] [33] [54] [72] [76] [80] [129]. Thus, higher temperatures are required to form anatase and rutile when titania is coated on silica. While more energy is needed for the formation of anatase from amorphous titania on a support, the anatase that is then formed has a higher thermal stability. It has even been reported that the anatase-rutile transformation only happens in some composites with a high temperature of 1000°C [54] [129]. The increase in temperature required for the crystalline transformations depends on the thickness of the titania, since a thicker layer is less influenced by the support [54]. For the titania-silica composites, the crystal growth by calcination can cause shrinkage stress when the titania structure shrinks due to the density increase and removal of chemically and physically adsorbed water. As the silica works like an anchor against the shrinkage, stress is produced on the structure which can lead to the breakage of some Ti-O-Si bonds [150].

4. The Influence of Silica on Photocatalytic Titania in Low Titania Content Composites

Titania-silica composites have more different properties than pure titania than simply a higher stability and a higher specific surface area, especially when the titania content is very low. Many researchers have studied the low titania composites because of these different properties. Anpo *et al.* [5] were one of the first who studied them. Using the CVD method on a porous silica glass, they found some interesting results which include: 1) below three layers, no anatase could be measured with X-ray diffraction, while it could still be present; 2) the band-gap became larger (4.1 eV) for just a monolayer titania; 3) the titanium was 4-coordinated in a tetrahedral structure instead of the 6-coordinated octahedral structure in pure anatase or rutile; 4) the tetrahedral titania with a large band gap catalyzed different reactions like the decomposition of N_2O as will be explained in Section 4.1; and 5) the photocatalytic efficiency per amount of catalyst was much higher for low titania content composites, which will be explained in Section 4.3.

4.1. The Larger Band Gap and Its Influence on the Photocatalytic Activity

The band gap of titania increases when going from bulk anatase to the tetrahedral titania. The normal band gap for crystalline titania is around 3.0 - 3.2 eV, but the measureable band gap from a very low amount of titania on the surface of silica can be much larger [5] [21] [68] [69] [151] [152]. There are two effects responsible for this increase. The first is the quantum size effect, which increases the band gap with decreasing crystal size, when the size is below 2 nm [68]. The second effect is caused by the difference in energy levels of the energy bands from silica and the energy bands from titania, close to the titania-silica interface [1] [21] [68]. Band gaps up to 4.1 eV [5] [69] could be measured due to these two effects. When the band gap becomes larger, electrons require more energy to be excited to the conduction band. For the applications that use sun-light or normal indoor light as the light source, this larger band gap is a disadvantage, since even less of the light spectra can then be absorbed.

On the other hand, the energy that is absorbed is used more efficiently because of the larger band gap. A larger band gap lowers the chance for recombination and increases the redox potentials of the excited electrons and holes. This higher redox potential increases the efficiency of the formation of the radical hydroxyl and superoxides molecules and enables the titania to catalyze different reactions [5] [17]-[20]. For example, Yamashita *et al.* [18] measured that pure titania transformed NO mostly into oxidized species, while NO decomposed to N_2 and O_2 in the presence of composites prepared with an ion-exchange method, in which titanium ions replaced silicon ions. In the same system [19] and similar systems with other zeolites [20], the same observation was made with the reaction of CO_2 and H_2O. With the ion-exchange composites, methanol was mostly produced while methane was produced by the titania samples. Another example of the difference in catalytic reactions taking place is from a study by Gao *et al.* [17] who observed that in the presence of tetrahedral titania, methanol reacted to methyl formate ($C_2H_4O_2$) and formaldehyde (CH_2O) while in the presence of octahedral titania, methanol reacted to dimethyl ether (C_2H_6O). While photocatalytic titania has some potential in reducing the amount of greenhouse gasses in air [153], these reactions show that the composites have an even greater potential to be useful against climate change.

4.2. Higher Density of Hydroxyl Groups

Binary metal oxides often have better catalytic properties than single metal oxides because they have extra acid sites on their surface in the form of hydroxyl groups [1] [154]. The titania-silica composites are one of those binary systems, and an increase in acid sites has been measured in several different studies [1] [18] [37] [56] [74] [80] [83]. The increase in hydroxyl groups is important for the photocatalytic activity and hydrophilicity as these depend on the amount of hydroxyl groups on the surface. Tanabe *et al.* [154] made the hypothesis that this increase in hydroxyl groups is caused by the difference in coordination numbers. The coordination number for silicon atoms in silica is 4 and for titanium atoms in crystalline titania it is 6. So when titanium atoms are introduced in, or on silica in low amounts and form the tetrahedral structure, an excess of negative −2 charge per titanium atom is created. This excess charge causes Brönsted acidity on the surface after absorbing enough protons to compensate the charge. Walter *et al.* [82] showed, using neutron diffraction, that the number of hydroxyl groups is indeed affected when titanium atoms are introduced into the silica structure by the difference in coordination number. They showed an increase in hydroxyl groups mainly caused by the increase in strain in the structure. Liu *et al.* [74] and Doolin *et al.* [80] both used the sol-gel method to make titania-rich and titania-poor composites and compared them to pure titania and silica. They measured indeed an increase in Brönsted acidity in the composites especially where there were Ti-O-Si bonds, which was in agreement with Tanabe. However, for the titania-poor composites, the increase in acidity was lower than for titania-rich composites, which is in disagreement with the model of Tanabe. So far, no model has been proposed yet, that explains the extra hydroxyl groups better than the model of Tanebe *et al.*, but these studies about the extra hydroxyl groups do show that the mechanism is related with the Ti-O-Si bond [1].

4.3. Higher Photocatalytic Efficiency of Low Titania Content Composites

Other researchers [17] [35] [62] [67] [81] [83] found similar results as Anpo *et al.* [5] on different low titania content composites and these researchers often observed an increase in photocatalytic efficiency per amount of catalyst compared to pure titania. The high efficiencies of these low titania content composites, which do not even have enough titania for a full monolayer, are caused by: 1) the high specific surface area of the silica supports used; 2) the ability of the silica to adsorb many molecules for longer times than titania, especially with the extra hydroxyl groups; 3) the fact that the titania is used more efficiently since all the titania is at the surface; 4) the higher redox potentials of the electrons and holes; and 5) the fact that silica can scatter the light to the titania without being able to absorb its energy. In addition, during the photocatalytic measurements in these studies, UV-light was used. The measurements using UV-light might not represent the applications which use sun-light as the light source, since the decrease of possible light absorption caused by the increase in band gap is less with UV-light.

5. Conclusions

The titania-silica composites are interesting materials because they have the potential to make photocatalytic materials more cost-effective. For the same level of photocatalytic activity, fewer resources have to be invested with the titania-silica composites than with pure titania. The titania-silica composites can, with less and cheaper material, have the same photocatalytic efficiency as pure titania for a longer time since the composites can have a higher photocatalytic efficiency, lower production costs and increased durability. The applications of the photocatalytic material including the applications that degrade pollutants, become then more attractive for companies to produce on a larger scale which can eventually lead to an overall improvement of the quality of air and water.

To obtain this cost-effective photocatalytic material, the titania-silica composites need to be synthesized with a method that has low production costs but still produces composites which have a high photocatalytic efficiency.

- For a high efficiency, the method needs to deposit an anatase layer with thickness of maximum 5 nm on a large specific surface area. Any layer larger than 5 nm will have titania, which does not contribute to the photocatalysis, since it is too far away from the surface.

- If some of the crystal structure is rutile instead of anatase, it can have some increase in photocatalytic efficiency because of the separation of holes and electrons at the interface of the two forms. However, the amount of rutile should not be too high, as the lifetime and conductivity of excited electrons and holes in rutile are less favorable than in anatase.

- When the crystal size is too small, the titania may have an increase in band gap. While it has been reported that such an increase in band gap can cause higher photocatalytic efficiencies, it is important to note that it will make the titania absorb less visible light.

- The titania should be chemically bonded to a silica substrate which has a large specific surface area, high mechanical and thermal stability as well as low production costs.

- The most promising methods for low cost photocatalytic composites are the ones that involve hydrolysis (precipitation and sol-gel methods), as these methods ensure that more than one layer of titania can form without the need of expensive materials. However, the hydrolysis of titania precursors can be hard to control. The most important parameters on which the hydrolysis rate is dependent are the pH, temperature, concentration of water and precursor, the speed at which these parameters are changed during the reaction (e.g. by addition of water) and the type of precursor used. How much influence each parameter has on the hydrolysis rate depends on the method used. It is important that the reaction speed of the hydrolysis should be slow enough so that the condensation of titania monomers on the substrate's surface is more likely to happen than polymerization between monomers.

- For the transformation of amorphous titania to anatase, calcination or hydrothermal treatment can be applied. The temperature and time required to obtain anatase crystals from amorphous titania depend on the mobility of the titania molecules, which can be influenced by: the chemical bonds to the silica, any nucleated crystals already present, and other chemicals present (e.g. adsorbed polymers, salts). While having more crystalline anatase is beneficial for the photocatalytic activity, crystallization does not always produce materials with a higher photocatalytic efficiency, since during the growth of the crystals, the specific surface area is reduced and anatase can transform into rutile at high temperatures.

When all these points are fulfilled, the resulting titania-silica composites will have the required properties to be a cost-effective material which can compete with pure titania in photocatalytic applications. Even with the increased complexity, the composites are an excellent alternative to pure titania nanoparticles.

References

[1] Gao, X. and Wachs, I.E. (1999) Titania-Silica as Catalysts: Molecular Structural Characteristics and Physico-Chemical Properties. *Catalysis Today*, **51**, 233-254. http://dx.doi.org/10.1016/S0920-5861(99)00048-6

[2] Hüsken, G., Hunger, M. and Brouwers, H.J.H. (2009) Experimental Study of Photocatalytic Concrete Products for Air Purification. *Building and environment*, **44**, 2463-2474. http://dx.doi.org/10.1016/j.buildenv.2009.04.010

[3] Fujishima, A., Zhang, X. and Tryk, D.A. (2008) TiO$_2$ Photocatalysis and Related Surface Phenomena. *Surface Science Reports*, **63**, 515-582. http://dx.doi.org/10.1016/j.surfrep.2008.10.001

[4] Hashimoto, K., Irie, H. and Fujishima, A. (2005). TiO$_2$ Photocatalysis: A Historical Overview and Future Prospects. *Japanese Journal of Applied Physics*, **44**, 8269. http://dx.doi.org/10.1143/JJAP.44.8269

[5] Anpo, M., Aikawa, N., Kubokawa, Y., Che, M., Louis, C. and Giamello, E. (1985) Photoluminescence and Photocatalytic Activity of Highly Dispersed Titanium Oxide Anchored onto Porous Vycor Glass. *The Journal of Physical Chemistry*, **89**, 5017-5021. http://dx.doi.org/10.1021/j100269a025

[6] Fujishima, A., and Zhang, X. (2006) Titanium Dioxide Photocatalysis: Present Situation and Future Approaches. *Comptes Rendus Chimie*, **9**, 750-760. http://dx.doi.org/10.1016/j.crci.2005.02.055

[7] Turchi, C.S. and Ollis, D.F. (1990) Photocatalytic Degradation of Organic Water Contaminants: Mechanisms Involving Hydroxyl Radical Attack. *Journal of catalysis*, **122**, 178-192. http://dx.doi.org/10.1016/0021-9517(90)90269-P

[8] Ballari, M.M. and Brouwers, H.J.H. (2013) Full Scale Demonstration of Air-Purifying Pavement. *Journal of hazardous materials*, **254**, 406-414. http://dx.doi.org/10.1016/j.jhazmat.2013.02.012

[9] Hüsken, G., Hunger, M., and Brouwers, H.J. (2007) Comparative Study on Cementitious Products Containing Titanium Dioxide as Photo-Catalyst. *Proceedings of the International RILEM Symposium on Photocatalysis, Environment and Construction Materials—TDP*, Florence, 8-9 October 2007, 147-154.

[10] Hunger, M., Hüsken, G. and Brouwers, H.J.H. (2010) Photocatalytic Degradation of Air Pollutants—From Modeling to Large Scale Application. *Cement and Concrete Research*, **40**, 313-320. http://dx.doi.org/10.1016/j.cemconres.2009.09.013

[11] Linsebigler, A.L., Lu, G. and Yates Jr., J.T. (1995) Photocatalysis on TiO_2 Surfaces: Principles, Mechanisms, and Selected Results. *Chemical Reviews*, **95**, 735-758. http://dx.doi.org/10.1021/cr00035a013

[12] Anpo, M., Aikawa, N., Kodama, S. and Kubokawa, Y. (1984) Photocatalytic Hydrogenation of Alkynes and Alkenes with Water over Titanium Dioxide. Hydrogenation Accompanied by Bond Fission. *The Journal of Physical Chemistry*, **88**, 2569-2572. http://dx.doi.org/10.1021/j150656a028

[13] Sirikawinkobkul, N., Kalambaheti, C., Jiemsirilers, S., Kashima, D.P. and Jinawath, S. (2009) Synthesis, Characterization and Photocatalytic Activity of Visible-Light Titania/Silica Photocatalyst. *18th International Conference on Composite Materials*, Edinburgh, 27-31 July 2009.

[14] Montes, M., Getton, F.P., Vong, M.S.W. and Sermon, P.A. (1997) Titania on Silica. A Comparison of Sol-Gel Routes and Traditional Methods. *Journal of Sol-Gel Science and Technology*, **8**, 131-137. http://dx.doi.org/10.1007/BF02436830

[15] Huang, C.H., Bai, H., Liu, S.L., Huang, Y.L. and Tseng, Y.H. (2011) Synthesis of Neutral SiO_2/TiO_2 Hydrosol and Its Photocatalytic Degradation of Nitric Oxide Gas. *Micro & Nano Letters*, **6**, 646-649. http://dx.doi.org/10.1049/mnl.2011.0331

[16] Ding, Z., Lu, G.Q. and Greenfield, P.F. (2000) A Kinetic Study on Photocatalytic Oxidation of Phenol in Water by Silica-Dispersed Titania Nanoparticles. *Journal of Colloid and Interface Science*, **232**, 1-9. http://dx.doi.org/10.1006/jcis.2000.7154

[17] Gao, X., Bare, S.R., Fierro, J.L.G., Banares, M.A. and Wachs, I.E. (1998) Preparation and *In-Situ* Spectroscopic Characterization of Molecularly Dispersed Titanium Oxide on Silica. *The Journal of Physical Chemistry B*, **102**, 5653-5666. http://dx.doi.org/10.1021/jp981423e

[18] Yamashita, H., Ichihashi, Y., Anpo, M., Hashimoto, M., Louis, C. and Che, M. (1996) Photocatalytic Decomposition of NO at 275 K on Titanium Oxides Included within Y-Zeolite Cavities: The Structure and Role of the Active Sites. *The Journal of Physical Chemistry*, **100**, 16041-16044. http://dx.doi.org/10.1021/jp9615969

[19] Anpo, M., Yamashita, H., Ichihashi, Y., Fujii, Y. and Honda, M. (1997) Photocatalytic Reduction of CO_2 with H_2O on Titanium Oxides Anchored within Micropores of Zeolites: Effects of the Structure of the Active Sites and the Addition of Pt. *The Journal of Physical Chemistry B*, **101**, 2632-2636. http://dx.doi.org/10.1021/jp962696h

[20] Anpo, M., Yamashita, H., Ikeue, K., Fujii, Y., Zhang, S.G., Ichihashi, Y. and Tatsumi, T. (1998) Photocatalytic Reduction of CO_2 with H_2O on Ti-MCM-41 and Ti-MCM-48 Mesoporous Zeolite Catalysts. *Catalysis Today*, **44**, 327-332. http://dx.doi.org/10.1016/S0920-5861(98)00206-5

[21] Fernández, A., Caballero, A. and González-Elipe, A.R. (1992) Size and Support Effects in the Photoelectron Spectra of Small TiO_2 Particles. *Surface and Interface Analysis*, **18**, 392-396. http://dx.doi.org/10.1002/sia.740180604

[22] Anpo, M., Yamashita, H., Ichihashi, Y. and Ehara, S. (1995) Photocatalytic Reduction of CO_2 with H_2O on Various Titanium Oxide Catalysts. *Journal of Electroanalytical Chemistry*, **396**, 21-26. http://dx.doi.org/10.1016/0022-0728(95)04141-A

[23] Ding, Z., Hu, X., Lu, G.Q., Yue, P.L. and Greenfield, P.F. (2000) Novel Silica Gel Supported TiO_2 Photocatalyst Synthesized by CVD Method. *Langmuir*, **16**, 6216-6222. http://dx.doi.org/10.1021/la000119l

[24] Yamashita, H., Ichihashi, Y., Harada, M., Stewart, G., Fox, M.A. and Anpo, M. (1996) Photocatalytic Degradation of 1-Octanol on Anchored Titanium Oxide and on TiO_2 Powder Catalysts. *Journal of Catalysis*, **158**, 97-101. http://dx.doi.org/10.1006/jcat.1996.0010

[25] Sayilkan, F., Asilturk, M., Sener, S., Erdemoglu, S., Erdemoglu, M. and Sayilkan, H. (2007) Hydrothermal Synthesis, Characterization and Photocatalytic Activity of Nanosized TiO_2 Based Catalysts for Rhodamine B Degradation. *Turkish Journal of Chemistry*, **31**, 211-221.

[26] Chuan, X.Y., Hirano, M. and Inagaki, M. (2004) Preparation and Photocatalytic Performance of Anatase-Mounted Natural Porous Silica, Pumice, by Hydrolysis under Hydrothermal Conditions. *Applied Catalysis B: Environmental*, **51**, 255-260. http://dx.doi.org/10.1016/j.apcatb.2004.03.004

[27] Hirano, M. and Ota, K. (2004) Preparation of Photoactive Anatase-Type TiO_2/Silica Gel by Direct Loading Anatase-Type TiO_2 Nanoparticles in Acidic Aqueous Solutions by Thermal Hydrolysis. *Journal of Materials Science*, **39**, 1841-1844. http://dx.doi.org/10.1023/B:JMSC.0000016199.85213.0b

[28] Hirano, M. and Ota, K. (2004) Direct Formation and Photocatalytic Performance of Anatase (TiO_2)/Silica (SiO_2) Composite Nanoparticles. *Journal of the American Ceramic Society*, **87**, 1567-1570. http://dx.doi.org/10.1111/j.1551-2916.2004.01567.x

[29] Kim, E.Y., Whang, C.M., Lee, W.I. and Kim, Y.H. (2006) Photocatalytic Property of SiO_2/TiO_2 Nanoparticles Pre-

pared by Sol-Hydrothermal Process. *Journal of Electroceramics*, **17**, 899-902.
http://dx.doi.org/10.1007/s10832-006-9071-5

[30] Fu, X., Clark, L.A., Yang, Q. and Anderson, M.A. (1996) Enhanced Photocatalytic Performance of Titania-Based Binary Metal Oxides: TiO_2/SiO_2 and TiO_2/ZrO_2. *Environmental Science & Technology*, **30**, 647-653.
http://dx.doi.org/10.1021/es950391v

[31] Anderson, C. and Bard, A.J. (1995) An Improved Photocatalyst of TiO_2/SiO_2 Prepared by a Sol-Gel Synthesis. *The Journal of Physical Chemistry*, **99**, 9882-9885. http://dx.doi.org/10.1021/j100024a033

[32] Cheng, P., Zheng, M.P., Huang, Q., Jin, Y.P. and Gu, M.Y. (2003) Enhanced Photoactivity of Silica-Titania Binary Oxides Prepared by Sol-Gel Method. *Journal of Materials Science Letters*, **22**, 1165-1168.
http://dx.doi.org/10.1023/A:1025187330150

[33] Smitha, V.S., Manjumol, K.A., Baiju, K.V., Ghosh, S., Perumal, P. and Warrier, K.G.K. (2010) Sol-Gel Route to Synthesize Titania-Silica Nano Precursors for Photoactive Particulates and Coatings. *Journal of Sol-Gel Science and Technology*, **54**, 203-211. http://dx.doi.org/10.1007/s10971-010-2178-9

[34] Guo, X.C. and Dong, P. (1999) Multistep Coating of Thick Titania Layers on Monodisperse Silica Nanospheres. *Langmuir*, **15**, 5535-5540. http://dx.doi.org/10.1021/la990220u

[35] Kamaruddin, S. and Stephan, D. (2014) Sol-Gel Mediated Coating and Characterization of Photocatalytic Sand and Fumed Silica for Environmental Remediation. *Water, Air, & Soil Pollution*, **225**, 1948.
http://dx.doi.org/10.1007/s11270-014-1948-3

[36] Shan, A.Y., Ghazi, T.I.M. and Rashid, S.A. (2010) Immobilisation of Titanium Dioxide onto Supporting Materials in Heterogeneous Photocatalysis: A Review. *Applied Catalysis A: General*, **389**, 1-8.
http://dx.doi.org/10.1016/j.apcata.2010.08.053

[37] Guan, K. (2005) Relationship between Photocatalytic Activity, Hydrophilicity and Self-Cleaning Effect of TiO_2/SiO_2 Films. *Surface and Coatings Technology*, **191**, 155-160. http://dx.doi.org/10.1016/j.surfcoat.2004.02.022

[38] Jung, K.Y. and Park, S.B. (2000) Enhanced Photoactivity of Silica-Embedded Titania Particles Prepared by Sol-Gel Process for the Decomposition of Trichloroethylene. *Applied Catalysis B: Environmental*, **25**, 249-256.
http://dx.doi.org/10.1016/S0926-3373(99)00134-4

[39] Ismail, A.A., Ibrahim, I.A., Ahmed, M.S., Mohamed, R.M. and El-Shall, H. (2004) Sol-Gel Synthesis of Titania-Silica Photocatalyst for Cyanide Photodegradation. *Journal of Photochemistry and Photobiology A: Chemistry*, **163**, 445-451.
http://dx.doi.org/10.1016/j.jphotochem.2004.01.017

[40] Xie, C., Xu, Z., Yang, Q., Xue, B., Du, Y. and Zhang, J. (2004) Enhanced Photocatalytic Activity of Titania-Silica Mixed Oxide Prepared via Basic Hydrolyzation. *Materials Science and Engineering: B*, **112**, 34-41.
http://dx.doi.org/10.1016/j.mseb.2004.05.011

[41] Zhang, X., Zhang, F. and Chan, K.Y. (2005) Synthesis of Titania-Silica Mixed Oxide Mesoporous Materials, Characterization and Photocatalytic Properties. *Applied Catalysis A: General*, **284**, 193-198.
http://dx.doi.org/10.1016/j.apcata.2005.01.037

[42] Yang, J., Zhang, J., Zhu, L., Chen, S., Zhang, Y., Tang, Y. and Li, Y. (2006) Synthesis of Nano Titania Particles Embedded in Mesoporous SBA-15: Characterization and Photocatalytic Activity. *Journal of Hazardous Materials*, **137**, 952-958. http://dx.doi.org/10.1016/j.jhazmat.2006.03.017

[43] Maggos, T., Plassais, A., Bartzis, J.G., Vasilakos, C., Moussiopoulos, N. and Bonafous, L. (2008) Photocatalytic Degradation of NO_x in a Pilot Street Canyon Configuration Using TiO_2-Mortar Panels. *Environmental Monitoring and Assessment*, **136**, 35-44. http://dx.doi.org/10.1007/s10661-007-9722-2

[44] Strini, A., Cassese, S. and Schiavi, L. (2005) Measurement of Benzene, Toluene, Ethylbenzene and *o*-Xylene Gas Phase Photodegradation by Titanium Dioxide Dispersed in Cementitious Materials Using a Mixed Flow Reactor. *Applied Catalysis B: Environmental*, **61**, 90-97. http://dx.doi.org/10.1016/j.apcatb.2005.04.009

[45] Ângelo, J., Andrade, L. and Mendes, A. (2014) Highly Active Photocatalytic Paint for NO_x Abatement under Real-Outdoor Conditions. *Applied Catalysis A: General*, **484**, 17-25. http://dx.doi.org/10.1016/j.apcata.2014.07.005

[46] Chen, J. and Poon, C.S. (2009) Photocatalytic Construction and Building Materials: From Fundamentals to Applications. *Building and Environment*, **44**, 1899-1906. http://dx.doi.org/10.1016/j.buildenv.2009.01.002

[47] Paz, Y. (2010) Application of TiO_2 Photocatalysis for Air Treatment: Patents' Overview. *Applied Catalysis B: Environmental*, **99**, 448-460. http://dx.doi.org/10.1016/j.apcatb.2010.05.011

[48] Yu, Q.L. and Brouwers, H.J.H. (2009) Indoor Air Purification Using Heterogeneous Photocatalytic Oxidation. Part I: Experimental Study. *Applied Catalysis B: Environmental*, **92**, 454-461. http://dx.doi.org/10.1016/j.apcatb.2009.09.004

[49] Yu, Q.L. and Brouwers, H.J.H. (2013) Design of a Novel Photocatalytic Gypsum Plaster: With the Indoor Air Purification Property. *Advanced Materials Research*, **651**, 751-756. http://dx.doi.org/10.4028/www.scientific.net/AMR.651.751

[50] Poon, C.S. and Cheung, E. (2007) NO Removal Efficiency of Photocatalytic Paving Blocks Prepared with Recycled Materials. *Construction and Building Materials*, **21**, 1746-1753. http://dx.doi.org/10.1016/j.conbuildmat.2006.05.018

[51] de Melo, J.V.S., Trichês, G., Gleize, P.J.P. and Villena, J. (2012) Development and Evaluation of the Efficiency of Photocatalytic Pavement Blocks in the Laboratory and after One Year in the Field. *Construction and Building Materials*, **37**, 310-319. http://dx.doi.org/10.1016/j.conbuildmat.2012.07.073

[52] Castillo, R., Koch, B., Ruiz, P. and Delmon, B. (1994) Influence of Preparation Methods on the Texture and Structure of Titania Supported on Silica. *Journal of Materials Chemistry*, **4**, 903-906. http://dx.doi.org/10.1039/jm9940400903

[53] Morrison, C. and Kiwi, J. (1989) Preparation and Characterization of TiO$_2$-SiO$_2$ Aerosil Colloidal Mixed Dispersions. *Journal of the Chemical Society, Faraday Transactions* 1: *Physical Chemistry in Condensed Phases*, **85**, 1043-1048. http://dx.doi.org/10.1039/f19898501043

[54] Hsu, W.P., Yu, R. and Matijević, E. (1993) Paper Whiteners: I. Titania Coated Silica. *Journal of Colloid and Interface Science*, **156**, 56-65. http://dx.doi.org/10.1006/jcis.1993.1080

[55] Galan-Fereres, M., Mariscal, R., Alemany, L.J., Fierro, J.L.G. and Anderson, J.A. (1994) Ternary V-Ti-Si Catalysts and Their Behaviour in the CO + NO Reaction. *Journal of the Chemical Society, Faraday Transactions*, **90**, 3711-3718. http://dx.doi.org/10.1039/ft9949003711

[56] Galan-Fereres, M., Alemany, L.J., Mariscal, R., Banares, M.A., Anderson, J.A. and Fierro, J.L. (1995) Surface Acidity and Properties of Titania-Silica Catalysts. *Chemistry of Materials*, **7**, 1342-1348. http://dx.doi.org/10.1021/cm00055a011

[57] Choi, H.H., Park, J. and Singh, R.K. (2005) Nanosized Titania Encapsulated Silica Particles Using an Aqueous TiCl$_4$ Solution. *Applied Surface Science*, **240**, 7-12. http://dx.doi.org/10.1016/j.apsusc.2004.06.147

[58] Sun, Z., Bai, C., Zheng, S., Yang, X. and Frost, R.L. (2013) A Comparative Study of Different Porous Amorphous Silica Minerals Supported TiO$_2$ Catalysts. *Applied Catalysis A*: *General*, **458**, 103-110. http://dx.doi.org/10.1016/j.apcata.2013.03.035

[59] Srinivasan, S., Datye, A.K., Hampden-Smith, M., Wachs, I.E., Deo, G., Jehng, J.M., Turek, A.M. and Peden, C.H.F. (1991) The Formation of Titanium Oxide Monolayer Coatings on Silica Surfaces. *Journal of Catalysis*, **131**, 260-275. http://dx.doi.org/10.1016/0021-9517(91)90343-3

[60] Srinivasan, S., Datye, A.K., Smith, M.H. and Peden, C.H.F. (1994) Interaction of Titanium Isopropoxide with Surface Hydroxyls on Silica. *Journal of Catalysis*, **145**, 565-573. http://dx.doi.org/10.1006/jcat.1994.1068

[61] Mariscal, R., Palacios, J.M., Galan-Fereres, M. and Fierro, J.L.G. (1994) Incorporation of Titania into Preshaped Silica Monolith Structures. *Applied Catalysis A*: *General*, **116**, 205-219. http://dx.doi.org/10.1016/0926-860X(94)80290-4

[62] Salama, T.M., Tanaka, T., Yamaguchi, T. and Tanabe, K. (1990) EXAFS/XANES Study of Titanium Oxide Supported on SiO$_2$: A Structural Consideration on the Amorphous State. *Surface Science*, **227**, L100-L104. http://dx.doi.org/10.1016/0039-6028(90)90379-m

[63] Ellestad, O.H. and Blindheim, U. (1985) Reactions of Titanium Tetrachloride with Silica Gel Surfaces. *Journal of Molecular Catalysis*, **33**, 275-287. http://dx.doi.org/10.1016/0304-5102(85)85001-X

[64] Aronson, B.J., Blanford, C.F. and Stein, A. (1997) Solution-Phase Grafting of Titanium Dioxide onto the Pore Surface of Mesoporous Silicates: Synthesis and Structural Characterization. *Chemistry of Materials*, **9**, 2842-2851. http://dx.doi.org/10.1021/cm970180k

[65] Huang, Y.Y., Zhao, B.Y. and Xie, Y.C. (1998) A Novel Way to Prepare Silica Supported Sulfated Titania. *Applied Catalysis A*: *General*, **171**, 65-73. http://dx.doi.org/10.1016/S0926-860X(98)00071-4

[66] Morrow, B.A. and McFarlan, A.J. (1990) Chemical Reactions at Silica Surfaces. *Journal of Non-Crystalline Solids*, **120**, 61-71. http://dx.doi.org/10.1016/0022-3093(90)90191-N

[67] Haukka, S., Lakomaa, E.L. and Root, A. (1993) An IR and NMR Study of the Chemisorption of Titanium Tetrachloride on Silica. *The Journal of Physical Chemistry*, **97**, 5085-5094. http://dx.doi.org/10.1021/j100121a040

[68] Nakayama, T., Onisawa, K., Fuyama, M. and Hanazono, M. (1992) TiO$_2$/SiO$_2$ Multilayer Insulating Films for ELDs. *Journal of the Electrochemical Society*, **139**, 1204-1206. http://dx.doi.org/10.1149/1.2069367

[69] Lassaletta, G., Fernandez, A., Espinos, J.P. and Gonzalez-Elipe, A.R. (1995) Spectroscopic Characterization of Quantum-Sized TiO$_2$ Supported on Silica: Influence of Size and TiO$_2$-SiO$_2$ Interface Composition. *The Journal of Physical Chemistry*, **99**, 1484-1490. http://dx.doi.org/10.1021/j100005a019

[70] Nakayama, T. (1994) Structure of TiO$_2$/SiO$_2$ Multilayer Films. *Journal of the Electrochemical Society*, **141**, 237-241. http://dx.doi.org/10.1149/1.2054690

[71] Hayashi, T., Yamada, T. and Saito, H. (1983) Preparation of Titania-Silica Glasses by the Gel Method. *Journal of Materials Science*, **18**, 3137-3142. http://dx.doi.org/10.1007/BF00700798

[72] Li, Z., Hou, B., Xu, Y., Wu, D., Sun, Y., Hu, W. and Deng, F. (2005) Comparative Study of Sol-Gel-Hydrothermal and Sol-Gel Synthesis of Titania-Silica Composite Nanoparticles. *Journal of Solid State Chemistry*, **178**, 1395-1405. http://dx.doi.org/10.1016/j.jssc.2004.12.034

[73] Fu, X. and Qutubuddin, S. (2001) Synthesis of Titania-Coated Silica Nanoparticles Using a Nonionic Water-in-Oil Microemulsion. *Colloids and Surfaces A: Physicochemical and Engineering Aspects*, **179**, 65-70. http://dx.doi.org/10.1016/S0927-7757(00)00723-8

[74] Liu, Z.F., Tabora, J. and Davis, R.J. (1994) Relationships between Microstructure and Surface Acidity of Ti-Si Mixed Oxide Catalysts. *Journal of Catalysis*, **149**, 117-126. http://dx.doi.org/10.1006/jcat.1994.1277

[75] Mine, E., Hirose, M., Kubo, M., Kobayashi, Y., Nagao, D. and Konno, M. (2006) Synthesis of Submicron-Sized Titania-Coated Silica Particles with a Sol-Gel Method and Their Application to Colloidal Photonic Crystals. *Journal of Sol-Gel Science and Technology*, **38**, 91-95. http://dx.doi.org/10.1007/s10971-006-5855-y

[76] Lee, D.W., Ihm, S.K. and Lee, K.H. (2005) Mesostructure Control Using a Titania-Coated Silica Nanosphere Framework with Extremely High Thermal Stability. *Chemistry of Materials*, **17**, 4461-4467. http://dx.doi.org/10.1021/cm050485w

[77] Lee, J.W., Kong, S., Kim, W.S. and Kim, J. (2007) Preparation and Characterization of SiO_2/TiO_2 Core-Shell Particles with Controlled Shell Thickness. *Materials Chemistry and Physics*, **106**, 39-44. http://dx.doi.org/10.1016/j.matchemphys.2007.05.019

[78] Do Kim, K., Bae, H.J. and Kim, H.T. (2003) Synthesis and Characterization of Titania-Coated Silica Fine Particles by Semi-Batch Process. *Colloids and Surfaces A: Physicochemical and Engineering Aspects*, **224**, 119-126. http://dx.doi.org/10.1016/S0927-7757(03)00252-8

[79] Rupp, W., Hüsing, N. and Schubert, U. (2002) Preparation of Silica-Titania Xerogels and Aerogels by Sol-Gel Processing of New Single-Source Precursors. *Journal of Materials Chemistry*, **12**, 2594-2596. http://dx.doi.org/10.1039/b204956b

[80] Doolin, P.K., Alerasool, S., Zalewski, D.J. and Hoffman, J.F. (1994) Acidity Studies of Titania-Silica Mixed Oxides. *Catalysis Letters*, **25**, 209-223. http://dx.doi.org/10.1007/BF00816302

[81] Hanprasopwattana, A., Srinivasan, S., Sault, A.G. and Datye, A.K. (1996) Titania Coatings on Monodisperse Silica Spheres (Characterization Using 2-Propanol Dehydration and TEM). *Langmuir*, **12**, 3173-3179. http://dx.doi.org/10.1021/la950808a

[82] Walters, J.K., Rigden, J.S., Dirken, P.J., Smith, M.E., Howells, W.S. and Newport, R.J. (1997) An Atomic-Scale Study of the Role of Titanium in TiO_2:SiO_2 Sol-Gel Materials. *Chemical Physics Letters*, **264**, 539-544. http://dx.doi.org/10.1016/S0009-2614(96)01359-0

[83] Klein, S., Weckhuysen, B.M., Martens, J.A., Maier, W.F. and Jacobs, P.A. (1996) Homogeneity of Titania-Silica Mixed Oxides: On UV-DRS Studies as a Function of Titania Content. *Journal of Catalysis*, **163**, 489-491. http://dx.doi.org/10.1006/jcat.1996.0350

[84] Liu, G., Liu, Y., Yang, G., Li, S., Zu, Y., Zhang, W. and Jia, M. (2009) Preparation of Titania-Silica Mixed Oxides by a Sol-Gel Route in the Presence of Citric Acid. *The Journal of Physical Chemistry C*, **113**, 9345-9351. http://dx.doi.org/10.1021/jp900577c

[85] Fujishima, A., Rao, T.N. and Tryk, D.A. (2000) Titanium Dioxide Photocatalysis. *Journal of Photochemistry and Photobiology C: Photochemistry Reviews*, **1**, 1-21. http://dx.doi.org/10.1016/S1389-5567(00)00002-2

[86] Carp, O., Huisman, C.L. and Reller, A. (2004) Photoinduced Reactivity of Titanium Dioxide. *Progress in Solid State Chemistry*, **32**, 33-177. http://dx.doi.org/10.1016/j.progsolidstchem.2004.08.001

[87] Kitano, M., Matsuoka, M., Ueshima, M. and Anpo, M. (2007) Recent Developments in Titanium Oxide-Based Photocatalysts. *Applied Catalysis A: General*, **325**, 1-14. http://dx.doi.org/10.1016/j.apcata.2007.03.013

[88] Gaya, U.I. and Abdullah, A.H. (2008) Heterogeneous Photocatalytic Degradation of Organic Contaminants over Titanium Dioxide: A Review of Fundamentals, Progress and Problems. *Journal of Photochemistry and Photobiology C: Photochemistry Reviews*, **9**, 1-12. http://dx.doi.org/10.1016/j.jphotochemrev.2007.12.003

[89] Macwan, D.P., Dave, P.N. and Chaturvedi, S. (2011) A Review on Nano-TiO_2 Sol-Gel Type Syntheses and Its Applications. *Journal of Materials Science*, **46**, 3669-3686. http://dx.doi.org/10.1007/s10853-011-5378-y

[90] Simonsen, M.E., Li, Z. and Søgaard, E.G. (2009) Influence of the OH Groups on the Photocatalytic Activity and Photoinduced Hydrophilicity of Microwave Assisted Sol-Gel TiO_2 Film. *Applied Surface Science*, **255**, 8054-8062. http://dx.doi.org/10.1016/j.apsusc.2009.05.013

[91] Yu, J., Jimmy, C.Y., Ho, W. and Jiang, Z. (2002) Effects of Calcination Temperature on the Photocatalytic Activity and Photo-Induced Super-Hydrophilicity of Mesoporous TiO_2 Thin Films. *New Journal of Chemistry*, **26**, 607-613. http://dx.doi.org/10.1039/b200964a

[92] Wang, R., Hashimoto, K., Fujishima, A., Chikuni, M., Kojima, E., Kitamura, A. and Watanabe, T. (1997) Light-Induced Amphiphilic Surfaces. *Nature*, **388**, 431-432. http://dx.doi.org/10.1038/41233

[93] Mezhenny, S., Maksymovych, P., Thompson, T.L., Diwald, O., Stahl, D., Walck, S.D. and Yates, J.T. (2003) STM Studies of Defect Production on the $TiO_2(110)$-(1×1) and $TiO_2(110)$-(1×2) Surfaces Induced by UV Irradiation. *Chemical Physics Letters*, **369**, 152-158. http://dx.doi.org/10.1016/S0009-2614(02)01997-8

[94] White, J.M., Szanyi, J. and Henderson, M.A. (2003) The Photon-Driven Hydrophilicity of Titania: A Model Study Using $TiO_2(110)$ and Adsorbed Trimethyl Acetate. *The Journal of Physical Chemistry B*, **107**, 9029-9033. http://dx.doi.org/10.1021/jp0345046

[95] Miyauchi, M., Nakajima, A., Fujishima, A., Hashimoto, K. and Watanabe, T. (2000) Photoinduced Surface Reactions on TiO_2 and $SrTiO_3$ Films: Photocatalytic Oxidation and Photoinduced Hydrophilicity. *Chemistry of Materials*, **12**, 3-5. http://dx.doi.org/10.1021/cm990556p

[96] Luttrell, T., Halpegamage, S., Tao, J., Kramer, A., Sutter, E. and Batzill, M. (2014) Why Is Anatase a Better Photocatalyst than Rutile?—Model Studies on Epitaxial TiO_2 Films. *Scientific Reports*, **4**, Article No.: 4043. http://dx.doi.org/10.1038/srep04043

[97] Mattioli, G., Filippone, F., Alippi, P. and Bonapasta, A.A. (2008) *Ab Initio* Study of the Electronic States Induced by Oxygen Vacancies in Rutile and Anatase TiO_2. *Physical Review B*, **78**, Article ID: 241201. http://dx.doi.org/10.1103/PhysRevB.78.241201

[98] Xu, M., Gao, Y., Moreno, E.M., Kunst, M., Muhler, M., Wang, Y., Idriss, H. and Wöll, C. (2011) Photocatalytic Activity of Bulk TiO_2 Anatase and Rutile Single Crystals Using Infrared Absorption Spectroscopy. *Physical Review Letters*, **106**, Article ID: 138302. http://dx.doi.org/10.1103/PhysRevLett.106.138302

[99] Ohtani, B., Ogawa, Y. and Nishimoto, S.I. (1997) Photocatalytic Activity of Amorphous-Anatase Mixture of Titanium (IV) Oxide Particles Suspended in Aqueous Solutions. *The Journal of Physical Chemistry B*, **101**, 3746-3752. http://dx.doi.org/10.1021/jp962702+

[100] Bickley, R.I., Gonzalez-Carreno, T., Lees, J.S., Palmisano, L. and Tilley, R.J. (1991) A Structural Investigation of Titanium Dioxide Photocatalysts. *Journal of Solid State Chemistry*, **92**, 178-190. http://dx.doi.org/10.1016/0022-4596(91)90255-G

[101] Scanlon, D.O., Dunnill, C.W., Buckeridge, J., Shevlin, S.A., Logsdail, A.J., Woodley, S.M. and Sokol, A.A. (2013) Band Alignment of Rutile and Anatase TiO_2. *Nature Materials*, **12**, 798-801. http://dx.doi.org/10.1038/nmat3697

[102] Mogyorósi, K., Farkas, A., Dékány, I., Ilisz, I. and Dombi, A. (2002) TiO_2-Based Photocatalytic Degradation of 2-Chlorophenol Adsorbed on Hydrophobic Clay. *Environmental Science & Technology*, **36**, 3618-3624. http://dx.doi.org/10.1021/es015843k

[103] Mogyorosi, K., Dekany, I. and Fendler, J.H. (2003) Preparation and Characterization of Clay Mineral Intercalated Titanium Dioxide Nanoparticles. *Langmuir*, **19**, 2938-2946. http://dx.doi.org/10.1021/la025969a

[104] Kun, R., Mogyorósi, K. and Dékány, I. (2006) Synthesis and Structural and Photocatalytic Properties of TiO_2/Montmorillonite Nanocomposites. *Applied Clay Science*, **32**, 99-110. http://dx.doi.org/10.1016/j.clay.2005.09.007

[105] Kibanova, D., Trejo, M., Destaillats, H. and Cervini-Silva, J. (2009) Synthesis of Hectorite-TiO_2 and Kaolinite-TiO_2 Nanocomposites with Photocatalytic Activity for the Degradation of Model Air Pollutants. *Applied Clay Science*, **42**, 563-568. http://dx.doi.org/10.1016/j.clay.2008.03.009

[106] Matthews, R.W. (1991) Photooxidative Degradation of Coloured Organics in Water Using Supported Catalysts. TiO_2 on Sand. *Water Research*, **25**, 1169-1176. http://dx.doi.org/10.1016/0043-1354(91)90054-T

[107] Matthews, R.W. and McEvoy, S.R. (1992) Photocatalytic Degradation of Phenol in the Presence of Near-UV Illuminated Titanium Dioxide. *Journal of Photochemistry and Photobiology A: Chemistry*, **64**, 231-246. http://dx.doi.org/10.1016/1010-6030(92)85110-G

[108] European Commission (2007) Reference Document on Best Available Techniques for the Manufacture of Large Volume Inorganic Chemicals—Solids and Other Industry

[109] Iler, R.K. (1979) The Chemistry of Silica: Solubility, Polymerization, Colloid and Surface Properties, and Biochemistry. Wiley, New York.

[110] Stöber, W., Fink, A. and Bohn, E. (1968) Controlled Growth of Monodisperse Silica Spheres in the Micron Size Range. *Journal of Colloid and Interface Science*, **26**, 62-69. http://dx.doi.org/10.1016/0021-9797(68)90272-5

[111] Lazaro, A., Brouwers, H.J.H., Quercia, G. and Geus, J.W. (2012) The Properties of Amorphous Nano-Silica Synthesized by the Dissolution of Olivine. *Chemical Engineering Journal*, **211**, 112-121. http://dx.doi.org/10.1016/j.cej.2012.09.042

[112] Lazaro, A., Van de Griend, M.C., Brouwers, H.J.H. and Geus, J.W. (2013) The Influence of Process Conditions and Ostwald Ripening on the Specific Surface Area of Olivine Nano-Silica. *Microporous and Mesoporous Materials*, **181**,

254-261. http://dx.doi.org/10.1016/j.micromeso.2013.08.006

[113] Lazaro, A., Quercia, G., Brouwers, H. and Geus, J. (2013) Synthesis of a Green Nano-Silica Material Using Beneficiated Waste Dunites and Its Application in Concrete. *World Journal of Nano Science and Engineering*, **3**, 41-51. http://dx.doi.org/10.4236/wjnse.2013.33006

[114] Lazaro, A., Benac-Vegas, L., Brouwers, H.J.H., Geus, J.W. and Bastida, J. (2015) The Kinetics of the Olivine Dissolution under the Extreme Conditions of Nano-Silica Production. *Applied Geochemistry*, **52**, 1-15. http://dx.doi.org/10.1016/j.apgeochem.2014.10.015

[115] Gu, W. and Tripp, C.P. (2005) Role of Water in the Atomic Layer Deposition of TiO_2 on SiO_2. *Langmuir*, **21**, 211-216. http://dx.doi.org/10.1021/la047811r

[116] Sugimoto, T., Zhou, X. and Muramatsu, A. (2002) Synthesis of Uniform Anatase TiO_2 Nanoparticles by Gel-Sol Method: 1. Solution Chemistry of $Ti(OH)_n^{(4-n)+}$ Complexes. *Journal of Colloid and Interface Science*, **252**, 339-346. http://dx.doi.org/10.1006/jcis.2002.8454

[117] Lee, G.H. and Zuo, J.M. (2004) Growth and Phase Transformation of Nanometer-Sized Titanium Oxide Powders Produced by the Precipitation Method. *Journal of the American Ceramic Society*, **87**, 473-479. http://dx.doi.org/10.1111/j.1551-2916.2004.00473.x

[118] Wang, T.H., Navarrete-López, A.M., Li, S., Dixon, D.A. and Gole, J.L. (2010) Hydrolysis of $TiCl_4$: Initial Steps in the Production of TiO_2. *The Journal of Physical Chemistry A*, **114**, 7561-7570. http://dx.doi.org/10.1021/jp102020h

[119] Yoldas, B.E. (1986) Hydrolysis of Titanium Alkoxide and Effects of Hydrolytic Polycondensation Parameters. *Journal of Materials Science*, **21**, 1087-1092. http://dx.doi.org/10.1007/BF01117399

[120] Zhang, X.T., Sato, O., Taguchi, M., Einaga, Y., Murakami, T. and Fujishima, A. (2005) Self-Cleaning Particle Coating with Antireflection Properties. *Chemistry of Materials*, **17**, 696-700. http://dx.doi.org/10.1021/cm0484201

[121] Ryu, D.H., Kim, S.C., Koo, S.M. and Kim, D.P. (2003) Deposition of Titania Nanoparticles on Spherical Silica. *Journal of Sol-Gel Science and Technology*, **26**, 489-493. http://dx.doi.org/10.1023/A:1020791130557

[122] Bischoff, B.L. and Anderson, M.A. (1995) Peptization Process in the Sol-Gel Preparation of Porous Anatase (TiO_2). *Chemistry of Materials*, **7**, 1772-1778. http://dx.doi.org/10.1021/cm00058a004

[123] Mahshid, S., Askari, M. and Ghamsari, M.S. (2007) Synthesis of TiO_2 Nanoparticles by Hydrolysis and Peptization of Titanium Isopropoxide Solution. *Journal of Materials Processing Technology*, **189**, 296-300. http://dx.doi.org/10.1016/j.jmatprotec.2007.01.040

[124] Takahashi, Y. and Matsuoka, Y. (1988) Dip-Coating of TiO_2 Films Using a Sol Derived from Ti(O-*i*-Pr)$_4$-Diethanol-lamine-H$_2$O-*i*-PrOH System. *Journal of Materials Science*, **23**, 2259-2266. http://dx.doi.org/10.1007/BF01115798

[125] Kato, K., Tsuzuki, A., Taoda, H., Torii, Y., Kato, T. and Butsugan, Y. (1994) Crystal Structures of TiO_2 Thin Coatings Prepared from the Alkoxide Solution via the Dip-Coating Technique Affecting the Photocatalytic Decomposition of Aqueous Acetic Acid. *Journal of Materials Science*, **29**, 5911-5915. http://dx.doi.org/10.1007/BF00366875

[126] Imoberdorf, G.E., Irazoqui, H.A., Cassano, A.E. and Alfano, O.M. (2005) Photocatalytic Degradation of Tetrachloroethylene in Gas Phase on TiO_2 Films: A Kinetic Study. *Industrial & Engineering Chemistry Research*, **44**, 6075-6085. http://dx.doi.org/10.1021/ie049185z

[127] Negishi, N., Takeuchi, K. and Ibusuki, T. (1998) Preparation of the TiO_2 Thin Film Photocatalyst by the Dip-Coating Process. *Journal of Sol-Gel Science and Technology*, **13**, 691-694. http://dx.doi.org/10.1023/A:1008640905357

[128] Negishi, N. and Takeuchi, K. (2001) Preparation of TiO_2 Thin Film Photocatalysts by Dip Coating Using a Highly Viscous Solvent. *Journal of Sol-Gel Science and Technology*, **22**, 23-31. http://dx.doi.org/10.1023/A:1011204001482

[129] Kim, D.J., Hahn, S.H., Oh, S.H. and Kim, E.J. (2002) Influence of Calcination Temperature on Structural and Optical Properties of TiO_2 Thin Films Prepared by Sol-Gel Dip Coating. *Materials Letters*, **57**, 355-360. http://dx.doi.org/10.1016/S0167-577X(02)00790-5

[130] Sonawane, R.S., Hegde, S.G. and Dongare, M.K. (2003) Preparation of Titanium (IV) Oxide Thin Film Photocatalyst by Sol-Gel Dip Coating. *Materials Chemistry and Physics*, **77**, 744-750. http://dx.doi.org/10.1016/S0254-0584(02)00138-4

[131] Daoud, W.A. and Xin, J.H. (2004) Low Temperature Sol-Gel Processed Photocatalytic Titania Coating. *Journal of Sol-Gel Science and Technology*, **29**, 25-29. http://dx.doi.org/10.1023/B:JSST.0000016134.19752.b4

[132] Crepaldi, E.L., Soler-Illia, G.J.D.A., Grosso, D., Cagnol, F., Ribot, F. and Sanchez, C. (2003) Controlled Formation of Highly Organized Mesoporous Titania Thin Films: From Mesostructured Hybrids to Mesoporous Nanoanatase TiO_2. *Journal of the American Chemical Society*, **125**, 9770-9786. http://dx.doi.org/10.1021/ja030070g

[133] Kajihara, K., Nakanishi, K., Tanaka, K., Hirao, K. and Soga, N. (1998) Preparation of Macroporous Titania Films by a Sol-Gel Dip-Coating Method from the System Containing Poly(ethylene glycol). *Journal of the American Ceramic Society*, **81**, 2670-2676. http://dx.doi.org/10.1111/j.1151-2916.1998.tb02675.x

[134] Park, H.K., Kim, D.K. and Kim, C.H. (1997) Effect of Solvent on Titania Particle Formation and Morphology in Thermal Hydrolysis of TiCl₄. *Journal of the American Ceramic Society*, **80**, 743-749. http://dx.doi.org/10.1111/j.1151-2916.1997.tb02891.x

[135] Jiang, X., Herricks, T. and Xia, Y. (2003) Monodispersed Spherical Colloids of Titania: Synthesis, Characterization, and Crystallization. *Advanced Materials*, **15**, 1205-1209. http://dx.doi.org/10.1002/adma.200305105

[136] Sun, J. and Gao, L. (2002) pH Effect on Titania-Phase Transformation of Precipitates from Titanium Tetrachloride Solutions. *Journal of the American Ceramic Society*, **85**, 2382-2384. http://dx.doi.org/10.1111/j.1151-2916.2002.tb00467.x

[137] Matthews, A. (1976) The Crystallization of Anatase and Rutile from Amorphous Titanium Dioxide under Hydrothermal Conditions. *American Mineralogist*, **61**, 419-424.

[138] Nam, H.D., Lee, B.H., Kim, S.J., Jung, C.H., Lee, J.H. and Park, S. (1998) Preparation of Ultrafine Crystalline TiO₂ Powders from Aqueous TiCl₄ Solution by Precipitation. *Japanese Journal of Applied Physics*, **37**, 4603-4608. http://dx.doi.org/10.1143/JJAP.37.4603

[139] Pedraza, F. and Vazquez, A. (1999) Obtention of TiO₂ Rutile at Room Temperature through Direct Oxidation of TiCl₃. *Journal of Physics and Chemistry of Solids*, **60**, 445-448. http://dx.doi.org/10.1016/S0022-3697(98)00315-1

[140] Terabe, K., Kato, K., Miyazaki, H., Yamaguchi, S., Imai, A. and Iguchi, Y. (1994) Microstructure and Crystallization Behaviour of TiO₂ Precursor Prepared by the Sol-Gel Method Using Metal Alkoxide. *Journal of Materials Science*, **29**, 1617-1622. http://dx.doi.org/10.1007/BF00368935

[141] Sun, J., Gao, L. and Zhang, Q. (2003) Synthesizing and Comparing the Photocatalytic Properties of High Surface Area Rutile and Anatase Titania Nanoparticles. *Journal of the American Ceramic Society*, **86**, 1677-1682. http://dx.doi.org/10.1111/j.1151-2916.2003.tb03539.x

[142] Zhang, Q., Gao, L. and Guo, J. (2000) Effects of Calcination on the Photocatalytic Properties of Nanosized TiO₂ Powders Prepared by TiCl₄ Hydrolysis. *Applied Catalysis B: Environmental*, **26**, 207-215. http://dx.doi.org/10.1016/S0926-3373(00)00122-3

[143] Chen, J., Gao, L., Huang, J. and Yan, D. (1996) Preparation of Nanosized Titania Powder via the Controlled Hydrolysis of Titanium Alkoxide. *Journal of Materials Science*, **31**, 3497-3500.

[144] Banfield, J. (1998) Thermodynamic Analysis of Phase Stability of Nanocrystalline Titania. *Journal of Materials Chemistry*, **8**, 2073-2076. http://dx.doi.org/10.1039/a802619j

[145] Yanagisawa, K. and Ovenstone, J. (1999) Crystallization of Anatase from Amorphous Titania Using the Hydrothermal Technique: Effects of Starting Material and Temperature. *The Journal of Physical Chemistry B*, **103**, 7781-7787. http://dx.doi.org/10.1021/jp990521c

[146] Bavykin, D.V., Dubovitskaya, V.P., Vorontsov, A.V. and Parmon, V.N. (2007) Effect of TiOSO₄ Hydrothermal Hydrolysis Conditions on TiO₂ Morphology and Gas-Phase Oxidative Activity. *Research on Chemical Intermediates*, **33**, 449-464. http://dx.doi.org/10.1163/156856707779238702

[147] Wang, C.C. and Ying, J.Y. (1999) Sol-Gel Synthesis and Hydrothermal Processing of Anatase and Rutile Titania Nanocrystals. *Chemistry of Materials*, **11**, 3113-3120. http://dx.doi.org/10.1021/cm990180f

[148] Penn, R.L. and Banfield, J.F. (1999) Morphology Development and Crystal Growth in Nanocrystalline Aggregates under Hydrothermal Conditions: Insights from Titania. *Geochimica et Cosmochimica Acta*, **63**, 1549-1557. http://dx.doi.org/10.1016/S0016-7037(99)00037-X

[149] Zhang, H. and Banfield, J.F. (2002) Kinetics of Crystallization and Crystal Growth of Nanocrystalline Anatase in Nanometer-Sized Amorphous Titania. *Chemistry of Materials*, **14**, 4145-4154. http://dx.doi.org/10.1021/cm020072k

[150] Kumar, S.R., Suresh, C., Vasudevan, A.K., Suja, N.R., Mukundan, P. and Warrier, K.G.K. (1999) Phase Transformation in Sol-Gel Titania Containing Silica. *Materials Letters*, **38**, 161-166. http://dx.doi.org/10.1016/S0167-577X(98)00152-9

[151] Zhang, J., Hu, Y., Matsuoka, M., Yamashita, H., Minagawa, M., Hidaka, H. and Anpo, M. (2001) Relationship between the Local Structures of Titanium Oxide Photocatalysts and Their Reactivities in the Decomposition of NO. *The Journal of Physical Chemistry B*, **105**, 8395-8398. http://dx.doi.org/10.1021/jp012080e

[152] Yoneyama, H., Haga, S. and Yamanaka, S. (1989) Photocatalytic Activities of Microcrystalline Titania Incorporated in Sheet Silicates of Clay. *The Journal of Physical Chemistry*, **93**, 4833-4837. http://dx.doi.org/10.1021/j100349a031

[153] Caillol, S. (2011) Fighting Global Warming: The Potential of Photocatalysis against CO₂, CH₄, N₂O, CFCs, Tropospheric O₃, BC and Other Major Contributors to Climate Change. *Journal of Photochemistry and Photobiology C: Photochemistry Reviews*, **12**, 1-19. http://dx.doi.org/10.1016/j.jphotochemrev.2011.05.002

[154] Tanabe, K., Sumiyoshi, T., Shibata, K., Kiyoura, T. and Kitagawa, J. (1974) A New Hypothesis Regarding the Surface Acidity of Binary Metal Oxides. *Bulletin of the Chemical Society of Japan*, **47**, 1064-1066. http://dx.doi.org/10.1246/bcsj.47.1064

15

Sol-Gel Synthesis Using Novel Chelating Agent and Electrochemical Characterization of Binary Doped LiMn$_2$O$_4$ Spinel as Cathode Material for Lithium Rechargeable Batteries

Ramasamy Thirunakaran[1*], Gil Hwan Lew[2], Won-Sub Yoon[2*]

[1]CSIR-Central Electrochemical Research Institute, Karaikudi, India
[2]Department of Energy Science, Sungkyunkwan University, Suwon, Republic of Korea
Email: [*]rthirunakaran@yahoo.com, [*]wsyoon@skku.edu

Abstract

LiMn$_2$O$_4$ and LiCu$_x$Cr$_y$Mn$_{2-x-y}$O$_4$ (x = 0.50; y = 0.05 - 0.50) powders have been synthesized via sol-gel method for the first time using Myristic acid as chelating agent. The synthesized samples have been taken to physical and electrochemical characterization such as thermo gravimetric analysis (TG/DTA), X-ray diffraction (XRD), Fourier transform infrared spectroscopy (FT-IR), field-emission scanning electron microscopy (FESEM), transmission electron microscopy (TEM) and electrochemical characterization *viz.*, electrochemical galvanostatic cycling studies, electrochemical impedance spectroscopy (EIS) and differential capacity curves (dQ/dE). XRD patterns of LiMn$_2$O$_4$ and LiCu$_x$Cr$_y$Mn$_{2-x-y}$O$_4$ confirm high degree of crystallinity with good phase purity. FESEM image of undoped pristine spinel lucidly depicts cauliflower morphology with good agglomerated particle size of 50 nm while 0.5-Cu doped samples depict the pebbles morphology. TEM images of the spinel LiMn$_2$O$_4$ and LiCu$_{0.5}$Cr$_{0.05}$Mn$_{1.45}$O$_4$ authenticate that all the synthesized particles via sol-gel method are nano-sized (100 nm) with spherical surface and cloudy particles morphology. The LiMn$_2$O$_4$ samples calcined at 850°C deliver the high discharge capacity of 130 mA·h/g with cathodic efficiency of 88% corresponds to 94% columbic efficiency in the first cycle. Among all four compositions studied, LiCu$_{0.5}$Cr$_{0.05}$Mn$_{1.45}$O$_4$ delivers 124 mA·h/g during the first cycle and shows stable performance with a low capacity fade of 1.1 mA·h/g cycle over the investigated 10 cycles.

Keywords

Multi-Doping, Sol-Gel Method, Myristic Acid, Differential Capacity, Spinel Cathode

---

[*]Corresponding authors.

1. Introduction

Spinel $LiMn_2O_4$ has been zeroed in attractive and promising cathode materials for lithium-ion batteries owing to its high voltages, proper Mn^{3+}/Mn^{4+} redox potential, high energy densities and high power densities. Lithium-ion batteries are much dependent on cathode material insertion-deinsertion processes. $LiMn_2O_4$, $LiCoO_2$ are used as commercial positive cathode materials for various applications and high-voltage spinel cathode materials are also used for various energy-storage devices. Among all the positive materials used for lithium-ion battery applications, $LiMn_2O_4$ is an apt cathode material for rechargeable lithium-ion batteries, owing to its low cost, easy availability, environmentally benign nature, non-toxicity, and ease of synthesis while comparing with other layered oxides such as $LiCoO_2$ and $LiNiO_2$ [1]-[3]. However, specific capacity of undoped $LiMn_2O_4$ decreases gradually upon repeated cycling carried out at elevated temperature [4] [5]. It is well known that the capacity fading is caused due to several factors such as Jahn-Teller distortion, two-phase unstable reaction [2], slow dissolution of manganese into the electrolyte [6], lattice instability [7], and particle size distribution [8] [9]. In order to overcome the problem of Jahn-Teller distortion for obtaining the high capacity retention, several researchers have investigated earlier lithium rich spinels with various divalent, trivalent and tetravalent-doped ions such as Cr, Fe, Zn, Cu, Ga, Co, Al, Ni and Ti [10]. Ohzuku et al. [8] and Lee et al. [11] showed that partial doping of cations is effective in suppressing the capacity fade upon cycling. Moreover, the capacity fade of $LiMn_2O_4$ often happens much in 3 V region which can be completely suppressed by doping selenium with $LiMn_2O_4$ [12]. Low temperature synthesis methods viz., sol-gel [13] [14], chemical precipitation [15], hydrothermal and pechini process [16] have been used to obtain cathode materials with expected physical and electrochemical properties to use in lithium-ion batteries. Spinel $LiMn_2O_4$ and Zn, Co, Ni and substituted $LiMn_2O_4$ synthesized via facile sol-gel method to improve the electrochemical and structural properties of $LiMn_2O_4$ spinel based on electrode materials for Li-ion batteries [17]. The present work highlights that efforts have been taken to synthesize physical and electrochemical characterization of $LiMn_2O_4$ and $LiCu_xCr_yMn_{2-x-y}O_4$ (x = 0.50; y = 0.05 - 0.50) via sol-gel method using Myristic acid as new chelating agent for the exploration of this facile synthesis method with good electrochemical performance.

2. Experimental

$LiMn_2O_4$ and $LiCu_xCr_yMn_{2-x-y}O_4$ (x = 0.50; y = 0.05 - 0.50) powders have been synthesized via sol-gel method using Myristic acid as chelating agent. **Figure 1** depicts the flow chart of the synthesis procedure. Stoichiometric amounts of acetates of lithium, manganese, and the dopant salts such as acetate of copper and chromium were dissolved independently in triple-distilled water. The chelating agent (Myristic acid) is dissolved slowly in heating ethyl alcohol gently. After dissolving all the salts gently, it will be mixed with chelating agent. A small amount of precursor has been taken for TG/DTA analysis to under the thermal behaviour and the rest of the sample is calcined at 850°C for 6 h. For thermal analysis, the precursors are heated in air atmosphere at 10°C/min to 850°C. All the calcined samples are subjected for physical characterization using thermo gravimetric/differential thermal analysis (TG/DTA-Seiko Exstart 6000, Japan), X-ray diffraction (XRD-Bruker D2 Phaser desktop, Cu source, AXS, Karisruhe, Germany), Fourier-transform infrared spectroscopy (FTIR-Tensor 27, Bruker (Germany)), field emission scanning electron microscopy (FESEM-JEOL, Model-JSM-7000F (Japan)), transmission electron microscopy (TEM-JEOL, JEM-ARM 200F, Japan), energy dispersive X-ray analysis (EDAX-JEOL, Model-JSM-7000F, Japan) and galvanostatic charge-discharge cycling studies (WonATech-Model, WBCS 3000, South Korea), electrochemical impedance spectroscopy (EIS-Potentiostat/Galva-nostat, Model-273A, USA) and differential capacity curves (dQ/dE).

Electrochemical Studies

Coin cells of 2032 configuration have been assembled in an argon filled glove box (MBraun, Germany) using lithium foil as anode, Celgard 2400 as separator, 1 M solution of $LiPF_6$ and 0.3 M $LiBF_4$ in 3:7 (v/v) volume mixture of ethylene carbonate (EC) and diethyl carbonate (DEC) as electrolyte and the synthesized material has been as cathode. The cathode is prepared by slurry coating procedure from a mix comprising synthesized compound, Super P carbon black as conducting material and poly (vinylidene fluoride) 5% PVdF binder in n-methyl-2-pyrrolidone (NMP) solution mixed in the percentage ratio of 80:10:10 so as to form slurry. The slurry has been coated over aluminum foil and vacuum dried at 110°C for 2 h. Electrode blanks of 18 mm diameter

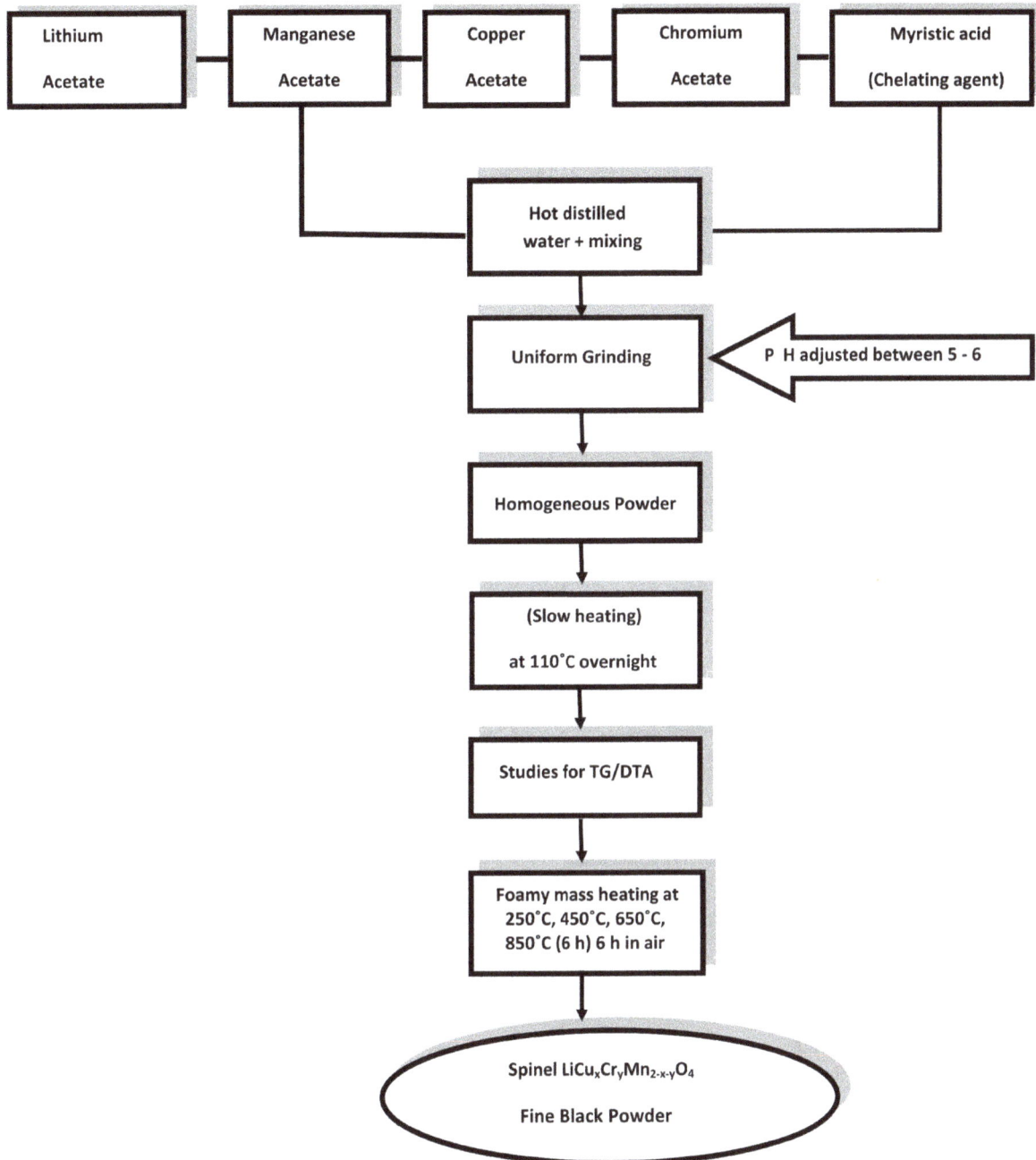

Figure 1. Flow chart for synthesis of $LiCu_xCr_yMn_{2-x-y}O_4$ by sol-gel method using Myristic acid as chelating agent.

are punched out and used as cathode in the coin cell. The fabricated 2032 coin cells have been cycled at a constant current of C/10 rate between 3.0 to 4.5 V using an in-house battery cycling unit.

3. Results and Discussion

3.1. Thermal Studies

The TG/DTA curve of $LiMn_2O_4$ precursor is depicted in **Figure 2(a)**. The TGA curve clearly illustrates the three weight loss regions. *Ab initio*, the low weight loss of 5% is seen up to 100°C owing to removal of water. Further, another two regions are observed between 100 V and 350°C extending with maximum weight loss of 45% may be assigned to the decomposition of chelating agent (Myristic acid) and acetate precursors. It is well known that

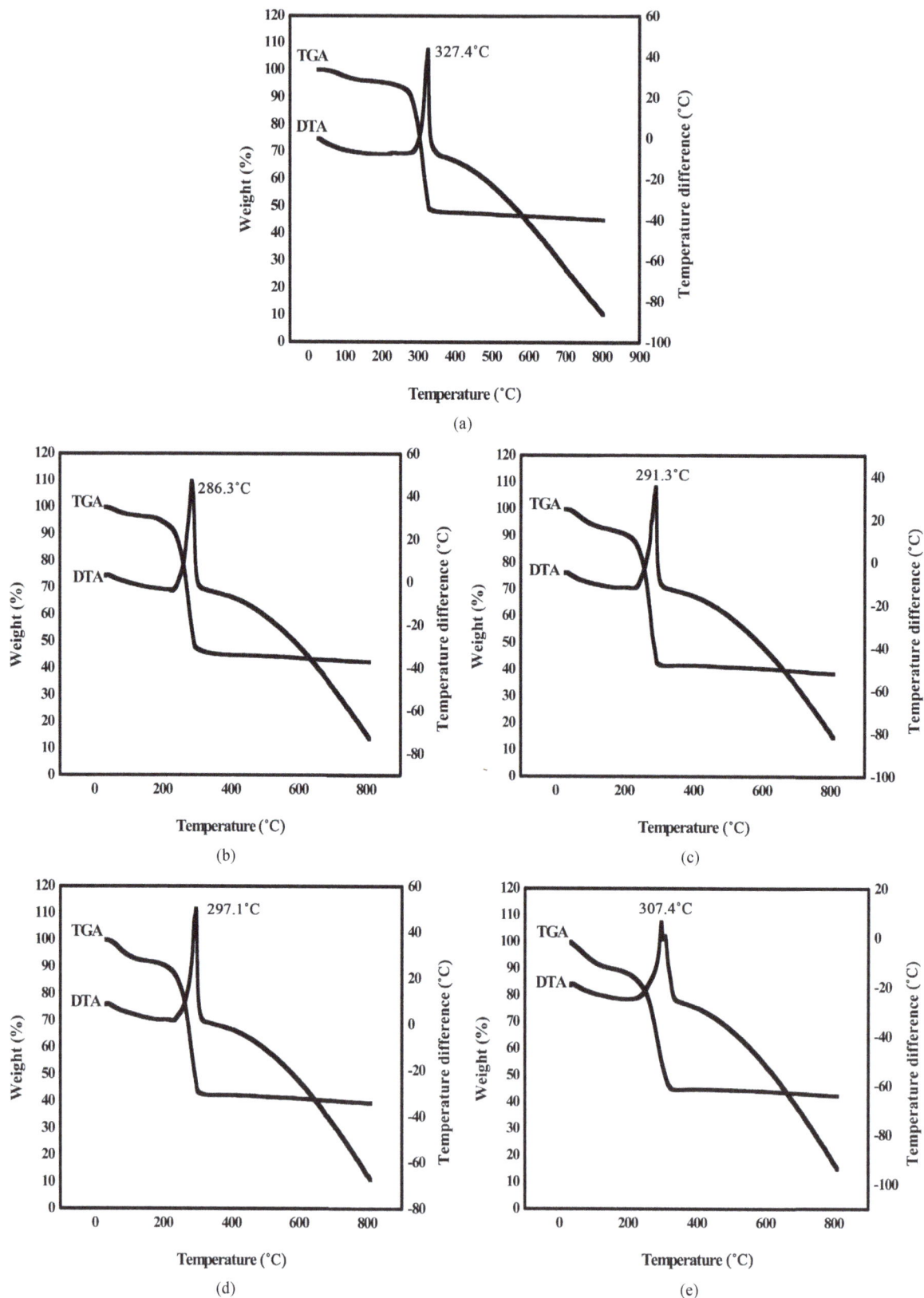

Figure 2. Thermo gravimetric and differential thermal analysis (TG/DTA) of $LiMn_2O_4$ and $LiCu_xCr_yMn_{2-x-y}O_4$ spinel precursor. (a) Undoped; (b) Cu: 0.50; (c) Cu: 0.50; Cr: 0.05; (d) Cu: 0.50; Cr: 0.10 and (e) Cu: 0.50; Cr: 0.50.

in the case of DTA curve, it depicts lucidly a well defined exothermic peak at 327.4°C indicating the formation of the spinel compound which is mirrored in the XRD pattern corresponding to high degree of crystallinity for the sample calcined at 850°C. Furthermore, all the peak reflections of $LiMn_2O_4$ have been shown unclouded when the precursor is calcined at 250°C and 450°C. Also, the TGA curve shows that the temperature above 350°C stops totally thermal events without any further thermal reaction after the formation of spinel compounds.

TG/DTA curves of spinel $LiCu_xCr_yMn_{2-x-y}O_4$ (x = 0.50; y = 0.05 - 0.50) precursors have been synthesized via sol-gel method using Myristic acid as chelating agent are presented in **Figures 2(b)-(e)**. TGA curve of copper doped spinel precursor shows three characteristic weight loss zones (**Figure 2(b)**). Initially, the low weight loss of 5% up to 100°C is an account of removal of moisture. The second and third weight loss zone of 40% up to 300°C is attributed to the decomposition of acetate precursor of copper. DTA curve shows a well defined an exothermic peak at 286.3°C suggesting the formation of spinel product. DTA curve corroborates no further thermal reactions are taking place beyond 300°C.

Moreover, in the case of dual doped spinel ($LiCu_xCr_yMn_{2-x-y}O_4$) precursor, $LiCu_{0.5}Cr_{0.05}Mn_{1.45}O_4$ shows the same three weight loss zones with weight loss of 5% compared to other all dual dopant concentration of the spinel precursors. This low amount in weight loss has been reflected in charge-discharge study to deliver the high discharge capacity of the spinel. Among all dual doped spinel precursors, equal doping ratio or Cu-Cr high doping of $LiCu_{0.5}Cr_{0.5}MnO_4$, shows a well defined exothermic peak (see **Figure 2(e)**) at 307.4°C with higher formation temperature owing to higher specific heat of Cr than Cu and Mn (Cr = 0.46 kJ/kg K; Cu = 0.39 kJ/kg K; Mn = 0.48 kJ/kg K).

Hence, in all cases, the weight loss zones complete up to 100°C may be attributed to the removal of moisture and resulting in the formation of spinel at around 307.4°C - 327.4°C. This temperature difference (20°C) is depicted in the formation temperature of all doped precursors which leads the reaction to begin much earlier and involving higher heat energy (307.4°C) as indicated in the DTA curve (**Figure 2(e)**). All precursors show lucidly thermally inactive regions beyond 300°C indicating the closure of thermal events.

3.2. X-Ray Diffraction Studies

Figures 3(a)-(e) depicts the XRD patterns of $LiMn_2O_4$ samples calcined at different temperatures: (a) as synthesized; (b) 250°C; (c) 450°C; (d) 650°C; (e) 850°C. It is evident that as synthesized and the samples calcined at 250°C exhibits broad and indistinct reflections indicating the nebulous nature of the compounds. Moreover, the sample calcined at 450°C and 650°C exhibits an additional impurity peaks may be attributed to the formation of α-Mn_2O_3 and $LiMn_2O_3$. Similarly, the high intensity peaks such as (111), (311), (222), (400), (331), (551), (440) and (351) have been obtained for the samples calcined at 850°C for 6 h confirms the high degree of crystallinity with better phase purity. These planes are in good agreement with the results obtained for the spinel compound synthesized via sol-gel method or solid-state method [17]-[26]. It is well known that the spinel compound has an Fd3m space group wherein lithium occupies at 8a tetrahedral sites, manganese and dopant ions occupy the

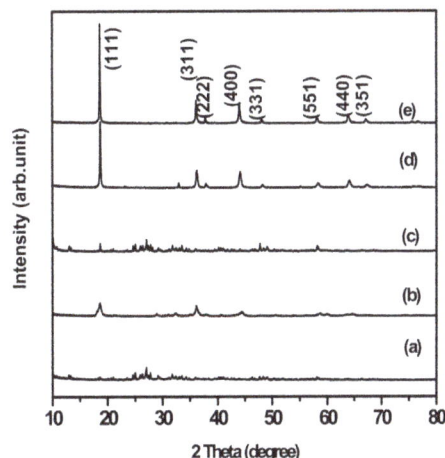

Figure 3. XRD patterns of sol-gel-synthesized undoped $LiMn_2O_4$ samples calcined at different temperatures *viz.*, (a) As synthesized; (b) 250°C; (c) 450°C; (d) 650°C; and (e) 850°C.

16d sites and oxygen at 32e sites. All the XRD peak reflections for synthesized spinel match perfectly with Joint committee on Powder Diffraction Standard (JCPDS card No. 35-782).

The XRD patterns of $LiCu_xCr_yMn_{2-x-y}O_4$ are shown in **Figures 4(a)-(d)** with different stoichiometric amounts of divalent and trivalent metal cations *viz.*, Cu and Cr (Cu-0.5; Cr-0.05-0.5), synthesized via sol-gel method calcined at 850°C. $LiMn_2O_4$ and $LiCu_xCr_yMn_{2-x-y}O_4$ depict the formation of phase pure crystalline spinels in substantiating the first order reaction. Moreover, it is clear that all the peaks corresponding to (111), (311), (222), (400), (331), (551), (440) and (351) which are in good accordance with other researchers [24]. All the XRD peak reflections for this pristine spinel perfectly match with Joint committee on Powder Diffraction Standard (JCPDS card No. 35-782).

3.3. FTIR Spectroscopy

The FT-IR spectra of sol-gel synthesized $LiMn_2O_4$ powders calcined at different temperatures *viz.*, 250°C, 450°C, 650°C and 850°C are shown in **Figures 5(a)-(d)**. It is clear that the broad peaks appear for the sample calcined at low temperature (250°C, 450°C and 650°C) for the sake of the water molecules (O-H bond). FT-IR studies on $LiAl_xMn_{2-x}O_4$ have been investigated through solid-state combustion synthesis method in which all the frequencies showing striking similarity with our present investigation [27]. It is ratified that in the case of black coloured spinel compound obtained via sol-gel method exhibiting the two IR spectral bands at wavelengths between 513 - 516 cm^{-1} and 607 - 610 cm^{-1} for the sample calcined at 850°C could be assignable to the Li-O bending vibration mode and Li-Mn-O stretching vibration band. The sample calcined at low temperature (250°C) depicting the low wave number while comparing to undoped spinel calcined at 850°C. The wave number corresponding to the $LiMn_2O_4$ and $LiCu_xCr_yMn_{2-x-y}O_4$ spinels are given in **Table 1** and **Table 2**. Also, the sample calcined at 850°C depicts the IR band at around 610 cm^{-1} which is slightly shifted towards the lower wave number.

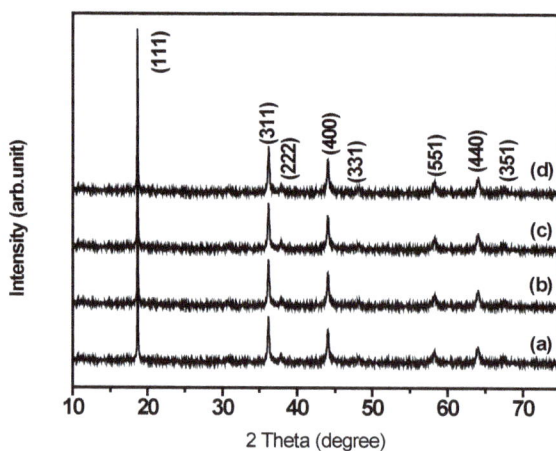

Figure 4. XRD patterns of $LiCu_xCr_yMn_{2-x-y}O_4$ powders calcined at 850°C. (a) Cu: 0.50; (b) Cu: 0.50; Cr: 0.05; (c) Cu: 0.50; Cr: 0.10 and (d) Cu: 0.50; Cr: 0.50.

Table 1. FTIR frequencies for the peaks observed for $LiMn_2O_4$.

No	Temperature	Wave number (cm⁻¹)	Assignments
1	250	513	Li-O
		607	Li-Mn-O
2	450	505	Li-O
		613	Li-Mn-O
3	650	516	Li-O
		619	Li-Mn-O
4.	850	516	Li-O
		610	Li-Mn-O

(a)

(b)

(c)

(d)

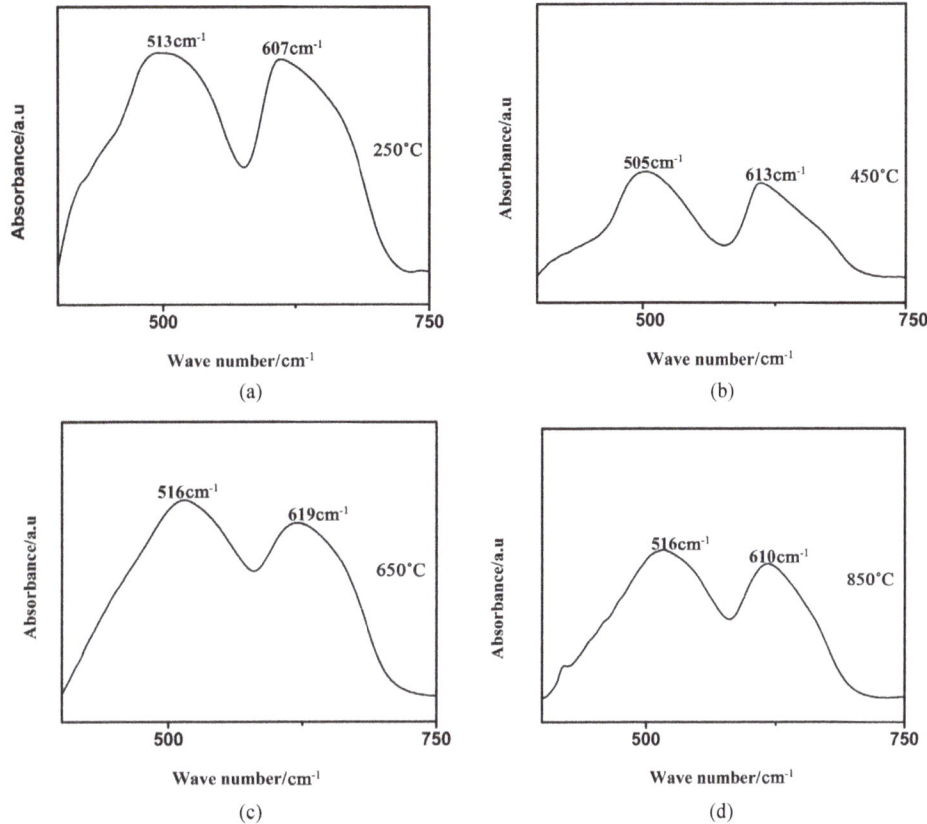

Figure 5. FT-IR spectra of spinel $LiMn_2O_4$ (a)-(d) particles calcined at different temperatures *viz.*, 250°C, 450°C, 650°C and 850°C.

Table 2. FTIR frequencies for the peaks observed for $LiCu_xCr_yMn_{2-x-y}O_4$.

No	Sample	Wave number (cm^{-1})	Assignments
1	$LiCu_{0.5}Mn_{1.5}O_4$	523	Li-O
		625	Li-Cu-Mn-O
2	$LiCu_{0.5}Cr_{0.05}Mn_{1.45}O_4$	505	Li-O
		621	Li-Cu-Cr-Mn-O
3	$LiCu_{0.5}Cr_{0.10}Mn_{1.40}O_4$	503	Li-O
		618	Li-Cu-Cr-Mn-O
4	$LiCu_{0.5}Cr_{0.5}Mn_{1.0}O_4$	525	Li-O
		625	Li-Cu-Cr-Mn-O

Figures 6(a)-(d) depict the FT-IR spectra of $LiCu_xCr_yMn_{2-x-y}O_4$ powders with varying amounts of Cu and Cr. It is clearly shown that the di and trivalent doped spinels exhibiting the two IR spectral bands between 523 - 525 cm^{-1} and 618 - 625 cm^{-1} may be attributed to the Li–O bending vibration at lower wave number and Li-Cu-Cr-Mn-O stretching vibration at higher wave number respectively. The FT-IR on $LiCr_xNi_yMn_{2-x-y}O_4$ reveals the synthesis process via sol-gel method [28]. These results are is in good agreement with our FTIR data. Similarly, $LiCr_xCu_yMn_{x-y}O_4$ has been synthesized via wet chemistry method and reveals the spectra of stretching and bending vibration [29]. Such reports show the striking similarity in our present investigation. Nevertheless, it is corroborated that there are no any significant impurity peak reflections in all samples were calcined at 850°C. High equal amount of dual doping spinel ($LiCu_{0.5}Cr_{0.5}MnO_4$) exhibits the spectral bands 525 cm^{-1} and 625 cm^{-1} which is similar to wave number of 0.5-Cu doped spinel.

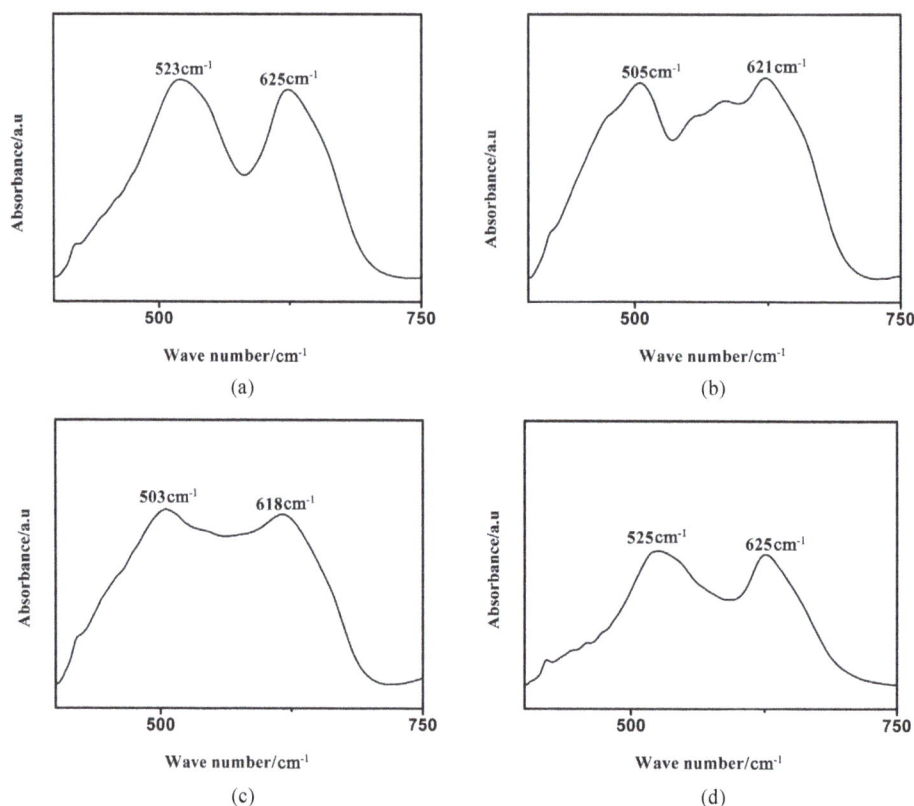

Figure 6. FT-IR spectra of LiCu$_x$Cr$_y$Mn$_{2-x-y}$O$_4$ particles with varying Cu-Cr doping calcined at 850˚C. (a) x = 0.50; (b) x = 0.50; y = 0.05; (c) x = 0.50; y = 0.10; (d) x = 0.50; y = 0.50.

3.4. FESEM Analysis

Figures 7(a)-(e) depict the Field Emission Scanning Electron Microscope images of undoped and LiCu$_x$Cr$_y$Mn$_{2-x-y}$O$_4$ spinel powders calcined at 850˚C. The undoped pristine spinel depicts cauliflower morphology with an average particle size of 50 nm (**Figure 7(a)**) with good agglomerated particles. Similarly in the case of Cu doped samples, it is lucidly seen pebbles morphology seems to be 2 μm. It is evident that 0.05-Cr (**Figure 7(c)**) doped spinel depicts spherical surface morphology with less particle size of 0.50 μm. The equal doping ratio of LiCu$_{0.50}$Cr$_{0.50}$Mn$_{1.0}$O$_4$ shows clearly that all the large particles are seen to be spherical grains haphazardly with an increased particle size of 2 μm as it possesses high amount of Cu-Cr doping which is reflected in electrochemical impedance spectroscopy (EIS) and cycling behavior of the doped spinel.

3.5. TEM Analysis

Figures 8(a)-(f) depict the TEM images of LiMn$_2$O$_4$ and LiCu$_x$Cr$_y$Mn$_{2-x-y}$O$_4$ particles calcined at 850˚C. **Figure 8(a)** illustrates the selected area of diffraction pattern (LiMn$_2$O$_4$) suggesting the diffuse hollow with multiple fringes. The undoped spinel shows (**Figure 8(b)**) uniform spherical morphology with particle size of (100 nm). In the case of LiCu$_{0.5}$Mn$_{1.5}$O$_4$, (**Figure 8(c)**) all the particles are seen like large spherical surface morphology. Moreover, LiCu$_{0.5}$Cr$_{0.05}$Mn$_{1.45}$O$_4$ (**Figure 8(d)**), depicts cloudy particles with good agglomerated particle size of 100 nm. This good agglomeration has been lucidly resulted to increase the high the electrochemical activity of the spinel among all dopants. Therefore, the high equal amount of dual doping spinel (LiCu$_{0.5}$Cr$_{0.5}$MnO$_4$) (**Figure 8(f)**) depicts the typical morphology with large particles since chromium content is higher than that of other dopant which is unclouded reflected in charge-discharge studies leading to rapid capacity fading upon the cycling.

3.6. EDAX Analysis

Figures 9(a)-(e) depict the EDAX peaks of Cu, Cr, Mn and O in LiMn$_2$O$_4$ and LiCu$_x$Cr$_y$Mn$_{2-x-y}$O$_4$ compounds.

(a)

(b)

(c)

(d)

(e)

Figure 7. FESEM images of spinel $LiMn_2O_4$ and $LiCu_xCr_yMn_{2-x-y}O_4$ particles calcined at 850°C. (a) Undoped; (b) Cu: 0.50; (c) Cu: 0.50; Cr: 0.05, (d) Cu: 0.50; Cr: 0.10 and (e) Cu: 0.50; Cr: 0.50.

All the EDAX peaks corroborate lucidly the actual compositions of the undoped and doped spinels to be pure without any additional impurities. **Table 3** depicts the EDAX compositions of various elements in $LiMn_2O_4$ and $LiCu_xCr_yMn_{2-x-y}O_4$.

3.7. Galvanostatic Charge-Discharge Studies

Figures 10(a)-(e) show the first cycle charge-discharge behavior of $LiMn_2O_4$ and $LiCu_xCr_yMn_{2-x-y}O_4$ with different stoichiometric concentration of Cu and Cr. The charge-discharge behavior of undoped and $LiCu_xCr_yMn_{2-x-y}O_4$ shows pellucid the de-intercalation/intercalation of lithium ions at the 8a tetrahedral sites from cubic spinel structure. At the beginning, $LiMn_2O_4$ has been synthesized via sol-gel method and calcined at 850°C delivers the high discharge capacity of 130 mA·h/g against the charging capacity of 138 mA·h/g with

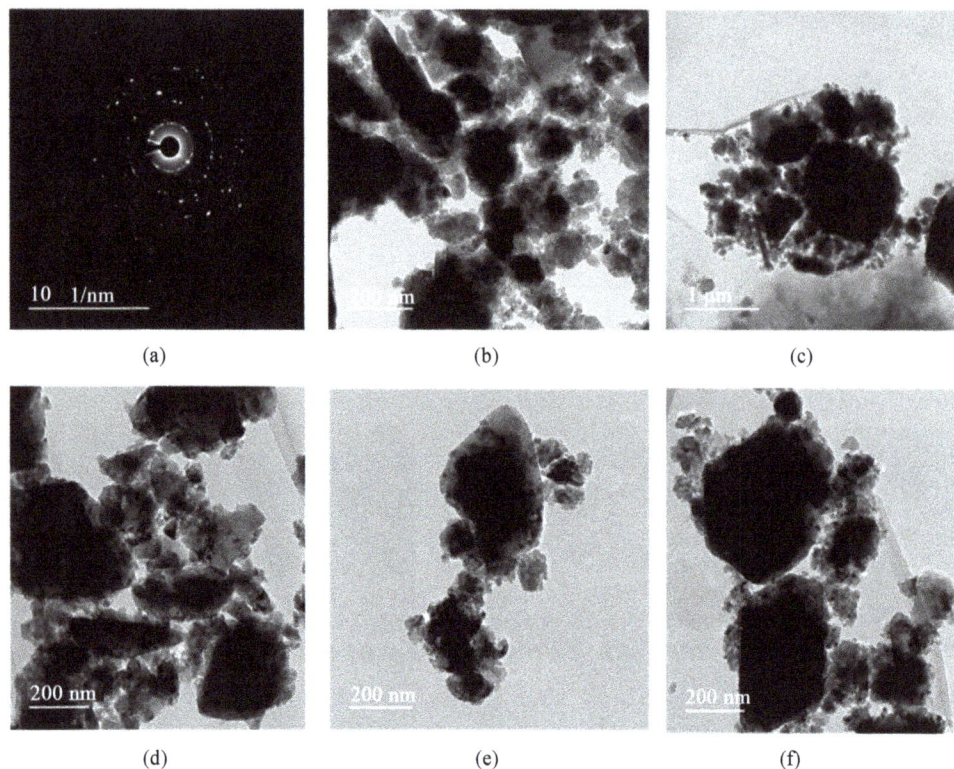

(a)   (b)   (c)

(d)   (e)   (f)

Figure 8. TEM images of LiMn$_2$O$_4$ and LiCu$_x$Cr$_y$Mn$_{2-x-y}$O$_4$ particles calcined at 850°C. (a) (b) Undoped; (c) Cu: 0.50; (d) Cu: 0.50; Cr: 0.05; (e) Cu: 0.50; Cr: 0.10 and (f) Cu: 0.50; Cr: 0.50.

Table 3. EDAX compositions of various elements in LiMn$_2$O$_4$ and LiCu$_x$Cr$_y$Mn$_{2-x-y}$O$_4$.

Element	Name of the compound	Weight %
	LiMn$_2$O$_4$	
O		34.02
Mn		65.98
	LiCr$_{0.5}$Mn$_{1.5}$O$_4$	
O		28.25
Mn		53.50
Cr		18.25
	LiCr$_{0.5}$Cu$_{0.05}$Mn$_{1.45}$O$_4$	
O		26.45
Mn		49.75
Cr		17.12
Cu		6.21
	LiCr$_{0.5}$Cu$_{0.10}$Mn$_{1.40}$O$_4$	
O		27.19
Mn		47.15
Cr		16.41
Cu		9.25
	LiCr$_{0.5}$Cu$_{0.50}$Mn$_{1.0}$O$_4$	
O		25.75
Mn		46.95
Cr		16.55
Cu		10.75

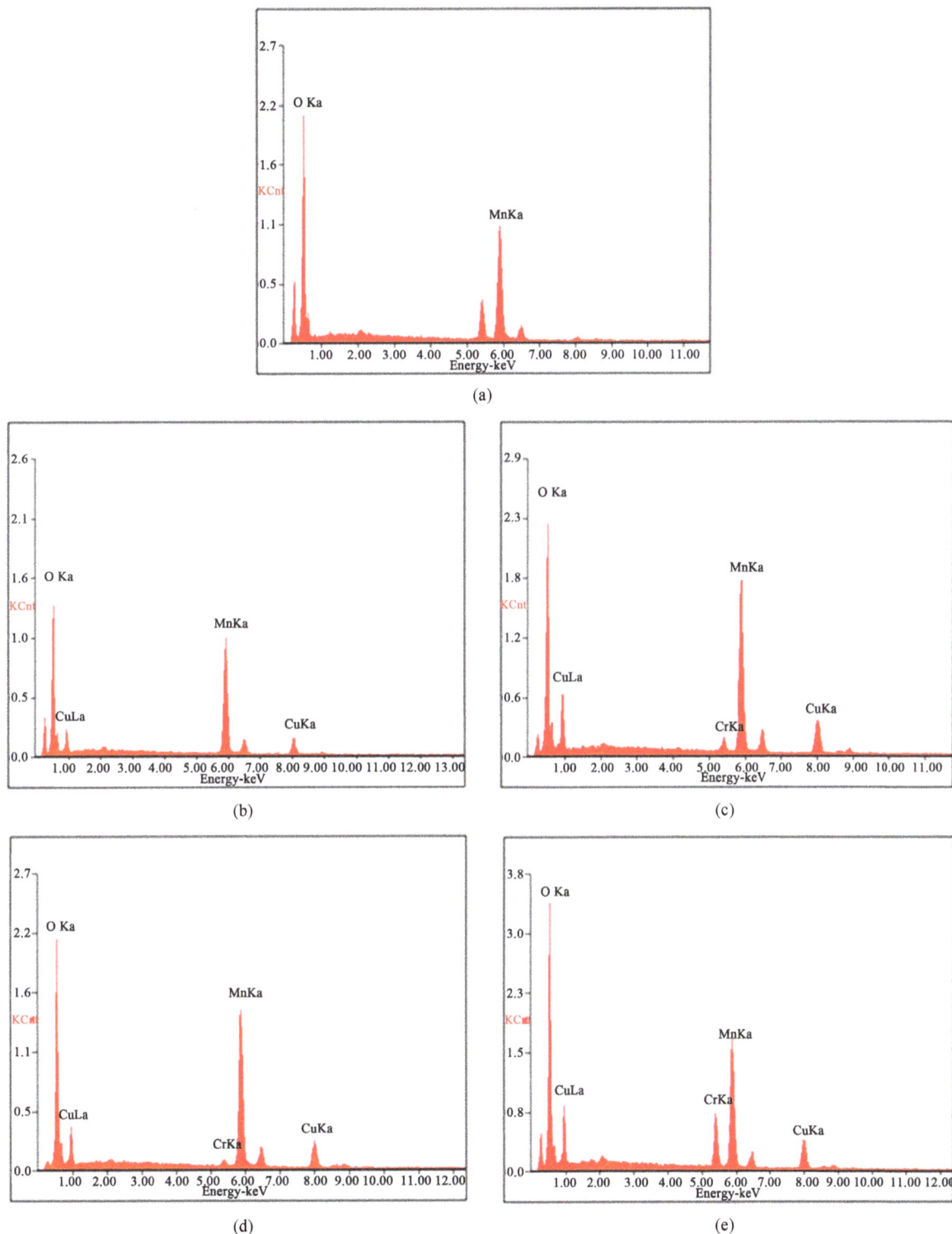

Figure 9. EDAX images of $LiMn_2O_4$ and $LiCu_xCr_yMn_{2-x-y}O_4$ powders calcined at 850°C. (a) Undoped; (b) x = 0.50; (c) x = 0.50; y = 0.05; (d) x = 0.50; y = 0.10; (e) x = 0.50; y = 0.50.

cathodic efficiency of 88% corresponds to 94% columbic efficiency during the first cycle. It is evident to note that the undoped spinel exhibits superior performance to other doped spinels ($LiCu_xCr_yMn_{2-x-y}O_4$). Moreover, undoped spinel has been derived via sol-gel method [30] and delivers the maximum discharge capacity of 125

(a)

(b)

(c)

(d)

(e)

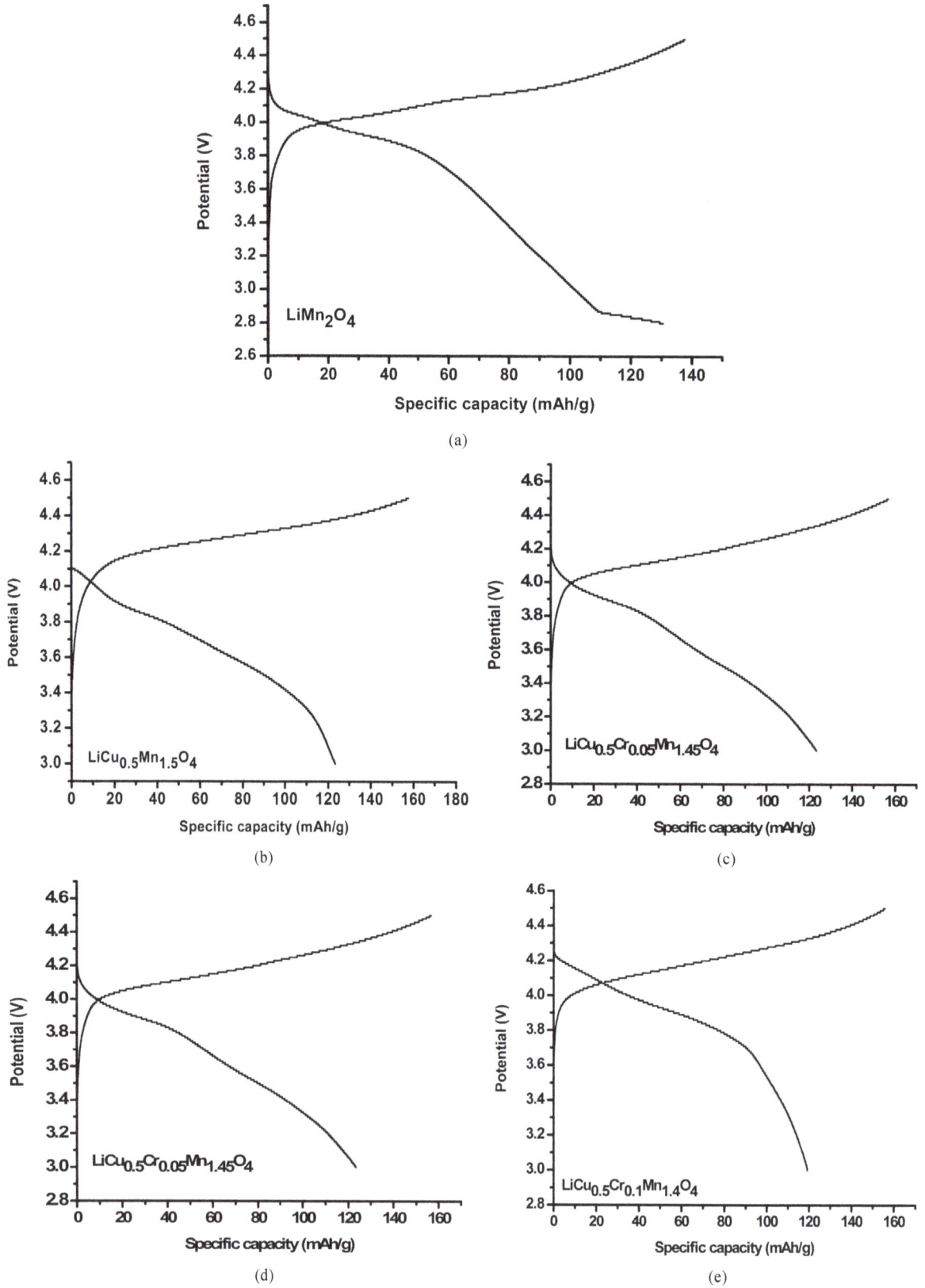

Figure 10. First cycle charge-discharge behaviour of $LiMn_2O_4$ and $LiCu_xCr_yMn_{2-x-y}O_4$. (a) Undoped; (b) $x = 0.50$; (c) $x = 0.50$; $y = 0.05$; (d) $x = 0.50$; $y = 0.10$; (e) $x = 0.50$; $y = 0.50$.

mA·h/g during the first cycle in which these results showing inferior performance to our present investigation. Similarly, another pristine LiMn$_2$O$_4$ has been synthesized through sol-gel method offers an initial reversible capacity of 128 mA·h/g [31] wherein it seems to be lower than our result. Also, copper doped spinel has been attempted to synthesize via sol-gel method [32] and delivers the maximum discharge of capacity of 128 mA·h/g during the first cycle. These experiment results are slightly lower than to our present report. Nevertheless, during sol-gel synthesis of spinel compounds, Myristic acid has been used as chelating agent to be favorable and to act as catalyst to fasten the reaction causing the multi-ligand chain between Mn-O and COO$^-$ resulting in the formation of pristine LiMn$_2$O$_4$ and LiCu$_x$Cr$_y$Mn$_{2-x-y}$O$_4$ for augmenting the high electrochemical stability of the spinel compound.

Figures 10(b)-(e) show the first cycle charge-discharge behavior of LiCu$_x$Cr$_y$Mn$_{2-x-y}$O$_4$ heated at 850˚C. Copper and copper-chromium doped spinels exhibit the discharge capacity of 122, 124, 120 and 120 mA·h/g corresponds to 77%, 78%, 77% and 76% columbic efficiency during the first cycle (**Figures 10(b)-(e)**). It is quite interesting to note that 0.5-Cu doped spinel has been experimented to synthesize through sol-gel route [32] and yields the maximum discharge capacity of 70 mA·h/ g in the first cycle. It shows the poor performance rather than our results. Similarly, 0.05-Cr^{3+} modified LiMn$_2$O$_4$ spinel intercalation cathodes through oxalic acid assisted sol-gel method for lithium rechargeable batteries has been reported [24] and exhibits an initial discharge capacity of 80 mA·h/g wherein these results are lower than that of our present performance. It is well known that the sample with low Cr-doping (LiCu$_{0.5}$Cr$_{0.05}$Mn$_{1.45}$O$_4$) exhibits the higher discharge capacity (124 mA·h/g) than the other all slightly high doping compositions. The increase in capacity may be due to the same oxidation state of chromium ions have smaller ionic radii than manganese ions, Cr^{3+} (0.615 Å), Mn^{3+} (0.68 Å), Cr^{4+} (0.58 Å), Mn^{4+} (0.60 Å). Also the stronger Cr-O bonds in the delithiated state (compare the binding energy of 1142 kJ·mol^{-1} for CrO$_2$ with 946 kJ mol^{-1} for α-MnO$_2$) may also be considered to contribute to stabilization energy of Cr^{3+} in the octahedral sites. Therefore, the equal doping ratio of spinel (LiCu$_{0.5}$Cr$_{0.5}$Mn$_{1.0}$O$_4$) delivers slightly lower performance (discharge capacity of 120 mA·h/g) than undoped and other dopant compositions may be owing to high order of cationic mixing of dopants, poor electronic conductivity, and higher electrochemical impedance has been developed inside the cell which is reflected in electrochemical impedance spectra (ECI).

The cycling behavior of undoped and doped spinels calcined at 850˚C over investigated 10 cycles with the corresponding columbic efficiencies (CE) is shown in **Figures 11(a)-(e)**. The undoped spinel exhibits the maximum discharge capacity of 130 mA·h/g in the first cycle. However, it besets from capacity fading drastically upon cycling up to 10 cycles and also it experiences a capacity fade of 1.4 mA·h/g cycle over the investigated 10 cycles with capacity retention of 89%. In the 5th and 10th cycle, the undoped spinel delivers a discharge capacity of 125, 116 mA·h/g or both same columbic efficiency of 99%. Similarly, the cells with copper and copper-chromim doped spinels deliver the discharge capacities of 123, 124, 120 and 120 mA·h/g during the first cycle and experiencing a capacity fade of 1.1, 1.1, 1.2 and 1.4 mA·h/g cycle with capacity retention of 92, 90, 91 and 88% for the Cu and Cu:Cr contents corresponding to 0.5, 0.5; 0.05, 0.5; 0.10 and 0.50; 0.50, respectively over the investigated 10 cycles. In the 5th and 10th cycle of doped spinels, the discharge capacities have been exhibited 118, 112; 115, 111; 112, 109; 114, 106 mA·h/g; with capacity fades of 0.1, 0.2; 0.1, 0.1; 0.1, 0.1; 0.1, 0.2 mA·h/g cycle corresponding to columbic efficiency of 99v, 98%; 99%, 99%; 99%, 99%; 99%, 98% for the Cu and Cu:Cr contents with regard to 0.5, 0.5:0.05, 0.5:0.10 and 0.50:0.50, respectively. Inter alia all dual doped compositions, the low Cr-doped spinels (LiCu$_{0.5}$Cr$_{0.05}$Mn$_{1.45}$O$_4$) offers better stable capacity retention up to 10 cycles. Several researchers have already reported [33] that 0.2-Cr doped LiMn$_2$O$_4$ has been synthesized by sol-gel method as cathodes in high-voltage lithium batteries and delivered the maximum discharge capacity of 78 mA·h/gin the 10th cycle. These results are not encouraging and inferior to our present investigation (109 mA·h/g). Also, 0.01-Cr doped spinel has been synthesized towards intercalation cathodes through oxalic acid assisted sol-gel method for lithium rechargeable batteries and delivers the inferior results of discharge capacity of 115 mA·h/g in the 10th cycle [24] when compare to our present investigation. Moreover, 0.01-Cr doped pristine spinel has also been prepared using Adipic acid as chelating agent via sol–gel route and exhibits the maximum discharge capacity of 117 mA·h/g in the 10th cycle [23] wherein these results are found to be somewhat lower in our earlier experiment rather than this work. In other words, 0.5-Cu doped spinel has been reported via sol-gel method [32] and delivers the maximum discharge capacity of 60 mA·h/g in the 10th cycle which is two times lower than our results. Moreover, 0.1-Al doped spinel has been experimented via traditional sol-gel method [34] exhibits 100 mA·h/g in the 10th cycle sans stable discharge capacity wherein these results are lower

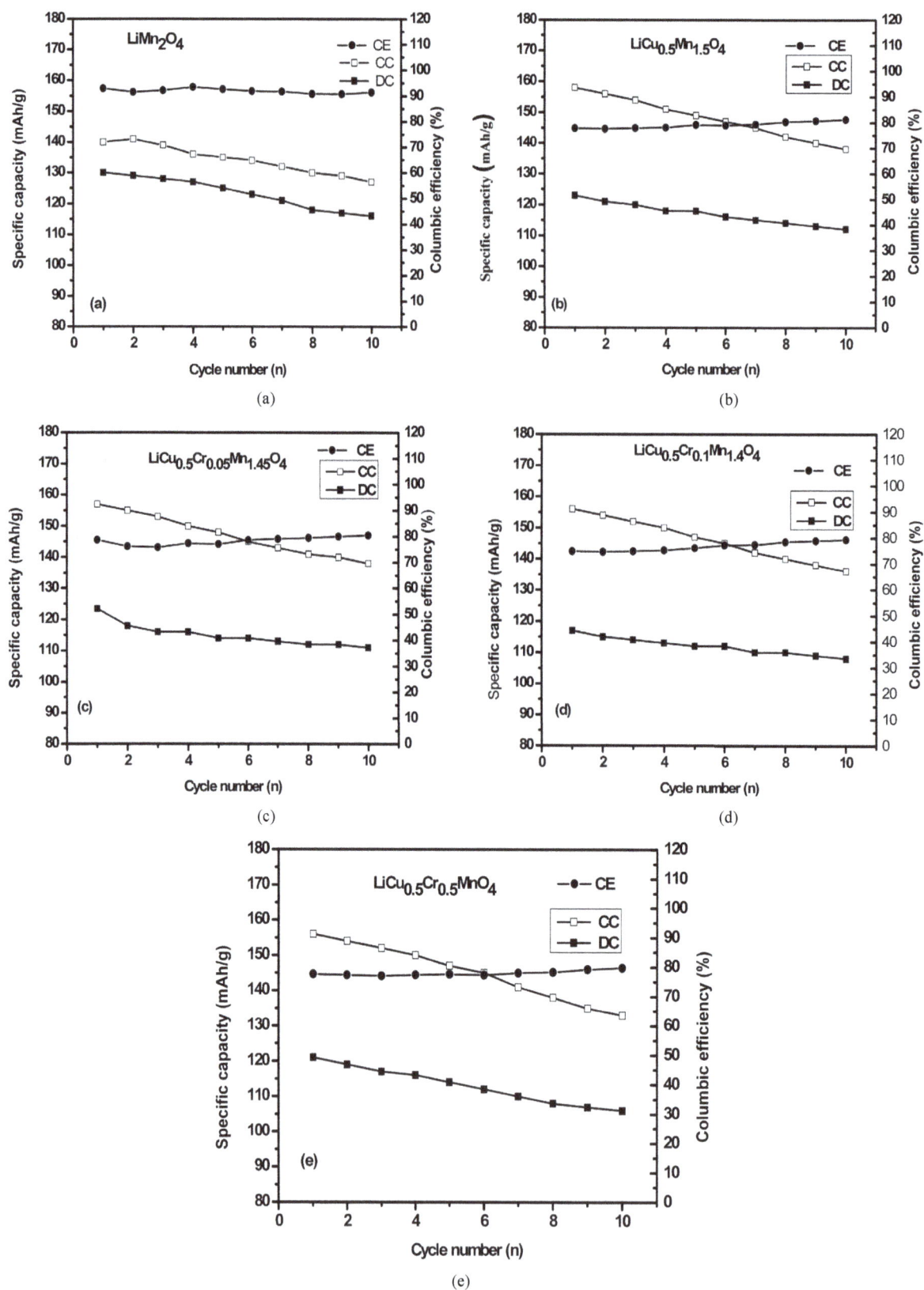

Figure 11. Cycling behaviour of $LiMn_2O_4$ and $LiCu_xCr_yMn_{2-x-y}O_4$. (a) Undoped; (b) x = 0.50; (c) x = 0.50; y = 0.05; (d) x = 0.50; y = 0.10; (e) x = 0.50; y = 0.50.

than our reported investigation (106 mA·h/g) with good capacity retention. It is quite interesting to know the discharge capacities of pristine spinels synthesized by different methods suggesting that 0.1-Cr doped spinel synthesizes via green chemistry method/sol-gel method and delivers the specific discharge capacity of 95 mA·h/g in the 10th cycle [24]. These results are not superior to our present work. Hence, it is concluded from the charge-discharge and cycling studies that among all dual doped spinels, the low amount of Cr-doped spinel (Li-$Cu_{0.5}Cr_{0.05}Mn_{1.45}O_4$) is an apt candidate to enhance the electrochemical stability of the spinel owing to higher octahedral stabilization energy of Cr^{3+} (1142 kJ·mole^{-1}) as compared to that of manganese (946 kJ·mole^{-1}) and effect of chromium will be more pronounced in reducing the capacity fade leads to decrease in cell volume to augment the stability of the structure during the deintercalation/intercalation process. Thus, low chromium doped spinel (LiCu$_{0.5}$Cr$_{0.05}$Mn$_{1.45}$O$_4$) stabilizes the Mn^{3+} structure for increasing good cycleability with better capacity retention. Also the stronger Cr-O bonds in the delithiated state (compare the binding energy of 1142 kJ·mol^{-1} for CrO$_2$ with 946 kJ·mol^{-1} for α-MnO$_2$) may also be considered to contribute to stabilization energy of Cr^{3+} in the octahedral sites.

3.8. Electrochemical Impedance Spectroscopy

Figures 12(a)-(e) show the Nyquist plot of electrochemical impedance spectra of (EIS) coin cells after 10 cycles. The Nyquist plots depict the single semicircle in the high frequency region and straight sloping line in the low frequency region. Such Nyquist plots are the relaxation frequency due to the interfacial polarization. All the Nyquist spectra depict the high-frequency semicircle with a maximum at a frequency of few kilo-ohms. This may be due to highly charged state and the typical behavior of the diffusion kinetic process inside the electrode materials. Also, having the high-frequency semicircle to the charge of the surface area of the spinel oxide positive electrodes. This real impedance belongs to both interparticle electronic contact and ionic diffusion via the passivation layer. It is clearly seen that among all the five compounds, LiCu$_{0.5}$Cr$_{0.05}$Mn$_{1.45}$O$_4$ shows low resistance of 190 Ohms while undoped spinel depicting the real impedance of 350 Ohms. Similarly, in the case of LiCu$_{0.5}$Mn$_{1.5}$O$_4$, LiCu$_{0.5}$Cr$_{0.1}$Mn$_{1.4}$O$_4$ provides higher resistance of 500 and 275 Ohms than that of LiCu$_{0.5}$Cr$_{0.05}$Mn$_{1.45}$O$_4$. Moreover, an equal amount of high doping pristine spinel (LiCu$_{0.5}$Cr$_{0.5}$Mn$_{1.0}$O$_4$) exhibits the high electrochemical polarization of 2000 Ohms which is seemed to be higher impedance relatively to all the dopants. During charging, Li/LiCu$_{0.5}$Cr$_{0.05}$Mn$_{1.45}$O$_4$ couple delivers very lower electrochemical polarization of 350 Ohms than that of undoped and all doped spinels for the sake of good kinetic reactions and good thermodynamics. Hence, among all dual doped spinels, the stable specific discharge capacity has been offered by the low Cr-doped spinel (LiCu$_{0.5}$Cr$_{0.05}$Mn$_{1.45}$O$_4$) may be due to the low order of electrochemical polarization/charge transfer resistance (R$_{ct}$) and better electronic conductivity with mitigating of Mn^{3+} from Mn^{4+} for high diffusion of lithium ions during intercalation and de-intercalation process. LiCu$_{0.5}$Cr$_{0.5}$Mn$_{1.0}$O$_4$ samples show higher electrochemical impedance as a consequence of increasing the charge transfer resistance and charge transfer of lithium ions on the surface of the electrodes which leads to capacity fading upon cycling (10 cycles). In view of the above, the impedance behaviour of 0.05-Cr doped spinel shows an excellent electrochemical performance.

3.9. dQ/dE vs. Potential Curves

The differential capacity curve of the parent LiMn$_2$O$_4$calcined at 850°C is shown in **Figure 13(a)**. This curve depicts lucidly two redox peaks respectively. The oxidative peaks correspond to the extraction of lithium ions and reductive peaks are insertion of lithium ions. The first two predominant anodic peaks are seen at about 4.01 and 4.15 V corresponds to Mn^{3+}/Mn^{4+} couples with an average oxidation state of 3.5 while another two well defined cathodic peaks appearing at around 3.98 and 4.1 V owing to the reductive behavior of the spinel. Moreover, these two anodic peaks of undoped spinel depict the high anodic peak currents at around 0.0018 and 0.00145 mA·h^{g-1}·mV^{-1} with the cathodic peak currents of −0.00195 and −0.00194 mA·h^{g-1}·mV^{-1}. Hence, it is vindicated that the anodic process is more facile than cathodic one.

Figure 13(b) depicts the differential capacity cure of LiCu$_{0.5}$Cr$_{0.05}$Mn$_{1.45}$O$_4$ calcined at 850°C. It is well known that two well defined anodic peaks are seen between 4.02 and 4.15 V attributed to the de-intercalation process with Mn^{3+}/Mn^{4+}, Cu^{2+}/Cu^{3+} and Cr^{3+}/Cr^{4+} couples respectively. Similarly, two anodic peaks of Li-Cu$_{0.5}$Cr$_{0.05}$Mn$_{1.45}$O$_4$ exhibit the high anodic peak currents at around 0.0065 and 0.002 mA·h^{g-1}·mV^{-1} with the cathodic peak currents of −0.0045 and −0.0021 mA·h^{g-1}·mV^{-1} at about 4.11 and 4 V. It is evident that the differential capacity curve of LiCu$_{0.5}$Cr$_{0.05}$Mn$_{1.45}$O$_4$ exhibits the better performance in both peak currents when

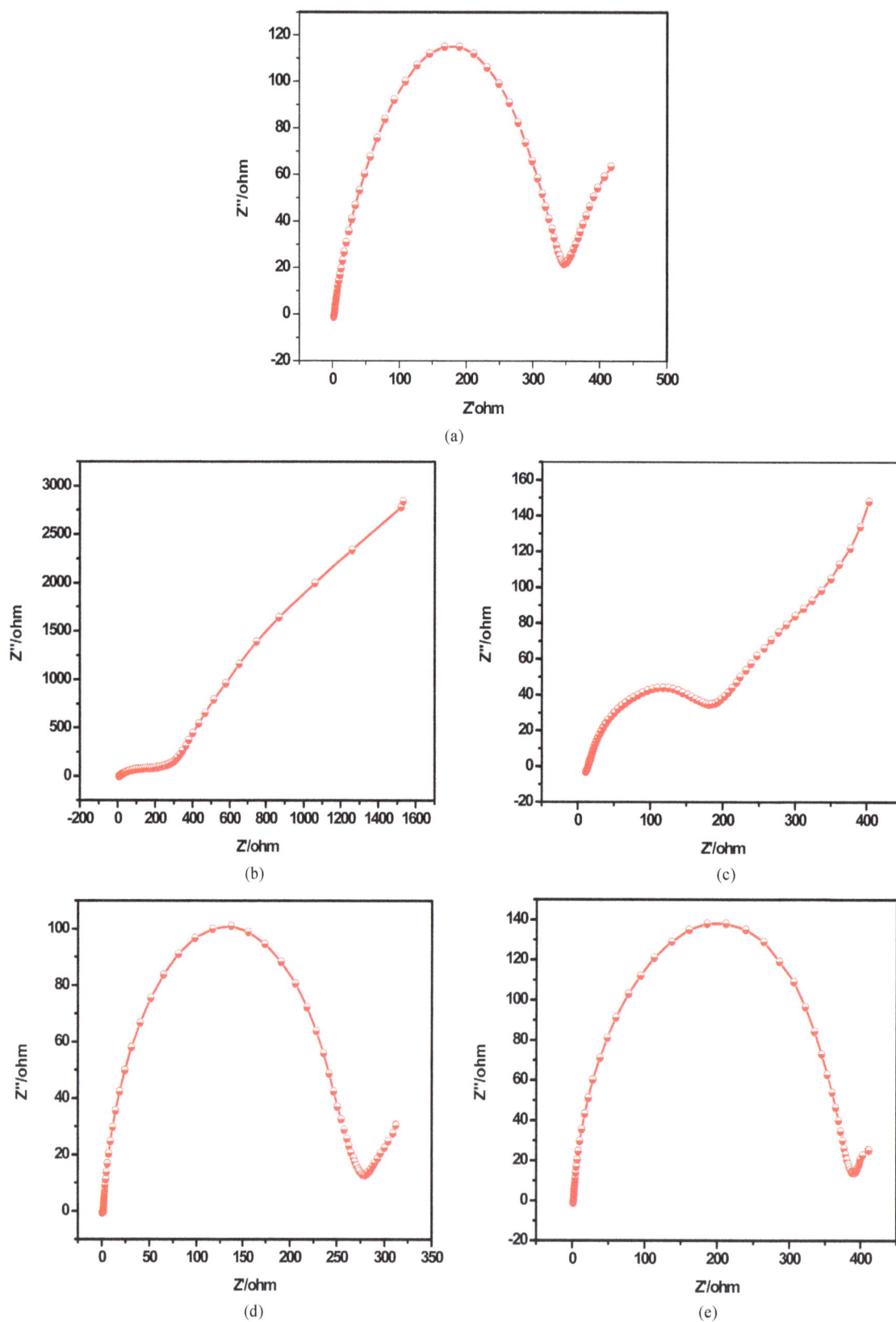

Figure 12. Electrochemical Impedance spectra (EIS) of the cells (LiMn$_2$O$_4$ and LiCu$_x$Cr$_y$Mn$_{2-x-y}$O$_4$) after 10 cycles. (a) Un-doped; (b) x = 0.50; (c) x = 0.50; y = 0.05; (d) x = 0.50; y = 0.10; (e) x = 0.50; y = 0.50.

(a)

(b)

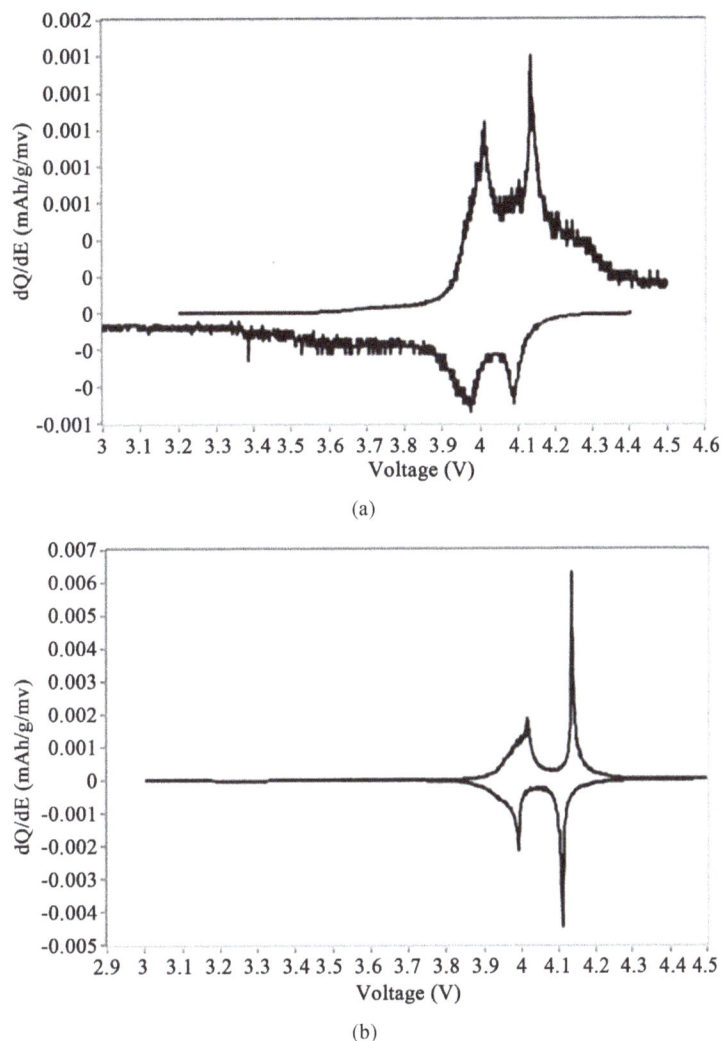

Figure 13. Differential capacity curves of (a) $LiMn_2O_4$; (b) $LiCu_{0.5}Cr_{0.05}Mn_{1.45}O_4$.

compared with undoped spinel. Hence, it is concluded that the low amount of chromium doped spinel ($LiCu_{0.5}Cr_{0.05}Mn_{1.45}O_4$) increases the electrochemical stability of the compound which has been reverberated in cycleability studies.

4. Conclusion

Pristine spinel $LiMn_2O_4$ and $LiCu_xCr_yMn_{2-x-y}O_4$ (x, y = Cu, Cr) (x = 0.50; y = 0.05 - 0.50) powders have been synthesized via sol-gel method for the first time using Myristic acid as the chelating agent to obtain micron sized particles for use as cathode materials in lithium rechargeable batteries. XRD patterns of $LiMn_2O_4$ and $LiCu_xCr_yMn_{2-x-y}O_4$ heated at 850°C corroborate the high degree of crystallinity with better phase purity of synthesized materials via sol-gel synthesis. FESEM image of undoped pristine spinel lucidly reveals cauliflower morphology with good agglomerated particle size of 50 nm while 0.5-Cu doped samples depict the pebbles morphology. TEM images of the spinel $LiMn_2O_4$ and $LiCu_{0.5}Cr_{0.05}Mn_{1.45}O_4$ substantiate that all the synthesized particles via sol-gel method are nano-sized (100 nm) with spherical surface and cloudy particles morphology. EDAX peaks confirm their actual composition of Cu, Cr, Mn and O in $LiMn_2O_4$ and $LiCu_xCr_yMn_{2-x-y}O_4$. Electrochemical impedance spectroscopy (EIS) measurements of spinel $LiMn_2O_4$ and $LiCu_{0.5}Cr_{0.05}Mn_{1.45}O_4$ exhibit the high and low order of polarization of (350 and 190 Ohms) after 10 cycles. Among all dopant concentrations attempted, $LiCu_{0.5}Cr_{0.05}Mn_{1.45}O_4$ sample exhibits the best performance (1st cycle discharge capacity: 124 mA·h/g) with low capacity fade of 1.1 mA·h/g cycle and capacity retention of 90%.

Acknowledgements

One of the authors Dr. R. Thirunakaran is thankful for the support given under the "Brain Pool Program of the Korean Federation of Science and Technology Societies" (KOFST), Republic of South Korea and also grateful to Prof. Won-Sub Yoon, for his help and valuable guidance. Further, Dr. R. Thirunakaran, thanks Dr. Vijaya-mohanan K. Pillai, Director, CSIR-CECRI for granting leave to avail the above fellowship. Also, this work was supported by the Human Resources Development Program (No. 20124010203270) of the Korea Institute of Energy Technology Evaluation and Planning (KETEP) grant funded by the Korea government Ministry of Trade, Industry and Energy. Many thanks are to the students of Professor for their co-operative helps and Cooperative Center for Research Facilities (CCRF) for completion of all physical characterization studies during my fellowship at Sungkyunkwan University (SKKU), South Korea.

References

[1] Tarascon, J.M., McKinnon, W.R., Coowar, F., Bowner, T.N., Amatucci, G. and Guyomard, D. (1994) Synthesis Conditions and Oxygen Stoichiometry Effects on Lithium Insertion into the Spinel $LiMn_2O_4$. *Journal of the Electrochemical Society*, **141**, 1421-1431. http://dx.doi.org/10.1149/1.2054941

[2] Gummow, R.J., de Kock, A. and Thackeray, M.M. (1994) Improved Capacity Retention in Rechargeable 4 V Lithium/Lithium-Manganese Oxide (Spinel) Cells. *Solid State Ionics*, **69**, 59-67. http://dx.doi.org/10.1016/0167-2738(94)90450-2

[3] Thackeray, M.M., de Kock, A., Rossouw, M.H., Liles, D., Bittihn, R. and Hoge, D. (1992) Spinel Electrodes from the Li-Mn-O System for Rechargeable Lithium Battery Applications. *Journal of the Electrochemical Society*, **139**, 363-366. http://dx.doi.org/10.1149/1.2069222

[4] Xia, Y., Zhou, Y. and Yoshio, M. (1997) Capacity Fading on Cycling of 4 V $Li/LiMn_2O_4$ Cells. *Journal of the Electrochemical Society*, **144**, 2593-2600. http://dx.doi.org/10.1149/1.1837870

[5] Pistoia, G., Antonini, A., Rosati, R. and Zane, D. (1996) Storage Characteristics of Cathodes for Li-Ion Batteries. *Electrochimca Acta*, **41**, 2683-2689. http://dx.doi.org/10.1016/0013-4686(96)00122-3

[6] Jang, D.H., Shin, J.Y. and Oh, S.M. (1996) Dissolution of Spinel Oxides and Capacity Losses in 4 V $Li/Li_xMn_2O_4$ Cells. *Journal of the Electrochemical Society*, **143**, 2204-2211. http://dx.doi.org/10.1149/1.1836981

[7] Yamada, A. (1996) Lattice Instability in Li $(Li_xMn_{2-x})O_4$. *Journal of Solid State Chemistry*, **122**, 160-165. http://dx.doi.org/10.1006/jssc.1996.0097

[8] Ohuzuku, T., Takeda, S. and Wakihara, M. (1999) Olivine Coated Spinel: 5 V System for High Energy Lithium Batteries. *Journal of Power Sources*, **90**, 81-82.

[9] Song, D., Ikuta, H., Uchida, T. and Wakihara, M. (1999) The Spinel Phases $LiAl_yMn_{2-y}O_4$ (y = 0, 1/12, 1/9, 1/6, 1/3) and Li $(Al, M)_{1/6} Mn_{11/6}O_4$ (M=Cr, Co) as the Cathode for Rechargeable Lithium Batteries. *Solid State Ionics*, **117**, 151-156. http://dx.doi.org/10.1016/S0167-2738(98)00258-6

[10] Iqbal, M.J. and Ahmad, Z. (2008) Electrical and Dielectric Properties of Lithium Manganate Nanomaterials Doped with Rare-Earth Elements. *Journal of Power Sources*, **179**, 763-769. http://dx.doi.org/10.1016/j.jpowsour.2007.12.115

[11] Lee, J.H., Hong, J.K., Jang, D.H., Sun, Y.K. and Oh, S.M. (2000) Degradation Mechanisms in Doped Spinels of $LiM_{0.05}Mn_{1.95}O_4$ (M = Li, B, Al, Co, and Ni) for Li Secondary Batteries. *Journal of Power Sources*, **89**, 7-14. http://dx.doi.org/10.1016/S0378-7753(00)00375-X

[12] Park, S.H., Park, K.S., Sun, Y.K. and Nahm, K.S. (2000) Synthesis and Characterization of a New Spinel, $Li_{1.02}Al_{0.25}Mn_{1.75}O_3S_{0.03}$, Operating at Potentials between 4.3 and 2.4 V. *Journal of Electrochemical Society*, **147**, 2116-2121. http://dx.doi.org/10.1149/1.1393494

[13] Bach, S., Henry, M., Baffier, N. and Livage, J. (1990) Sol-Gel Synthesis of Manganese Oxide. *Journal of Solid State Chemistry*, **88**, 325-333. http://dx.doi.org/10.1016/0022-4596(90)90228-P

[14] Perreira-Ramos, J.P. (1995) Electrochemical Properties of Cathodic Materials Synthesized by Low-Temperature Techniques. *Journal of Power Sources*, **54**, 120-126. http://dx.doi.org/10.1016/0378-7753(94)02051-4

[15] Barboux, P., Tarascon, J.M. and Shokoohi, F.K. (1991) The Use of Acetates as Precursors for the Low Temperature Synthesis of $LiMn_2O_4$ and $LiCoO_2$ Intercalation Compounds. *Journal of Solid State Chemistry*, **94**, 185-196. http://dx.doi.org/10.1016/0022-4596(91)90231-6

[16] Liu, W., Farrington, G.C., Chaput, F. and Dunn, B.J. (1996) Synthesis and Electrochemical Studies of Spinel Phase $LiMn_2O_4$ Cathode Materials Prepared by the Pechini Process. *Journal of Electrochemical Society*, **143**, 879-884. http://dx.doi.org/10.1149/1.1836552

[17] Thirunakaran, R., Kalaiselvi, N., Periasamy, P., RameshBabu, B., Renganathan, N.G. and Muniyandi, N. (2001) Signi-

ficance of Mg Doped $LiMn_2O_4$ Spinels as Attractive 4 V Cathode Materials for Use in Lithium Batteries. *Ionics*, **7**, 187-191. http://dx.doi.org/10.1007/BF02419227

[18]   Guo, S.H., Zhang, S.C., He, X.M., Pu, W.H., Jiang, C.Y. and Wan, C.R. (2008) Synthesis and Characterization of Sn-Doped $LiMn_2O_4$ Cathode Materials for Rechargeable Li-Ion Batteries. *Journal of the Electrochemical Society*, **155**, A760-A763. http://dx.doi.org/10.1149/1.2965635

[19]   Thirunakaran, R., Sivashanmugam, A., Gopukumar, S., Dunnill, C.W. and Gregory, D.H. (2008) Electrochemical Behaviour of Nano-Sized Spinel $LiMn_2O_4$ and $LiAl_xMn_{2-x}O_4$ (x = Al: 0.00 - 0.40) Synthesized via Fumaric Acid-Assisted Sol-Gel Synthesis for Use in Lithium Rechargeable Batteries. *Journal of Physics Chemistry of Solids*, **69**, 2082-2090. http://dx.doi.org/10.1016/j.jpcs.2008.03.009

[20]   Veluchamy, A., Ikuta, H. and Wakihara, M. (2001) Boron-Substituted Manganese Spinel Oxide Cathode for Lithium Ion Battery. *Solid State Ionics*, **143**, 161-171. http://dx.doi.org/10.1016/S0167-2738(01)00856-6

[21]   Fey, G.T.K., Lu, C.Z. and Prem Kumar, T. (2003) Solid-State Synthesis and Electrochemical Characterization of Li-$M_yCr_{0.5-y}Mn_{1.5}O_4$ (M = Fe or Al; $0.0 < y < 0.4$) Spinels. *Materials Chemistry and Physics*, **80**, 309-318. http://dx.doi.org/10.1016/S0254-0584(02)00522-9

[22]   Thirunakaran, R., Sivashanmugam, A., Gopukumar, S. and Rajalakshimi, R. (2009) Cerium and Zinc: Dual-Doped $LiMn_2O_4$ Spinels as Cathode Material for Use in Lithium Rechargeable Batteries. *Journal of Power Sources*, **187**, 565-574. http://dx.doi.org/10.1016/j.jpowsour.2008.10.134

[23]   Thirunakaran, R., Kim, K.T., Kang, Y.M., Seo, C.Y. and Lee, J.Y. (2004) Adipic Acid Assisted Sol-Gel Route for Synthesis of $LiCr_xMn_{2-x}O_4$ Cathode Material. *Journal of Power Sources*, **137**, 100-104. http://dx.doi.org/10.1016/j.jpowsour.2004.02.016

[24]   Thirunakaran, R., Kim, K.T., Kang, Y.M. and Lee, J.Y. (2005) Cr^{3+} Modified $LiMn_2O_4$ Spinel Intercalation Cathodes through Oxalic Acid Assisted Sol-Gel Method for Lithium Rechargeable Batteries. *Materials Research Bulletin*, **40**, 177-186. http://dx.doi.org/10.1016/j.materresbull.2004.08.013

[25]   Thirunakaran, R., Sivashanmugam, A., Gopukumar, S., Dunnill, C.W. and Gregory, D.H. (2008) Studies on Chromium/Aluminium Doped Managanese Spinel as Cathode Materials for Lithium-Ion Batteries—A Novel Chelated Sol-Gel Synthesis. *Journal of Materials Process & Technology*, **208**, 520-531. http://dx.doi.org/10.1016/j.jmatprotec.2008.01.017

[26]   Thirunakaran, R., Sivashanmugam, A., Gopukumar, S., Dunnill, C.W. and Gregory, D.H. (2008) Phthalic Acid Assisted Nano-Sized Spinel $LiMn_2O_4$ and $LiCr_xMn_{2-x}O_4$(x=0.00−0.40) via Sol-Gel Synthesis and Its Electrochemical Behaviour for Use in Li-Ion-Batteries. *Materials Research Bulletin*, **43**, 2119-2129. http://dx.doi.org/10.1016/j.materresbull.2007.09.021

[27]   Penga, C., Huanga, J., Guoa, Y., Lia, Q., Baia, H., Hea, Y., Sua, C. and Guao, J. (2015) Electrochemical Performance of Spinel $LiAl_xMn_{2-x}O_4$ Prepared Rapidly by Glucose-Assisted Solid-State Combustion Synthesis. *Vacuum*, **120**, 121-126. http://dx.doi.org/10.1016/j.vacuum.2015.07.001

[28]   Rajakumar, S., Thirunakaran, R., Sivashanmugam, A. and Gopukumar, S. (2010) Synthesis, Characterization, and Electrochemical Properties of $LiCr_xNi_yMn_{2-x-y}O_4$ Spinels as Cathode Material for 5 V Lithium Battery. *Journal of Electrochemical Society*, **157**, A333-A339. http://dx.doi.org/10.1149/1.3283015

[29]   Julien, C., Mangani, I.R., Selladurai, S. and Massot, M. (2002) Synthesis, Structure and Electrochemistry of $LiMn_{2-y}Cr_{y/2}Cu_{y/2}O_4$ ($0.0 \leq y \leq 0.5$) Prepared by Wet Chemistry. *Solid State Ionics*, **4**, 1031-1038. http://dx.doi.org/10.1016/s1293-2558(02)01357-2

[30]   Lu, C.H., Lin, Y.L. and Wang, H.C. (2003) Chromium-Ion Doped Spinel Lithium Manganate Nanoparticles Derived from the Sol-Gel Process. *Journal of Material Science Letters*, **22**, 615-618. http://dx.doi.org/10.1023/A:1023358731547

[31]   Sun, Y.K, Oh, I.H. and Kim, K.Y. (1997) Synthesis of Spinel $LiMn_2O_4$ by the Sol-Gel Method for a Cathode-Active Material in Lithium Secondary Batteries. *Industrial and Engineering Chemistry Research*, **36**, 4839-4846. http://dx.doi.org/10.1021/ie970227b

[32]   Murali, K.R., Saravanan, T. and Jeyachandran, M. (2008) Synthesis and Characterization of Copper Substituted Lithium Manganate Spinels. *Journal of Materials Science: Materials in Electronics*, **19**, 533-537. http://dx.doi.org/10.1007/s10854-007-9376-4

[33]   Hernán, L., Morales, J., Sánchez, L. and Santos, J. (1999) Use of Li-M-Mn-O [M=Co, Cr, Ti] Spinels Prepared by a Sol-Gel Method as Cathodes in High-Voltage Lithium Batteries. *Solid State Ionics*, **118**, 179-185. http://dx.doi.org/10.1016/S0167-2738(98)00449-4

[34]   Bao, S.J., Liang, Y.Y., Zhou, W.J., He, B.L. and Li, H.H. (2006) Synthesis and Electrochemical Properties of Li-$Al_{0.1}Mn_{1.9}O_4$ by Microwave-Assisted Sol-Gel Method. *Journal of Power Sources*, **154**, 239-245. http://dx.doi.org/10.1016/j.jpowsour.2005.03.220

16

Electrical Analysis of Indium Deep Levels Effects on Kink Phenomena of Silicon NMOSFETs

Abdelaali Fargi, Neila Hizem, Adel Kalboussi, Abdelkader Souifi

Department of Physics, Faculty of Sciences of Monastir, Monastir, Tunisia
Email: fargi.abdelaali@gmail.com

Abstract

Several methods of characterization of trap levels like I-V, C-V and transient spectroscopy (DLTS) were used to determine the accurate values of the activation energies of traps present in N+P junctions obtained after retrograde profile implantation of indium and boron on silicon. Four main traps located at Ev + 0.15 eV, Ev + 0.21 eV, Ev + 0.28 eV and Ev + 0.46 eV are reported. Shallow levels are also calculated from I-V characteristics. Concurrently, indium channel doped NMOSFETs are investigated showing the kink phenomenon. In order to discuss the relationship between the kink effect and the active indium trap level situated at 0.16 eV, the transient effects are studied by varying the integration time and the temperature. The effects of substrate polarization are also carried out showing the reduction of the kink with the bulk positive polarization.

Keywords

Electrical Characterization; MOSFET; Indium Deep Traps; Kink Effect; Impact Ionization

1. Introduction

Microelectronics' evolution implies the reduction of the components' size to improve the density of integration and the rapidity of circuits. The reduction of this size is particularly critical on the channel length under the gate, and leads to several parasitic effects when gate length goes down below 0.1 μm. The control of Short Channel Effects (SCE) becomes more challenging. This often requires a higher substrate doping concentration for effective SCE control. Thus, Retrograded Channel Profile (RCP) formation has been suggested, where a high substrate doping concentration is buried underneath the surface, leaving a lightly doped region at the surface [1]. The lightly doped Si surface results in less ionized impurity scattered in the channel, and this improves the electron mobility that is dominated by surface scattering. The highly doped region buried underneath the surface limits the gate depletion width, and therefore reduces the amount of SCE [2]. In order to build a NMOS Transistors with retrograde channel, indium seems to give the best profiles that are sufficiently fine and not very sensitive to annealing. It seems obvious that its heavy ion is less diffused allowing indium to give narrow profiles [3]. However, indium's mass, which is a quality for reaching retrograde profiles, will be a defect when indium becomes electronically active. Another problem arises: indium precipitates as soon as the concentrations exceed 10^{17} cm^{-3} [9]. Even with the advantages of indium doped silicon, when scaling down dimensions of MOSFETs, many parasitic effects appear such as the kink effect. This parasitic effect, which is characterized by a sharp increase in the drain-source current at a certain drain-to-source voltage ($V_{ds} = V_{kink}$), yields a high drain conductance. Many studies of the kink effect in field effect transistors such as MOSFETs (bulk and SOI) and HEMTs have been reported. However, due to its complex behavior, the origin of this effect is still a subject of interest. Some studies have established a link between the kink effect and the impact ionization phenomenon occurring in the channel [4], while other studies have correlated this effect with the presence of traps in the structure [5]. Other studies such as [6] have reported that the impact ionization depends on a hole trapping and detrapping phenomenon in HEMTs. The surface states can also affect the kink phenomenon as shown in GaAs MESFETS [7]. However, few works have been reported on how the deep trap affects the kink in Si MOSFETs. The aim of this work is to discuss the relative influence of the kink effect compared to the ionization of indium levels in silicon. A detailed study of the electrical properties of deep levels was performed in N+P junctions using I-V-T, C-V-Tandtransient spectroscopy (DLTS) techniques. In order to study the impact of deep trap levels on the kink effect, drain current measurements were carried out in the 100 - 350 K temperature ranges. The substrate-related kink dynamics are also analyzed.

2. Experimental Details

N+P well diodes and NMOSFET transistors were fabricated to study the effect of dopant redistribution. These devices were investigated in each case using a retrograde well indium and boron implantation as shown in **Figure 1** which was deduced from capacitance-voltage profiling. A P well of about 5×10^{17} cm^{-3} indium implanted, was formed in a P substrate at 100 nm under the interface. Boron implantation was performed to adjust the substrate doping to about 1017 cm^{-3}. The NMOSFET device had a gate oxide thickness of 3.2 nm. The surface of the active zone of the poly gate NMOSFETs was 100×100 μm^2 (L × W). N+P well diodes fabricated with 5×1016 cm^{-3} for both In and B dopants were used for testing the MOS devices. A variety of methods, DLTS and I-V-T, have been used to characterize the acceptor levels related to the presence of Indium in N+P well diodes and MOS transistors.

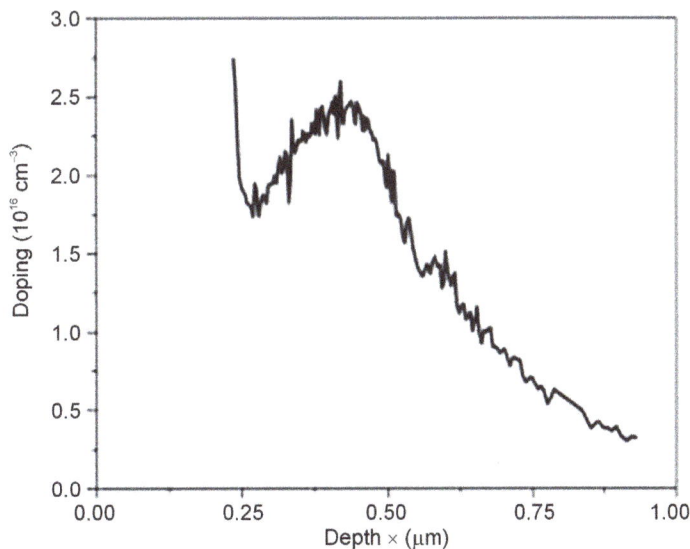

Figure 1. Retrograded doping profile showing indium and boron concentration deduced from Capacitance-Voltage profiling.

3. Diode Results

3.1. I-V-T Measurements

The current-voltage I-V characteristics of N+P well diodes have been measured over a wide range of temperature (80 - 300 K). Some intrinsic and contact properties such as barrier height $V_d = 0.7$ V, ideality factor n = 1.7, and series resistance Rs = 14 Ω were calculated from the forward current-voltage characteristics at room temperature. **Figure 2**, shows reverse current-voltage characteristic of a diode as a function of temperature. If the reverse current is related to traps generation from a deep level in the space charge region, the activation energy could be determined from $\ln(I/T5/2)$ versus 1000/T plot [8].

The analysis of the Arrhenius plots at different reverse polarizations reported in **Figure 3**, shows two straight lines; one for temperature higher than 170 K that gives activation energies between 0.15 - 0.28 eV (**Figure 3(a)**) and the second one gives activation energies between 66 - 23 meV for temperature lower than 170 K (**Figure 3(b)**). The activation energy values are reported in **Table 1**.

3.2. C-V-T Measurements

In order to investigate the freeze-out of carriers and the presence of the hole traps, **Figure 4**, shows the capacitance voltage as a function of temperature varying between 80 and 300 K at frequency 100 KHz. The dispersion of curves with temperature at zero bias is due to deep defect levels the density of which decreases when the depth increases.

Finally, we don't observe any anomaly in C-V curves even at a low temperature. In fact, all indium atoms are electrically active. Since Cerofolini *et al.* [9] observed that with an In concentration below 10^{17} cm^{-3} the junction is electrically active while with a concentration $> 10^{17}$ cm^{-3} evidence for precipitation as the mechanism responsible for inactivation was given.

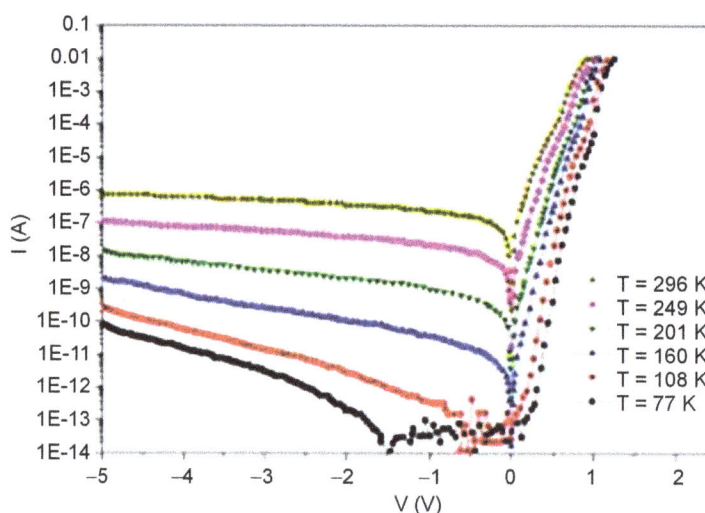

Figure 2. The reverse current-voltage characteristics at different temperature of N+P diode.

Table 1. Thermal activation energies of shallow and deep traps extracted from I-V-T measurements.

Polarization	Activation Energy	
	Shallowtraps (77 K < T< 160 K)	Deeptraps (200 K < T< 300 K)
0 V	23 ± 4 meV	0.28 ± 0.06 eV
−1 V	66 ± 16 meV	0.21 ± 0.02 eV
−2 V	55 ± 5 meV	0.21 ± 0.02 eV
−3 V	35 ± 6 meV	0.17 ± 0.02 eV
−4 V	33 ± 5 meV	0.15 ± 0.02 eV

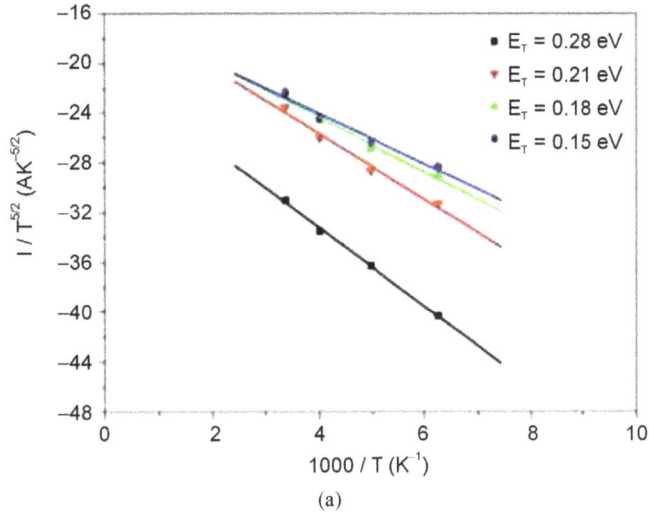

(a)

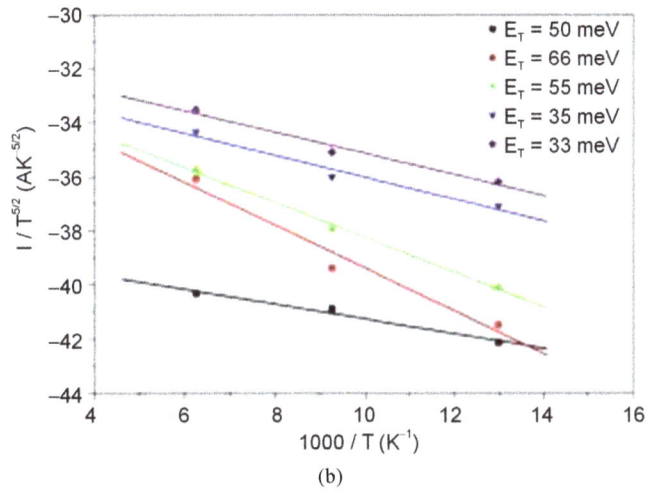

(b)

Figure 3. The analysis of the Arrhenius plots at different polarization: (a) For temperature upper than 170 K, (b) For temperature less than 170 K.

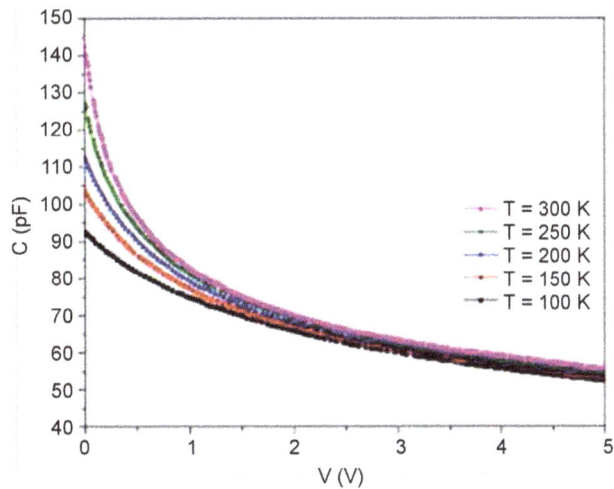

Figure 4. Capacitance-Voltage versus temperature of N+P junction at frequency f = 100 KHz.

3.3. DLTS Measurements

DLTS measurements, as shown in **Figure 5**, were performed in 40 - 300 K temperature range and made by lock-in amplifier technique (A-B) for different emission rates. Two main activation energy values 0.18 eV and 0.45 eV are extracted from the Arrhenius plot. A comparative study for the same temperature range of DLTS results with I-V measurements and later works revealed that the trap level located at 0.18 eV seems to be in good agreement with the 0.15 - 0.2 eV observed by I–V measurements and the 0.16 eV indium level measured by different characterization techniques: DLTS [10], Hall Effect [9], and Infrared Absorption [11]. The variety of values of in deep trap detected by different techniques is explained by the electric field effects due to coulomb traps [10].

4. MOSFET Results

4.1. I_{ds}-V_{ds} Characteristics

The **Figure 6** shows the static output characteristics (I_d-V_d) of the transistor at different gate voltages. We notice that the drain current continues to increase even in the saturation region. This unusual variation of the drain current is well known as the kink effect [12]. It consists of a sharp increase in the drain/source current at a certain drain-to-source voltage ($V_d = V_{kink}$). This variation of the drain current induces an increase of the drain/source output conductance (gds) and a decrease of the amplification factor. We observe that when the gate voltage increases, the amount of drain current and kink voltage (V_{kink}) increases too. The kink voltage seems to begate voltage dependent.

4.2. Kink Analysis

As seen in **Figure 7**, the pre-kink zone is highly temperature dependant. When the temperature decreases from 300 K the kink phenomena is more pronounced with a clear increase of the amount of drain current especially around 100 K where the In trap appears.

The trans conductance is maximal at the temperature range of indium deep trap appearance as shown in **Figure 8**. After the V_{kink} voltage the curves are identical. Therefore one concludes that the kink effect is related to the trapping and detrapping of the deep traps in the pre-kink region.

In order to get a better understanding of the influence of the deep traps were port in **Figure 9**, the changes of the I_d-V_d characteristics when the drain voltage is swept by long, medium and short speeds. When the integration time is short the kink tends to disappear and the drain current is higher than that of longer integration time.

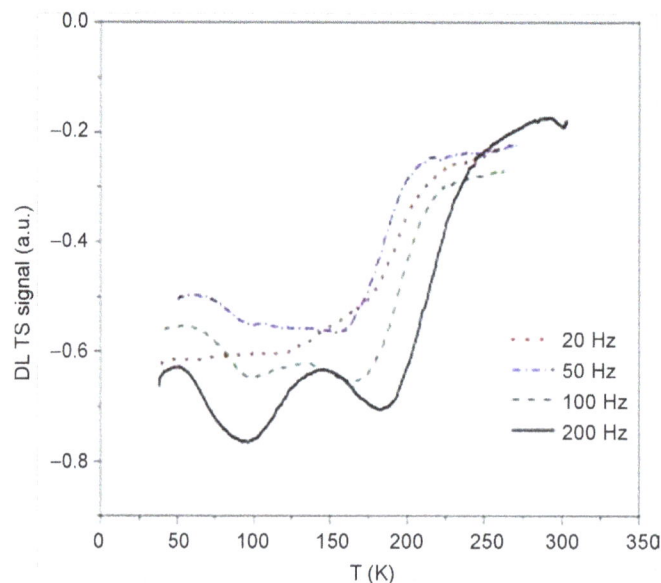

Figure 5. DLTS signal of N+P junction for reverse bias $V_r = -4$ V and pulse bias $V_p = 0$ V, with different emission rates.

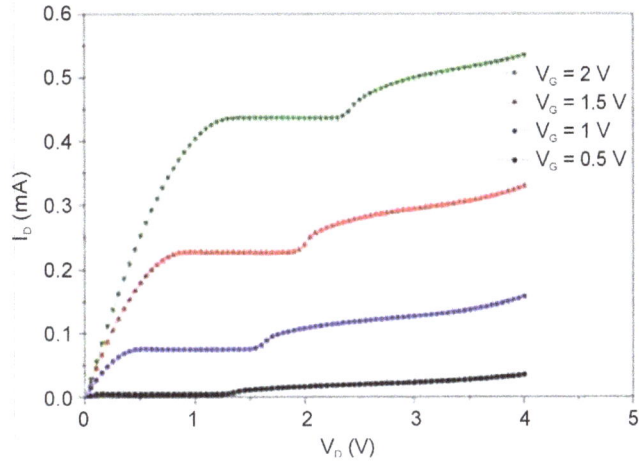

Figure 6. The output I_d-V_d characteristics at room temperature of NMOSFET transistor.

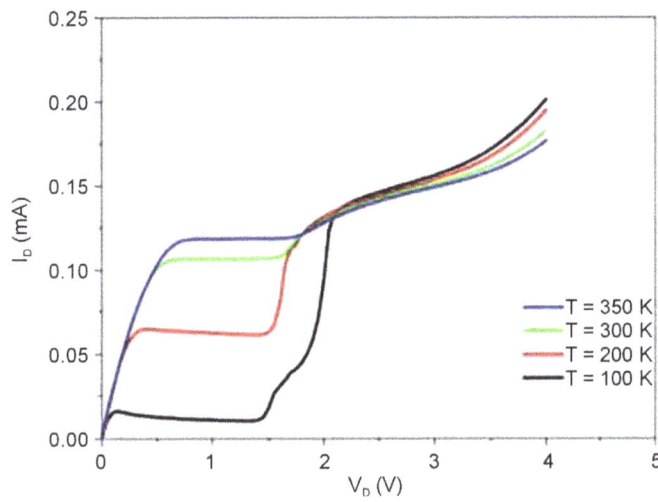

Figure 7. The output I_d-V_d characteristics in the temperature range between 100 and 350 K.

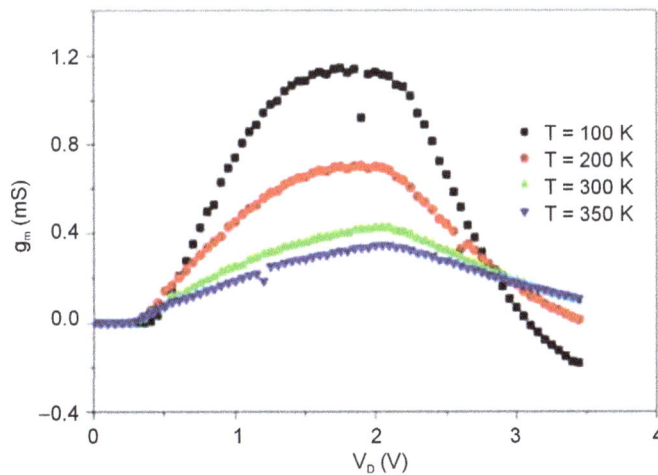

Figure 8. The temperature dependence of the transconductance-drain voltage characteristics.

This is because of the presence of hole traps that find enough time to capture injected electrons in thechannel when the integration time is longer resulting in are duced drain current. No changes in the characteristics for $V_d > V_{kink}$.

Next, we describe changes of the I_d-V_d characteristics when bulk polarization varies from -1 V to $+2$ V, at room temperature. **Figure 10**, shows the reduction of the kink with positive polarization of the bulk. If we compare the curves before and after the V_{kink} tension we will find that for $V_d < V_{kink}$ the drain current increases with the increasing of V_{sub}, but for $V_d > V_{kink}$ the drain current curves are confound with different bulk polarizations.

The inside shows that the pre-kink zone is sensitive to substrate polarization. The drain current rises and gets saturated at bulk positive polarization ($V_{sub} > +0.5$ V) for $V_d = 1.28$ V $< V_{kink}$, and $V_g = 1.5$ V. The drain current increase and the Kink effect reduction are related to the activation process of the indium deep level in the space charge region. We believe that at $V_g = 1.5$ V, for negative bulk polarization the space charge region be-comes larger and the capture process increases, a low drain current and a strong kink effect occur. For positive bulk polarization the space charge zone is reduced, the trapping phenomenon decreases and the kink is also reduced. This confirms the relation between the trap level and the kink effect.

The kink has also been reported in other devices such as SOI MOSFETs [13] [14], GaAs-based and InP-based HEMTs [15]-[17], etc. In the SOI MOSFETs, the kink is considered to occur due to the floating body effects and the parasitic bipolar effects originated from impact ionization of holes. In the HEMTs, impact ionization is

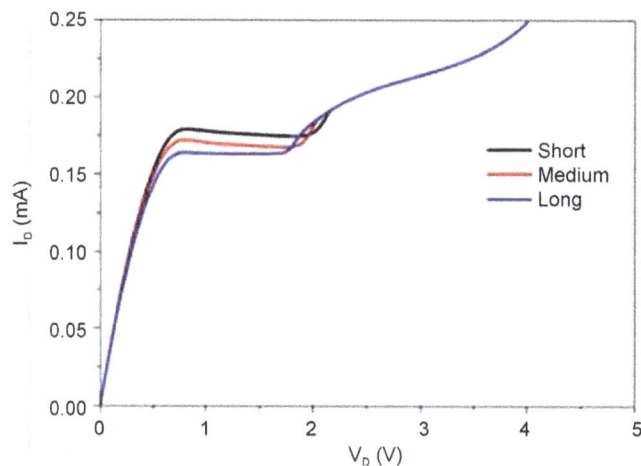

Figure 9. The changes of the transient effect with short long and medium integration time during data acquisitions.

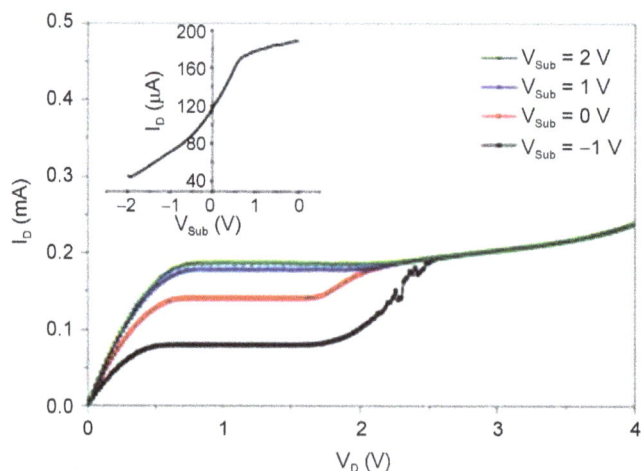

Figure 10. The effects of substrate polarization at 300 K. The inset shows the drain current-substrate voltage in the pre-kink zone ($V_d = 1.28$ V, $V_g = 1.5$ V).

believed to be a necessity for the kink, and some traps should be correlated to it. Some other works conclude that both impact ionization and trapping and detrapping by a deep level centre are responsible for kink effect [18] [19].

In this work, the analysis of the kink effect from the I_d-V_d characteristics leads to think that two different phenomena arise before and after V_{kink} voltage. Before V_{kink} we expect that the drain current behavior according to the temperature, the bulk polarization and integration time dependence is correlated to the incomplete ionization of the indium trap level. This trap can capture the electrons injected in the channel causing a reduction of the drain current. In contrast the drain current increase after the V_{kink} when V_d increases doesn't present any dependence on temperature or substrate polarization nor on integration time. This is probably correlated to the impact ionization effect. Indeed by increasing the voltage V_{ds} beyond the pinch-off voltage, the drain field becomes strong enough to ionize atoms in the space charge area channel at the drain side. These highly energetic electrons are able to extract other electron-hole pairs even at low temperature. The created electrons contribute strongly to the conduction channel. The holes are attracted towards the gate, but others are attracted to the substrate thus causing a gate-substrate current resulting in an increase of the drain current too [20]-[22].

5. Conclusion

Indium deep traps have been performed by I-V-T and DLTS measurements. It has been shown that indium induces two distinct levels: a shallow level at 20 meV and a deep one at 0.18 eV with electric field dependence. Electrical properties of NMOSFET transistors were characterized. The kink effect in MOSFET devices has been examined using temperature, bulk polarization and integration time. The incomplete ionization of indium could be the main origin of the kink effect. Depending on the temperature and the substrate polarization, the pre-kink region is changed, resulting in the change of the amount of the drain current. This behavior is mainly related to the presence of deep electron traps. Therefore, the non saturation of the drain current after the kink zone could be related to the impact ionization. This leads to concluding that both the impact ionization effect and trapping/detrapping by a deep level center are responsible for the increase of drain current. Transient and dynamic simulations have indicated that the kink phenomena originated from interaction between generated electrons and deep levels should be rather slow processes with long response times.

References

[1] De, I. and Osburn, C.M. (1999) Impact of Super-Steep-Retrograde Channel Doping Profiles on the Performance of Scaled Devices. *IEEE Transactions on Electronic Devices*, **46**, 1711-1717. http://dx.doi.org/10.1109/16.77716

[2] Skotnicki, T., Guérin, L., Mathiot, D., Gauneau, M., Grouillet, A., D'Anterroches, C., André, E., Bouillon, P. and Haond, M. (1994) Channel Engineering by Heavy Ion Implants. *24th ESSDRC*, **94**, 671-674.

[3] Ong, S.Y., Chor, E.F., Leung, Y.K., Lee, J., Li, W.S., See, A. and Chan, L. (2002) Steep Retrograde Indium Channel Profiling for High Performance nMOSFETs Device Fabrication. *Microelectronics Journal*, **33**, 55-60. http://dx.doi.org/10.1016/S0026-2692(01)00104-5

[4] Martinot, H. and Rossel, P. (1971) Carrier Multiplication in the Pinchoff Region of m.o.s. Transistors. *Electronics Letters*, **7**, 118-120. http://dx.doi.org/10.1049/el:19710078

[5] Ben Salem, M., Bouzgarrou, S., Sghaier, N., Kalboussi, A. and Souifi, A. (2006) Correlation between Static Characteristics and Deep Levels in InAlAs/InGaAs/InP HEMT'S. *Materials Science and Engeineering B*, **127**, 34-40. http://dx.doi.org/10.1016/j.mseb.2005.09.047

[6] Zaman, S. and Parker, A. (2007) Impact Ionization Dependence of Hole Trapping Phenomena in Hemts. *Journal of Science and Technology*, **2**, 8-12. http://dspace.daffodilvarsity.edu.bd:8080/bitstream/handle/123456789/460/Impact%20ionization.pdf?sequence=1

[7] Horio, K., Wakabayashi, A. and Yamada, T. (2000) Two-Dimensional Analysis of Substrate-Trap Effects on Turn-On Characteristics in GaAs MESFETs. *IEEE Transactions on Electronic Devices*, **47**, 2270-2276. http://dx.doi.org/10.1109/16.824738

[8] Lançon, R. and Marfaing, Y. (1969) Mécanisme de génération-recombinaison dans les jonctions p-n de tellurure de cadmium. *Le Journal de Physique*, **30**, 97-102. http://dx.doi.org/10.1051/jphys:0196900300109700

[9] Cerofolini, G.F., Pignatel, G.U., Mazzega, E. and Ottaviani, G. (1985) Supershallow Levels in Indium-Doped Silicon. *Journal of Applied Physics*, **58**, 2204-2207. http://dx.doi.org/10.1063/1.335988

[10] Jones, C.E. and Johnson, G.E. (1981) Deep Level Transient Spectroscopy Studies of Trapping Parameters Forcenters

in Indium-Doped Silicon. *Journal of Applied Physics*, **52** 5159-5163. http://dx.doi.org/10.1063/1.329416

[11] Parker, G.J., Brotherton, S.D., Gale, I. and Gill, A. (1983) Measurement of Concentration and Photoionization Cross Section of Indium in Silicon. *Journal of Applied Physics*, **54**, 3926-3929. http://dx.doi.org/10.1063/1.332566

[12] Nakahara, M., Iwasawa, H. and Yasutake, K. (1968) Anomalous Enhancement of Substrate Terminal Current beyond Pinch-Off in Silicon N-Channel MOS Transistors and Its Related Phenomena. *Proceedings of IEEE*, **56**, 2088-2090. http://dx.doi.org/10.1109/PROC.1968.6810

[13] Choi, J.Y. and Fossum, J.G. (1991) Analysis and Control of Floating-Body Bipolar Effects in Fully Depleted Submicrometer SOI MOSFET's. *IEEE Transactions on Electronic Devices*, **38**, 1384-1391. http://dx.doi.org/10.1109/16.81630

[14] Valdinoci, M., Colalongo, L., Baccarani, G., Fortunato, G., Pecora, A. and Policicchio, I. (1997) Floating Body Effects in Polysilicon Thin-Film Transistors. *IEEE Transactions on Electronic Devices*, **44**, 2234-2241. http://dx.doi.org/10.1109/16.644643

[15] Canali, C., Paccagnella, A., Pisoni, P., Tedesco, C., Telaroli, P. and Zanoni, E. (1991) Impact Ionization Phenomena in AlGaAs/GaAs HEMTs. *IEEE Transactions on Electronic Devices*, **38**, 2571-2573. http://dx.doi.org/10.1109/16.97428

[16] Hori, Y. and Kuzuhara, M. (1994) Improved Model for Kink Effect in AlGaAs/InGaAs Heterojunction FET's. *IEEE Transactions on Electronic Devices*, **41**, 2262-2267. http://dx.doi.org/10.1109/16.337437

[17] Suemitsu, T., Enoki, T., Sano, N., Tomizawa, M. and Ishii, Y. (1998) An Analysis of the Kink Phenomena in InAlAs/InGaAs HEMT's Using Two-Dimensional Device Simulation. *IEEE Transactions on Electronic Devices*, **45**, 2390-2399. http://dx.doi.org/10.1109/16.735714

[18] Bouzgarrou, S., Sghaier, Na., Ben Salem, M., Souifi, A. and Kalboussi, A. (2008) Influence of Interface States and Deep Levels on Output Characteristics of InAlAs/InGaAs/InP HEMTs. *Materials Science and Engineering C*, **28**, 676-679. http://dx.doi.org/10.1016/j.msec.2007.10.075

[19] Sornerville, M.H., Del Alamom J,A. and Hoke, W. (1996) Direct Correlation between Impact Ionization and the Kink Effect in InAlAs/InGaAs HEMTs. *IEEE Electron Device Letters*, **17**, 473-475. http://dx.doi.org/10.1109/55.537079

[20] Hafez, I.M., Ghibaudo, G. and Balestra, F. (1990) Reduction of Kink Effect in Short-Channel MOS Transistors. *IEEE Electron Device Letters*, **11**, 818-821. http://dx.doi.org/10.1109/55.46953

[21] Chen, S.S., Lin, S.C. and Kuo, J.B. (1996) Kink Effect on Subthreshold Current Conduction Mechanism for N-Channel Metal-Oxide-Silicon Devices. *Journal of Applied Physics*, **80**, 5821-5827. http://dx.doi.org/10.1063/1.363729

[22] Huang, C. and Gildenblat, S. (1990) Measurements and Modeling of the N-Channel MOSFET Inversion Layer Mobility and Device Characteristics in the Temperature Range 60 - 300 K. *IEEE Transactions on Electronic Devices*, **37**, 1289-1300. http://dx.doi.org/10.1109/16.108191

One-Step Microemulsion-Mediated Hydrothermal Synthesis of Nanocrystalline TiO$_2$

Xuyao Xu, Xiaosong Zhou*, Lin Ma, Miaoyan Mo, Cuifen Ren, Rongkai Pan

School of Chemistry Science & Technology, Institute of Physical Chemistry, and Development Center for New Materials Engineering & Technology in Universities of Guangdong, Zhanjiang Normal University, Zhanjiang, China
Email: *zxs801213@163.com

Abstract

Nanocrystalline TiO$_2$ powders with high photocatalytic activity were prepared by one-step microemulsion-mediated hydrothermal method using tetrabutylorthotitanate (TiO(C$_4$H$_9$)$_4$, TBOT) as precursor. The as-prepared TiO$_2$ powders were characterized by X-ray diffraction (XRD), transmission electron microscopy (TEM) and the Brunauer-Emmett-Teller (BET) specific surface area measurements. The effects of the oil/water ratio and hydrothermal temperature on the microstructures and photocatalytic activity of the TiO$_2$ powders were investigated. The results suggest that increasing the oil/water emulsion ratio significantly decreased the particle size of the as-prepared TiO$_2$ powders and improved the photocatalytic activity. With hydrothermal temperature increasing, the average crystallite size increased and the photocatalytic activities of TiO$_2$ powders decreased.

Keywords

Titanium Dioxide; Microemulsion; Hydrothermal Method; Photocatalytic Activity

1. Introduction

The excellent performance of semiconducting oxide materials in the photocatalytic realm has attracted scientific interest for ongoing research [1]. As an important n-type semiconductor material, nano-TiO$_2$ has recently stimulated increasing attention because of their promising applications in many fields, including sensors [2],

---

*Corresponding author.

self-cleaning photocatalytic surfaces and devices [3], dye-sensitized solar cells [4], environmental purification and hydrogen generation by water photoelectrolysis [5] [6]. Fundamental researches regarding the preparation of photocatalyst with highly photocatalytic activity, the immobilization of powder photocatalyst, and the improvement of photocatalyst performance are priorities to be considered [7].

Many methods, such as hydrolysis (chemical precipitation) [8], reverse micelles (microemulsion) [9] [10], sol-gel [11] [12] and hydrothermal crystallization [13]-[15], have been used to prepare TiO_2 nanocrystalline photocatalyst. These methods suffer the problems of high-temperature processing, long aging time, grain growth and vigorous stirring. The high processing temperatures and long reaction time result in the formation of large sized particles with a wide size distribution [16]. Consequently, it is in need of developing a novel and environmentally benign approach to preparing TiO_2 nanocrystalline photocatalyst. In this paper, a new one-step strategy by microemulsion-mediated low temperature hydrothermal route has been employed to synthesize crystalline nanosized TiO_2 powders. The main advantage of the proposed route is low reaction temperature and short processing time that prevents agglomeration in the formed particles. Varying oil/water microemulsion ratio, nano-sized TiO_2 particles with large surface area and a cuboid-liked shape were successfully synthesized at low temperature within short reaction period. The effects of hydrothermal processing temperature or time on the microstructures and photocatalytic behavior of the obtained TiO_2 powders were also investigated in detail.

2. Experimental Section

2.1. Preparation of the Catalysts

In a typical synthesis, 40 ml cyclohexane, 12 ml n-butanol, 8 ml polyoxyethylene (10) octylphenyl ether (OP-10), 4 ml tetrabutyl titanate and 4 ml HCl (0.1 M) were mixed together to form a dispersed mixture. After stirring for about 30 min, the mixture was finally transferred into a 100 ml Teflon-lined stainless steel autoclave. It was sealed tightly and maintained at 120°C for 1 h. After that, the autoclave was allowed to cool down naturally. The products were collected and washed three times using deionized water and absolute ethanol, then dried for 12 h under vacuum at 60°C. The as-synthesized samples were denoted as TX-Y (X denotes the oil/water ratio, and Y denotes the hydrothermal temperature).

2.2. Characterization of the Catalysts

XRD analysis was performed on a D/Max-2550 X-ray diffractometer with monochromatized CuKα radiation (λ = 0.1540562 nm). TEM was recorded on a transmission electron microscopy (TEM, JEOL JEM-200CX). The Brunauer-Emmett-Teller (BET) surface area (S_{BET}) was determined by nitrogen adsorption-desorption isotherm measurements at 77 K on a Micromeritics ASAP 2010 system. The samples were degassed in vacuum at 473 K until a pressure lower than 10^{-6} Torr before the actual measurements.

2.3. Photocatalyst Reaction

The photocatalytic reaction was conducted in the XPA-II photochemical reactor (Nanjing Xujiang Machine-electronic Plant). A 1000 W Xe lamp was used as the simulated solar light source, and a house-made filter was mounted on the lamp to eliminate infrared irradiation. MB (20 mg/L) was used as contamination. 20 mg photocatalyst powder was dispersed in 200 mL reaction solutions by ultrasonicating for 15 min, then the suspension was magnetically stirred in dark for 1 h. Air was blown into the reaction medium at a flow rate of 200 mL/min during the photocatalytic reaction. One 8 mL of the suspension was sampled and filtered. The concentration of the remaining MB was measured by a Hitachi UV-3010 spectrophotometer. The degradation ratio was calculated by $X = (A_0 - A)/A_0 \times 100\%$.

3. Results and Discussion

3.1. Morphology and Structures of the Samples

Figure 1(a) shows the XRD patterns of nano-sized TiO_2 powders prepared by one-step microemulsion-mediated hydrothermal route at 120°C for 1 h. All samples show pure anatase phase, which was not affected by the change in the oil/water ratio. The anatase crystal sizes were calculated by the Scherrer equation (**Table 1**). The XRD peaks of the obtained TiO_2 powders became broader with increasing the microemulsion oil/water ratio.

Table 1. Experimental conditions, SBET and particle size of the as-prepared samples.

Samples	R = oil/water (volume)	Hydrothermal temperature (°C)	S_{BET} ($m^2 \cdot g^{-1}$)	Particle size (nm)
T2-120	2	120	232	22.8
T5-120	5	120	294	14.3
T10-120	10	120	300	9.6
T10-140	10	140	287	16.5
T10-160	10	160	259	19.4

The broad diffraction peaks implied decrease in the particle size of TiO$_2$ powders.

Figures 1(b)-(d) show TEM micrographs of the as-prepared samples T2-120, T5-120 and T10-120. It is revealed that the morphology of the obtained TiO$_2$ particles was mono-dispersed. When the oil/water ratio was increased, the average particle sizes of the TiO$_2$ powders decreased, which were consistent with XRD result. It was reported that the size of the reverse micelle depends on the amount of water content in the microemulsion solution [17]. Once the oil/water ratio is increased, the size of the reverse micelle is correspondingly decreased. Since the grain growth of particles is limited by the size of the reverse micelles in the microemulsion hydrothermal process, increasing the oil/water ratio ultimately decreases the size of the TiO$_2$ particles formed therein.

Figure 2(a) depicts the XRD patterns of TiO$_2$ powders synthesized from one-step microemulsion-mediated solution after hydrothermally treated at 120°C, 140°C, and 160°C for 1 h. The XRD patterns indicate that anatase phases were formed in all samples. It was observed that with increasing hydrothermal processing temperatures, the peak intensities increased and the full width half maximum of the diffraction peak became narrower. This reveals that the crystallite size increased with increasing hydrothermal temperatures. This indicates the increase in the formation of larger crystallites and anatase phase crystallinity enhancement in the formed TiO$_2$ powders at elevated hydrothermal temperatures. Yu *et al.* also reported that the average crystallite sizes and degree of crystallization increase with increasing hydrothermal temperatures [18].

Figures 2(b)-(d) show TEM micrographs of the as-prepared samples hydrothermally treated at 120°C, 140°C and 160°C. It can be seen that after increasing the hydrothermal temperature to 160°C (**Figure 2(d)**), the TiO$_2$ particles retained the cuboid-like morphology. In addition, the particle size was found to be increased after the hydrothermal temperature was raised. With increasing hydrothermal temperature, the reverse micelles in the microemulsion will not be maintained. This will result in fast cluster nucleation oriented in random directions, thereby forming large particles [19].

3.2. Photocatalytic Activities of Samples

Figure 3 shows the comparison of photocatalytic activity of the TiO$_2$ powders synthesized in different hydrothermal conditions. The photocatalytic activity of Degussa P-25 powders, which is recognized as an excellent photocatalyst, was also measured as a reference to compare with that of the as-prepared catalysts. The BET specific surface areas of TiO$_2$ powders are shown in **Table 1**. It is revealed that the amount of methylene blue degraded by TiO$_2$ powders was increased when the oil/water ratio was raised. This is attributed to the increase in BET specific surface area of TiO$_2$ powders as reported by Yu *et al.* [20]. It can be seen that not all of the as-prepared TiO$_2$ powders had a greater photocatalytic activity than that of Degussa P-25. The photocatalytic activity should be a function of many physical parameters including the BET surface area, pore size and distribution, crystal size, crystalline, etc [18]. Therefore, the highest activities of the prepared TiO$_2$ powders at 120°C for 1 h could be attributed to the results of the above mentioned synergistic effects.

4. Conclusion

In summary, TiO$_2$ nano-particles with improved photocatalytic performance were successfully prepared via a one-step microemulsion hydrothermal process. The advantage of using this microemulsion mediated hydrothermal route is the significant reduction in reaction time and temperatures compared with the conventional hydrothermal process. Nanocrystalline TiO$_2$ powders with high photocatalytic activity were obtained after the hydrothermal treatment. The oil/water ratio and hydrothermal temperature obviously influenced the microstructures and photocatalytic activity of the as-prepared TiO$_2$ powders. Increasing the oil/water emulsion ratio

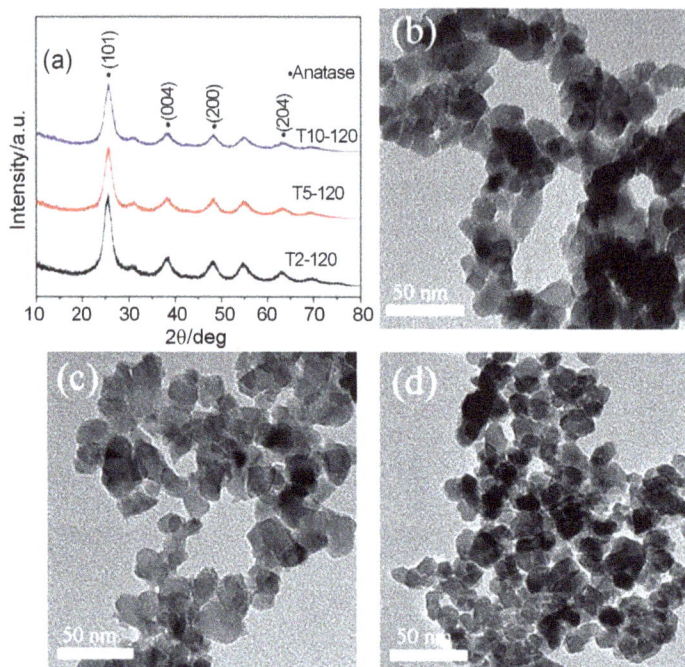

Figure 1. XRD pattern (a) and TEM images of the as-prepared samples: (b) T2-120; (c) T5-120 and (d) T10-120.

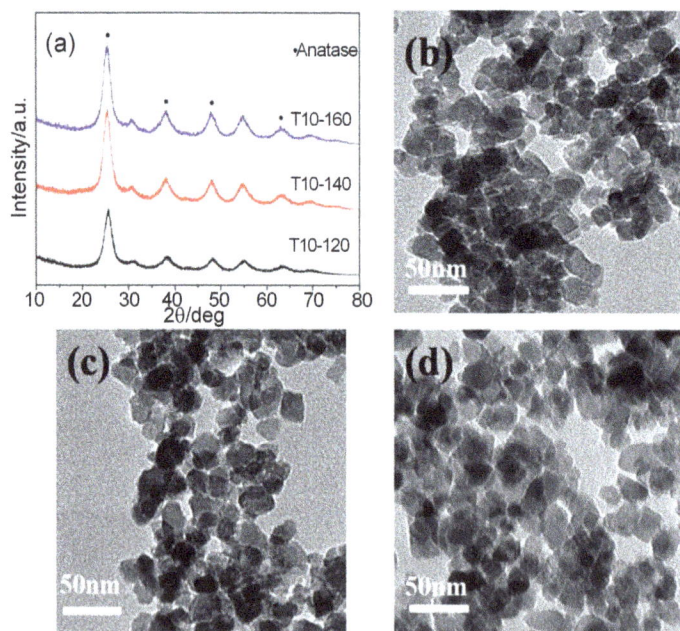

Figure 2. XRD pattern (a) and TEM images of the as-prepared samples: (b) T10-120; (c) T10-140 and (d) T10-160.

significantly decreased the particle size of the prepared TiO$_2$ powders and improved the photocatalytic activity. With increasing hydrothermal temperature, the average crystallite size increased. In contrast, the BET specific surface areas steadily decreased. Consequently, the photocatalytic activities of TiO$_2$ powders were reduced with hydrothermal temperature elevation. Our experimental results indicate that the photocatalytic activity of the TiO$_2$ powders prepared under an optimal hydrothermal condition (T10-120) exceeded that of Degussa P-25 powders.

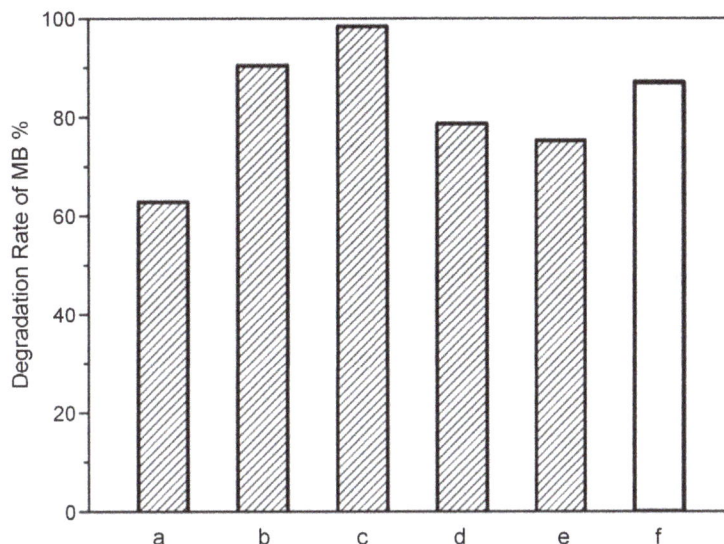

Figure 3. Comparison of degraded methylene blue rate of TiO_2 powders prepared in different hydrothermal treatment for 2.5 h: (a) T2-120; (b) T5-120; (c) T10-120; (d) T10-140; (e) T10-160 and (f) P-25.

Acknowledgements

This work was financially supported by Zhanjiang Science and Technology Research Projects (No. 2013B01152), Zhanjiang Normal University Natural Science Foundation for the Youth Program (No. L1202), Guangdong Natural Science Foundation for Ph.D Start-up Research Program (S2013040013755), and Colleges and Universities in Guangdong Province Science and Technology Innovation Project (2013KJCX0123).

References

[1]   Lu, N., Quan, X., Li, J.Y., Chen, S., Yu, H.T. and Chen, G.H. (2007) Fabrication of Boron-Doped TiO_2 Nanotube Array Electrode and Investigation of Its Photoelectrochemical Capability. *The Journal of Physical Chemistry C*, **111**, 11836-11842. http://dx.doi.org/10.1021/jp071359d

[2]   Gouma, P.I., Mills, M.J. and Sandhage, K.H. (2000) Fabrication of Free-Standing Titaniabased Gas Sensors by the Oxidation of Metallic Titanium Foils. *Journal of the American Ceramic Society*, **83**, 1007-1009. http://dx.doi.org/10.1111/j.1151-2916.2000.tb01320.x

[3]   Kamegawa, T., Shimizu, Y. and Yamashita H., (2012) Superhydrophobic Surfaces with Photocatalytic Self-Cleaning Properties by Nanocomposite Coating of TiO_2 and Polytetrafluoroethylene. *Advanced Materials*, **24**, 3697-3700. http://dx.doi.org/10.1002/adma.201201037

[4]   Tan, B. and Wu, Y. (2006) Dye Sensitized Solar Cells Based on Anatase TiO_2 Nanoparticle/Nanowire Composites. *The Journal of Physical Chemistry B*, **110**, 15932-15938. http://dx.doi.org/10.1021/jp063972n

[5]   Venkatachalam, N., Palanichamy, M. and Murugesan, V. (2007) Sol-Gel Preparation and Characterization of Alkaline Earth Metal Doped Nano TiO_2: Efficient Photocatalytic Degradation of 4-Chlorophenol. *Journal of Molecular Catalysis A*, **273**, 177-185. http://dx.doi.org/10.1016/j.molcata.2007.03.077

[6]   Allam, N.K., Alamgir, F. and El-Sayed, M.A. (2010) Enhanced Photoassisted Water Electrolysis Using Vertically Oriented Anodically Fabricated Ti-Nb-Zr-O Mixed Oxide Nanotube Arrays. *ACS Nano*, **4**, 5819-5826. http://dx.doi.org/10.1021/nn101678n

[7]   Yu, J.G., Su, Y.R., Cheng, B. and Zhou, M.H. (2006) Effects of pH on the Microstructures and Photocatalytic Activity of Mesoporous Nanocrystalline Titania Powders Prepared via Hydrothermal Method. *Journal of Molecular Catalysis A*, **258**, 104-112. http://dx.doi.org/10.1016/j.molcata.2006.05.036

[8]   Borse, P.H., Kankate, L.S., Dassenoy, F., Vogel, W., Urban, J. and Kulkarni, S.K. (2002) Synthesis and Investigations of Rutile Phase Nanoparticles of TiO_2. *Journal of Materials Science*, **13**, 553-559. http://dx.doi.org/10.1023/A:1019677730981

[9]   Hong, S.S., Lee, M.S. and Lee, G.D. (2003) Photocatalytic Decomposition of *p*-Nitrophenol over Titanium Dioxide Prepared by Reverse Microemulsion Method Using Nonionic Surfactants with Different Hydrophilic Groups. *Reaction*

Kinetics and Catalysis Letters, **80**, 145-151. http://dx.doi.org/10.1023/A:1026096628817

[10] Keswani, R.K., Ghodke, H., Sarkar, D., Khilar, K.C. and Srinivasa, R.S. (2010) Room Temperature Synthesis of Titanium Dioxide Nanoparticles of Different Phases in Water in Oil Microemulsion. *Colloid Surface A*, **369**, 75-81. http://dx.doi.org/10.1016/j.colsurfa.2010.08.001

[11] Pavasupree, S., Suzuki, Y., Art, S.P. and Yoshikaw, S. (2005) Preparation and Characterization of Mesoporous MO_2 (M = Ti, Ce, Zr, and Hf) Nanopowders by a Modified Sol-Gel Method. *Ceramics International*, **31**, 959-963. http://dx.doi.org/10.1016/j.ceramint.2004.10.009

[12] Venkatchalam, N., Palanichamy, M. and Murugesan, V. (2007) Sol-Gel Preparation and Characterization of Nanosize TiO_2: Its Photocatalytic Performance. *Materials Chemistry and Physics*, **104**, 454-459. http://dx.doi.org/10.1016/j.matchemphys.2007.04.003

[13] Yu, J.G., Zhang, L.J., Cheng, B. and Su, Y.R. (2007) Hydrothermal Preparation and Photocatalytic Activity of Hierarchically Sponge-Like Macro-/Mesoporous Titania. *The Journal of Physical Chemistry C*, **111**, 10582-10589. http://dx.doi.org/10.1021/jp0707889

[14] Kobayashi, M., Petrykin, V., Tomita, K. and Kakihana, M. (2011) Hydrothermal Synthesis of Brookite-Type Titanium Dioxide with Snowflake-Like Nanostructures Using a Water-Soluble Citratoperoxotitanate Complex. *Journal of Crystal Growth*, **337**, 30-37. http://dx.doi.org/10.1016/j.jcrysgro.2011.09.046

[15] Yin, H., Ding, G.Q., Gao, B., Huang, F.Q., Xie, X.M. and Jiang, M.H. (2012) Synthesis of Ultrafine Titanium Dioxide Nanowires Using Hydrothermal Method. *Materials Research Bulletin*, **47**, 3124-3128. http://dx.doi.org/10.1016/j.materresbull.2012.08.022

[16] Lu, C.H., Wu, W.H. and Kale, R.B. (2008) Microemulsion-Mediated Hydrothermal Synthesis of Photocatalytic TiO_2 Powders. *Journal of Hazardous Materials*, **154**, 649-654. http://dx.doi.org/10.1016/j.jhazmat.2007.10.074

[17] Lisiecki, I. and Pileni, M.P. (1993) Synthesis of Copper Metallic Clusters Using Reverse Micelles as Microreactors. *Journal of the American Chemical Society*, **115**, 3887-3896. http://dx.doi.org/10.1021/ja00063a006

[18] Yu, J., Wang, G., Cheng, B. and Zhou, M. (2007) Effects of Hydrothermal Temperature and Time on the Photocatalytic Activity and Microstructures of Bimodal Mesoporous TiO_2 Powders. *Applied Catalysis B*, **69**, 171-180. http://dx.doi.org/10.1016/j.apcatb.2006.06.022

[19] Zhang, P. and L. Gao, (2003) Synthesis and Characterization of CdS Nanorods via Hydrothermal Microemulsion. *Langmuir*, **19**, 208-210. http://dx.doi.org/10.1021/la0206458

[20] Yu, J., Su, Y. and Cheng, B. (2007) Template-Free Fabrication and Enhanced Photocatalytic Activity of Hierarchical Macro-/Mesoporous Titania. *Advanced Functional Materials*, **17**, 1984-1990. http://dx.doi.org/10.1002/adfm.200600933

18

Effect of Zirconium Oxide Nano-Fillers Addition on the Flexural Strength, Fracture Toughness, and Hardness of Heat-Polymerized Acrylic Resin

Mohamed Ashour Ahmed[1], Mohamed I. Ebrahim[2,3]

[1]Department of Prosthodontics, Faculty of Dental Medicine, Taif University, Taif, KSA
[2]Department of Restorative Dentistry, Taif University, Taif, KSA
[3]Department of Dental Biomaterial, Al-Azhar University, Cairo, Egypt
Email: drmohashour99@gmail.com

Abstract

Purpose: The mechanical strength of polymethyl methacrylate (PMMA) remains far from ideal for maintaining the longevity of denture. The purpose of this study was to evaluate the effect of Zirconium oxide (ZrO_2) nanofillers powder with different concentration (1.5%, 3%, 5% and 7%) on the flexural strength, fracture toughness, and hardness of heat-polymerized acrylic resin. **Materials and methods:** Zirconium oxide powders with different concentrations (1.5%, 3%, 5% and 7%) were incorporated into heat-cure acrylic resin (PMMA) and processed with optimal condition (2.5:1 Powder/monomer ratio, conventional packing method and water bath curing for 2 hours at 95°C) to fabricate test specimens of PMMA of dimensions (50 × 30 × 30 mm) for the flexural strength, fracture toughness, and (50 × 30 × 30 mm) were fabricated for measuring hardness. PMMA without additives was prepared as a test control. Three types of mechanical tests; flexural strength, fracture toughness and hardness were carried out on the samples. The recorded values of flexural strength in (MPa), fracture toughness in ($MPa.m^{1/2}$), and hardness (VHN) were collected, tabulated and statistically analyzed. One way analysis of variance (ANOVA) and Tukey's tests were used for testing the significance between the means of tested groups which are statistically significant when the P value ≤ 0.05. **Results:** Addition of Zirconium oxide nanofillers to PMMA significantly increased the flexural strength, fracture toughness and hardness. **Conclusion:** These results indicate that Zirconium oxide nanofillers added to PMMA has a potential as a reliable denture base material with increased flexural strength, fracture toughness, and hardness. According to the results of the present study, the best mechanical properties were achieved by adding 7%wt ZrO_2 concentration.

Keywords

Zirconium Oxide Nano-Fillers, Flexural Strength, Fracture Toughness, Hardness, Heat-Polymerized Acrylic Resin

1. Introduction

Acrylic resin polymethayl methacrylate (PMMA) has been the most popular material for the construction of dentures for many decades as it has many advantages such as good aesthetics, accurate fit, stability in the oral environment, easy laboratory and clinical manipulation, and inexpensive equipment's [1]. Although it is the most widely used in dentistry for fabrication of denture bases, this material is still insufficient to fulfill the perfect mechanical requirements for dental applications. This issue was attributed mainly to its low fracture resistance and plaque accumulation [2] [3]. In a survey to compare ten types of denture base resins it was found that nearly 70% of dentures had broken within the first 3 years of their delivery [2]. In a study evaluating the denture fracture, it was reported that 33% of the repairs were due to debonded/detached teeth, 29% of the repairs were because of midline fractures which were more commonly seen in the upper dentures and the rest were other types of fracture.

In another study the authors reported that the Mandibular partial denture was the most commonly needing repair [4]. So, the measuring of mechanical properties of the denture base materials is important to evaluate the effect of adding different strengthening materials [5].

Undoubtedly that, many trails were made to enhance mechanical properties of denture base materials either by adding chemical solutions such as a polyfunctional cross linking agent (polyethylene glycol dimethacrylate) [6] or by incorporating a rubber phase [7], metal fram [8], metal oxides [9], or fibers [10]. Despite these efforts to improve the fracture resistance of PMMA few have obtained promising results [11] [12]. The reinforcement of polymers used in dentistry with metal-composite systems has been a prime interest [12].

Zirconium oxidenano-particles powder has been selected to improve the properties of PMMA, as a bio-compatible material that possesses high fracture resistance, and to improve fracture toughness of ceramics by developing a new generation of ceramic-matrix composites [13] [14].

Since only limited amount of data regarding the effect of metal oxides on heat-cured PMMA are available in the literature, the purpose of this study was to investigate the influence of addition of metal oxides [zirconium oxide powder (ZrO_2)] on some mechanical properties of heat cured PMMA.

2. Materials and Methods

An *in vitro* study was conducted to evaluate the effect of Zirconium oxide nanofillers powder (ZrO_2) (5 - 15 nm) with different concentration (1.5%, 3%, 5% and 7%) on the flexural strength, fracture toughness, and hardness of heat-polymerized acrylic resin.

One type of heat-cure acrylic resin (PMMA) was used as the control (Acrostone (A), Anglo-Egyptian Company. Hegaz, Cairo, Egypt, Batch No.505/04), Zirconium oxide nanofillers powder (ZrO_2) (Sigma-Aldrich Germany, Trade 544,760) with different concentrations (1.5%, 3%, 5% and 7%) was added into heat-cure acrylic resin (PMMA) and processed with optimal condition (2.5:1 Powder/monomer ratio, conventional packing method and water bath curing for 2 hours at 95°C) 150 bar shapes specimenswere prepared to be used in this study. 50 specimenswere used for each test [flexural strength (group A), fracture toughness (group B), and hardness (group C)].

Grouping of the specimens:

Each group was further divided into five subgroups (1, 2, 3, 4 and 5) of 10 specimens each as shown in **Table 1**.

2.1. Flexural Strength

Specimens were tested by 3-point bend test on Lloyd universal testing machine (model LRX plus II, Fareham, England) at a cross head speed of 1 mm/min. For the 3 point bend test, a fixture was fabricated with the

Table 1. Classification and grouping of the specimens.

Groups	Subgroups	Description	No. of Specimens
	Group A1	Heat-cure acrylic resin (PMMA) without additives as control.	10 specimens
	Group A2	PMMA with 1.5% zirconium oxide nanofillers powder (ZrO_2).	10 specimens
Group A	Group A3	PMMA with 3% ZrO_2.	10 specimens
	Group A4	PMMA with 5% ZrO_2.	10 specimens
	Group A5	PMMA with 7% ZrO_2.	10 specimens
	Group B1	PMMA without additives as control.	10 specimens
	Group B2	PMMA with 1.5% ZrO_2.	10 specimens
Group B	Group B3	PMMA with 3% ZrO_2.	10 specimens
	Group B4	PMMA with 5% ZrO_2.	10 specimens
	Group B5	PMMA with 7% ZrO_2.	10 specimens
	Group C1	PMMA without additives as control.	10 specimens
	Group C2	PMMA with 1.5% ZrO_2.	10 specimens
Group C	Group C3	PMMA with 3% ZrO_2.	10 specimens
	Group C4	PMMA with 5% ZrO_2.	10 specimens
	Group C5	PMMA with 7% ZrO_2.	10 specimens
		Total	150 specimens

dimensions of $50 \times 30 \times 30$ mm [15]. On top of the fixture two plates were welded at a distance of 15 mm from the center on either side. A customized "T" shaped stress applicator rod with the dimension of 80×20 mm was fabricated, by which stress can be applied in the center of the specimen. The specimen was placed on the rollers in such a way that the center of the specimen coincided with the center of the distance between the two rollers. This whole unit was mounted on the lower jaw of the universal testing machine and the stress applicator rod was fixed on the upper jaw. A load was applied with "T" shaped rod on the center of the specimen until fracture occurred and peak force (F) values were recorded at this point in Newton [16].

The maximum force (F) necessary to produce fracture of the specimens was recorded in Newton. The flexural strength Q was calculated in (MPa) for all specimens from the "Equation (1)":

$$Q = \frac{3FI}{2BH^2} \tag{1}$$

"In this formula, "F" is the maximum load or force which is applied to the center of the specimen to fracture it (N); "I" is the distance between the two rests on the surface under the tensile force (mm); "B" is the width (mm) and "H" is the height of the specimen between the surfaces under the tensile and compressive forces (mm)."

2.2. Fracture Toughness

For fracture toughness testing, specimens were fabricated with the dimensions of $50 \times 30 \times 30$ mm [15]. After all specimens were stored in distilled water at 37°C for 24 hours, a notch was made in the middle of each specimen on one edge with 2.5 mm lengths using sand paper disk. Fracture toughness tests were performed on Lloyd universal testing machine (model LRX plus II, Fareham, England) with a cross-head speed of 1 mm/min, and peak load to fracture was recorded. The recorded data were used to determine the fracture toughness (K_{Ic}) in MPa.m$^{1/2}$ according to the "Equation (2)" [17]:

$$K_{ic} = pc/bw^{1/2} \cdot F\left(a/w\right) \tag{2}$$

where *pc* is the maximum load (kN) prior to crack advance, *b* is specimen thickness (cm), *w* is the width of the specimen (cm), *a* is crack length (cm) and *F* is calculated from the following Equation (3):

$$F_{(a/w)} = \frac{\left(2 + a/w\right)\left(0.886 + a/w - 13.32\,a^2/w^2 + a^3/w^3 - 5.6\,a^4/w^4\right)}{\left(1 - a/w\right)^{3/2}} \tag{3}$$

2.3. Hardness

For Hardness testing, specimens were fabricated with the dimensions of $15 \times 15 \times 5$ mm [18]. Surface hardness was determined using Digital Display Vickers Microhardness Tester (Model HVS-50, Laizhou Huayin Testing Instrument Co., Ltd. China) which is suitable for acrylic resin material. With a Vickers diamond indenter and a $20\times$ objective lens. A load of 20 gram was applied to the surface of the specimens for 15 sec. five indentations were equally placed over a specimen and not closer than 1 mm to the adjacent indentations or to the margin of the specimens were made on the surface of each specimen. The diagonal length of the indentations was measured by built in scaled microscope.

Surface microhardness calculation:

Vickers microhardness was obtained using the following Equation (4):

$$VHN = 1.854\,L/d^2 \tag{4}$$

where:

VHN: Vickers hardness in Kg/mm^2.

L: Load in Kg.

d: Length of the diagonals in mm.

The recorded values of flexural strength, fracture toughness, and hardness were collected, tabulated and statistically analyzed. One way analysis of variance (ANOVA) and Tukey's tests were used for testing the significance between the means of tested groups which are statistically significant when the P value ≤ 0.05.

3. Results

3.1. Flexural Strength

Both **Table 2** and **Figure 1** show a comparison between mean flexural strength in (MPa) of the tested groups of PMMA. ANOVA test showed statistically significant difference between all groups.

PMMA specimen with 7% zirconium oxide nanofillers (ZrO_2) (group E) showed significantly highest mean flexural strength followed by PMMA specimen with 5% (ZrO_2, group D) followed by PMMA specimen with 3% (ZrO_2, group C) then PMMA specimen with 1.5% (ZrO_2, group B). There were significant differences ($P < 0.05$) between studied groups. PMMA specimen without any additives (control group) showed significantly lowest mean flexural strength.

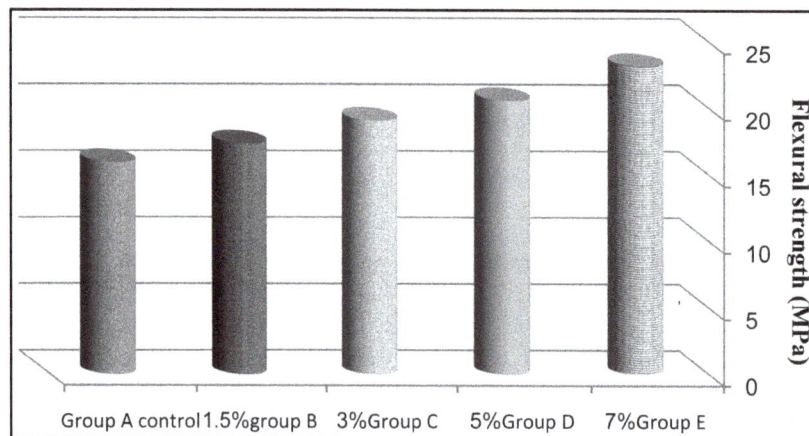

Figure 1. Bar chart of mean flexural strength (MPa) of the tested groups of PMMA.

Table 2. Comparison between mean flexural strength (MPa) of the tested groups of PMMA.

Group A1 Control group		Group A2 (1.5% ZrO_2)		Group A3 (3% ZrO_2)		Group A4 (5% ZrO_2)		Group A5 (7% ZrO_2)		P-value
Mean	SD	Mean	SD	Mean	SD	Mean	SD	Mean	SD	
85.54^e	1.145	95.18^d	4.46	105.24^c	3.63	116.04^b	3.028	123.72^a	1.96	0.000*

*Significant at $P \leq 0.05$, Means with different letters are significantly different according to Tukey's test.

3.2. Fracture Toughness

The tensile strength data showed there was significant improvement in the tested groups which were reinforced with zirconium oxide nanofillers (ZrO_2) (**Table 3** and **Figure 2**).

There was significant increase in the fracture toughness for groups reinforced with (1.5%, 3%, 5% and 7%) ZrO_2 when compared with control group.

3.3. Hardness

Both **Table 4** and **Figure 3** show the mean hardness of tested groups. All specimens showed hardness mean values higher than that control group. PMMA specimen with 7% zirconium oxide nanofillers (ZrO_2) (group E) showed significantly highest mean hardness followed by PMMA specimen with 5% (ZrO_2) (group D) followed by PMMA specimen with 3% (ZrO_2) (group C) then PMMA specimen with 1.5% (ZrO_2) (group B). There were significant differences ($P < 0.05$) between studies groups. PMMA specimen without any additives (control group) showed significantly lowest mean hardness.

4. Discussion

We principally aimed to assess possible improvements in the mechanical properties of PMMA, in particular, the FS, fracture toughness, and hardness, through incorporating of ZrO_2 Nano particles. There are three ways to improve the mechanical properties of PMMA: replacing PMMA with an alternative material; chemically modifying it; and reinforcing the PMMA with other materials [19] [20].

Addition of Zirconia Nano fillers to acrylic resin was found to improve mechanical properties. In addition to that ZrO_2 was used because it has excellent biocompatibility and white color which less likely to alter esthetic. The Nano-filler particles were used in this study as it yield a better dispersion, eliminate aggregation and improve its compatibility with organic polymer [21] [22]. Proper percentage range of zirconiumoxide Nano-fillers (Percentages of 1.5% - 7% by weight) was selected because percentages above 7% was leads to massive changes occurred in the color of acrylic [23].

Fractures in an acrylic denture base are a common clinical problem. Flexural strength of denture base resin was measured in this study because it is considered the primary mode of clinical failure [24]. Fatigue failure does not require strong biting forces as relatively small stresses caused by mastication over a period of time can eventually lead to the formation of a small crack, which propagates through the denture and results in a fracture. The maximal biting forces of a patient can reach up to 700 N, but these values are reduced (100 - 150 N) [25] with the removal of dentures. Denture fractures are essentially due to stress concentration and increased flexing [26]. Many authors found that the fracture toughness seems to be a suitable measurement to demonstrate the effects of resin modifications [27].

Hardness of the polymerized resin has been found to be sensitive to the residual monomer content in the resin material. Moreover, hardness measurement have been successfully used as an indirect method of evaluating polymerization depth of resin-based composite materials [28] and the degree of conversion of conventional heat-polymerizing and self-curing acrylic resins. In addition, hardness has been used to predict the wear resistance of dental materials [29].

The Results of the present study demonstrated a significant increase in flexural strength, fracture toughness, and hardness as the percentage of ZrO_2 fillers increased. This improvement in mechanical properties could be attributed to the high interfacial shear strength between the nanofiller and resin matrix as a result of formation of cross-links or supra molecular bonding which cover or shield the nanofillers which in turn prevent propagation of crack, also complete wetting of the nanofillers by resin lead to increase in flexural strength, fracture toughness, and hardness as volume of filler increased [30].

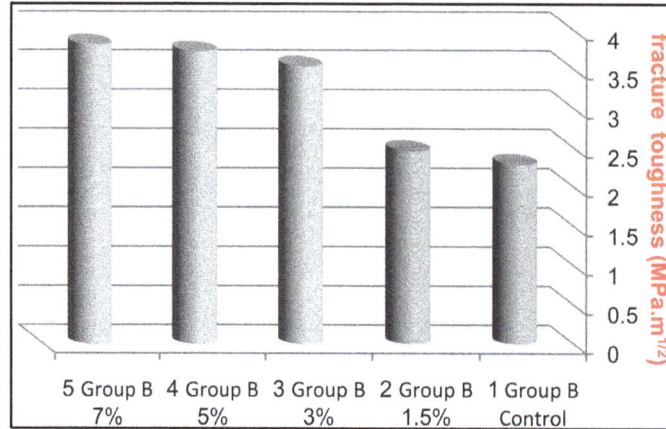

Figure 2. Bar chart of mean fracture toughness (MPa.m$^{1/2}$) of the tested groups of PMMA.

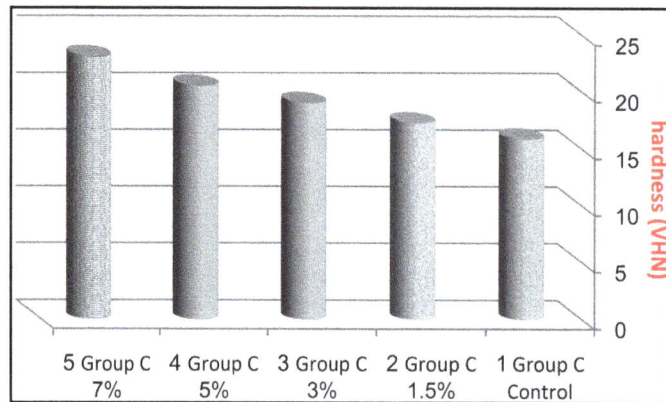

Figure 3. Bar chart of mean hardness (VHN) of the tested groups of PMMA.

Table 3. Comparison between mean fracture toughness (MPa.m$^{1/2}$) of the tested groups of PMMA.

Group B1 Control group		Group B2 (1.5% ZrO$_2$)		Group B3 (3% ZrO$_2$)		Group B4 (5% ZrO$_2$)		Group B5 (7% ZrO$_2$)		P-value
Mean	SD	Mean	SD	Mean	SD	Mean	SD	Mean	SD	
2.30[b]	0.158	2.47[b]	0.37	3.54[a]	0.08	3.73[a]	0.14	3.82[a]	0.16	0.000*

*Significant at P ≤ 0.05, Means with different letters are significantly different according to Tukey's test.

Table 4. Comparison between mean hardness (VHN) of the tested groups of PMMA.

Group C1 Control group		Group C2 (1.5% ZrO$_2$)		Group C3 (3% ZrO$_2$)		Group C4 (5% ZrO$_2$)		Group C5 (7% ZrO$_2$)		P-value
Mean	SD	Mean	SD	Mean	SD	Mean	SD	Mean	SD	
15.95[c]	0.96	17.35[c]	0.63	19.10[b]	1.07	20.60[b]	0.56	23.19[a]	1.05	0.000*

*Significant at P ≤ 0.05, Means with different letters are significantly different according to Tukey's test.

It is noted also from these results that concentration of ZrO$_2$ (3%wt) lead to the maximum value of fracture toughness. There is no significant improvement in fracture toughness values of the modified acrylic resin at the concentrations of ZrO$_2$ above that limit (5%wt and 7%wt). It is probably due to complete saturation of the polymer matrix with the ZrO$_2$ particles [31].

Improvement of hardness with the increase in concentration of ZrO_2 nanofillers may have be due to inherent characteristics of the ZrO_2 particles. ZrO_2 possesses strong ionic interatomic bonding, giving rise to its desirable material characteristics, that is, hardness and strength.

The results of this study are in good agreement with the findings reported by others who concluded that reinforcement of ceramics, dental restorative resins as well as acrylic resin with Zirconia nanoparticles could exhibit improvement in their mechanical properties [31]-[33]. The increase of mechanical properties was due to good bonding between nanofillers and resin matrix [34] [35].

5. Conclusions

Within the limitation of this study, we can conclude that:

Addition of zirconium oxide nanofillers to PMMA increased the flexural strength, fracture toughness, and hardness of heat polymerized acrylic resin. According to the results of the present study, the best result was got when using the concentration of 7%wt.

Further studies are needed to investigate its effect on other mechanical and physical properties with different concentrations.

References

[1]   Nejatian, T., Johnson, A. and Van Noort, R. (2006) Reinforcement of Denture Base Resin. *Advances in Science and Technology*, **49**, 124-129. http://dx.doi.org/10.4028/www.scientific.net/AST.49.124

[2]   Darbar, U.R., Huggett, R. and Harrison, A. (1994) Denture Fracture—A Survey. *British Dental Journal*, **176**, 342-345. http://dx.doi.org/10.1038/sj.bdj.4808449

[3]   John, J., Gangadhar, S.A. and Shah, I. (2001) Flexural Strength of Heat-Polymerized Polymethyl Methacrylate Denture Resin Reinforced with Glass, Aramid, or Nylon Fibers. *Journal of Prosthetic Dentistry*, **86**, 424-427. http://dx.doi.org/10.1067/mpr.2001.118564

[4]   El-Sheikh, A.M. (2006) SBA-Z: Causes of Denture Fracture: A Survey. *Saudi Dental Journal*, **18**, 149-154.

[5]   Vallittu, P.K., Alakuijala, P., Lassila, V.P., *et al.* (1994) *In Vitro* Fatigue Fracture of an Acrylic Resin-Based Partial Denture: An Exploratory Study. *The Journal of Prosthetic Dentistry*, **72**, 289-295. http://dx.doi.org/10.1016/0022-3913(94)90342-5

[6]   Kanie, T., Fujii, K., Arikawa, H., *et al.* (2000) Flexural Properties and Impact Strength of Denture Base Polymer Reinforced with Woven Glass Fibers. *Dental Materials*, **16**, 150-158. http://dx.doi.org/10.1016/S0109-5641(99)00097-4

[7]   Knott, N.J. (1989) The Durability of Acrylic Complete Denture Bases in Practice. *Quintessence International*, **20**, 341-343.

[8]   Balch, J.H., Smith, P.D., Marin, M.A., *et al.* (2013) Reinforcement of a Mandibular Complete Denture with Internal Metal Framework. *Journal of Prosthetic Dentistry*, **109**, 202-205. http://dx.doi.org/10.1016/S0022-3913(13)60045-1

[9]   Venkat,, R., Gopichander, N. and Vasantakumar, M. (2013) Comprehensive Analysis of Repair/Reinforcement Materials for Polymethyl Methacrylate Denture Bases: Mechanical and Dimensional Stability Characteristics. *The Journal of Indian Prosthodontic Society*, **13**, 439-449.

[10]   Xu, J., Li, Y., Yu, T., *et al.* (2013) Reinforcement of Denture Base Resin with Short Vegetable Fiber. *Dental Materials*, **29**, 1273-1279. http://dx.doi.org/10.1016/j.dental.2013.09.013

[11]   Franklin, P., Wood, D.J. and Bubb, N.L. (2005) Reinforcement of Poly(methyl methacrylate) Denture Base with Glass Flake. *Dental Materials*, **21**, 365-370. http://dx.doi.org/10.1016/j.dental.2004.07.002

[12]   Asar, N.V., Albayrak, H., Korkmaz, T., *et al.* (2013) Influence of Various Metal Oxides on Mechanical and Physical Properties of Heat-Cured Polymethyl Methacrylate Denture. *The Journal of Advanced Prosthodontics*, **5**, 241-247. http://dx.doi.org/10.4047/jap.2013.5.3.241

[13]   Mohamed, A. and Mohamed, A., Fallal, A. and Hawary, Y. (2007) Effect of Zirconium Oxide Reinforcement on Epithelial Oral Mucosa, Immunoglobulin and Surface Roughness of Complete Acrylic Heat-Cured Denture. *Egyptian Dental Journal*, **53**, 941-946.

[14]   Skukla, S. and Seal, S. (2003) Phase Stabilization in Nanocrystalline Zirconia. *Reviews on Advanced Materials Science*, **5**, 117-120.

[15]   Kamble, V.D., Parkhedkar, R.D. and Mowade, T.K. (2012) The Effect of Different Fiber Reinforcements on Flexural Strength of Provisional Restorative Resins: An *In-Vitro* Study. *The Journal of Advanced Prosthodontics*, **4**, 1-6. http://dx.doi.org/10.4047/jap.2012.4.1.1

[16] Dagar, S., Pakhan, A. and Tunkiwala, A. (2005) An *in Vitro* Evaluation of Flexural Strength of Direct and Indirect Provisionalization Materials. *Journal of Indian Prosthodontic Society*, **5**, 132-135. http://dx.doi.org/10.4103/0972-4052.17105

[17] Yilmaz, C. and Korkmaz, T. (2007) The Reinforcement Effect of Nano and Microfillers on Fracture Toughness of Two Provisional Resin Materials. *Materials and Design*, **28**, 2063-2070. http://dx.doi.org/10.1016/j.matdes.2006.05.029

[18] Osada, T., Ishimoto, T., Aoki, T., Suzuki, Y., Ohkubo, C. and Hosoi, T. (2010) Bending Strengths and Hardness of Autopolymerized Acrylic Resin. *International Chinese Journal of Dentistry*, **10**, 1-5.

[19] Kim, S.H. and Watts, D.C. (2004) The Effect of Reinforcement with Woven E-Glass Fibers on the Impact Strength of Complete Dentures Fabricated with High-Impact Acrylic Resin. *The Journal of Prosthetic Dentistry*, **91**, 274-280. http://dx.doi.org/10.1016/j.prosdent.2003.12.023

[20] Jagger, D.C., Harrison, A. and Jandt, K.D. (1999) The Reinforcement of Dentures. *Journal of Oral Rehabilitation*, **26**, 185-194. http://dx.doi.org/10.1046/j.1365-2842.1999.00375.x

[21] Suna, L.Y., Gibson, R.F., Gordaninejad, F. and Suhr, J. (2009) Energy Absorption Capability of Nanocomposites: A Review. *Composites Science and Technology*, **69**, 2392-2409. http://dx.doi.org/10.1016/j.compscitech.2009.06.020

[22] Ihab, N.S., Hasanayn, K.A. and Ali, N.A. (2012) Assessment of Zirconium Oxide Nano-Fillers Incorporation and Si-lanation on Impact, Tensile Strength and Color Alteration of Heat Polymerized Acrylic Resin. *Journal of Baghdad College of Dentistry*, **24**, 36-42.

[23] Shi, J.M., Bao, Y.Z., Huang, Z.M. and Weng, Z.X. (2004) Preparation of Poly (Methyl Methacrylate)/Nanometer Calcium Carbonate Composite by *in Situ* Emulsion Polymerization. *Journal of Zhejiang University Science A*, **5**, 709-713. http://dx.doi.org/10.1631/jzus.2004.0709

[24] Chitchumnong, P., Brooks, S.C. and Stafford, G.D. (1989) Comparison of Three- and Four-Point Flexural Strength Testing of Denture-Base Polymers. *Dental Materials*, **5**, 2-5. http://dx.doi.org/10.1016/0109-5641(89)90082-1

[25] Narva, K.K., Lassila, L.V.J. and Vallittu, P.K. (2005) Flexural Fatigue of Denture Base Polymer with Fiber-Reinforced Composite Reinforcement. *Composites Part A*, **36**, 1275-1281. http://dx.doi.org/10.1016/j.compositesa.2005.01.025

[26] Franklin, P., Wood, D.J. and Bubb, N.L. (2005) Reinforcement of Poly(Methyl Methacrylate) Denture Base with Glass Flake. *Dental Materials*, **21**, 365-370. http://dx.doi.org/10.1016/j.dental.2004.07.002

[27] Zappini, G., Kammann, A. and Wachter, W. (2003) Comparison of Fracture Tests of Denture Base Materials. *Journal of Prosthetic Dentistry*, **90**, 578-585. http://dx.doi.org/10.1016/j.prosdent.2003.09.008

[28] Dunn, W.J. and Bush, A.C. (2002) A Comparison of Polymerization by Light-Emitting Diode and Halogen-Based Light-Curing Units. *Journal of the American Dental Association*, **133**, 335-341. http://dx.doi.org/10.14219/jada.archive.2002.0173

[29] Lee, S.Y., Lai, Y.L. and Hsu, T.S. (2002) Influence of Polymerization Conditions on Monomer Elution and Micro-hardness of Autopolymerized Polymethyl Methacrylate Resin. *European Journal of Oral Sciences*, **110**, 179-183. http://dx.doi.org/10.1034/j.1600-0722.2002.11232.x

[30] Ayad, N.M., Badawi, M.F. and Fatah, A.A. (2008) Effect of Reinforcement of High-Impact Acrylic Resin with Zirconia on Some Physical and Mechanical Properties. *Revista de Clínica e Pesquisa Odontológica*, **4**, 145-151.

[31] Vojdani, M., Bagheri, R. and Khaledi, A.A.R. (2012) Effects of Aluminum Oxide Addition on the Flexural Strength, Surface Hardness, and Roughness of Heat-Polymerized Acrylic Resin. *Journal of Dental Sciences*, **7**, 238-244. http://dx.doi.org/10.1016/j.jds.2012.05.008

[32] Giordano II, R. (2000) A Comparison of All-Ceramic Restorative Systems: Part 2. *General Dentistry*, **48**, 38-40, 43-45.

[33] Furman, B., Rawls, H.R., Wellinghoff, S., Dixon, H., Lankford, J. and Nicolella, D. (2000) Metal-Oxide Nanoparticles for the Reinforcement of Dental Restorative Resins. *Critical Reviews in Biomedical Engineering*, **28**, 439-443. http://dx.doi.org/10.1615/CritRevBiomedEng.v28.i34.150

[34] Sun, L.Y., Gibson, R.F., Gordaninejad, F. and Suhr, J. (2009) Energy Absorption Capability of Nanocomposites: A Review. *Composites Science and Technology*, **69**, 2392-2409. http://dx.doi.org/10.1016/j.compscitech.2009.06.020

[35] Tinschert, J., Natt, G., Mautsch, W., Augthun, M. and Spiekermann, H. (2001) Fracture Resistance of Lithium Disilicate-, Alumina-, and Zirconia-Based Three-Unit Fixed Partial Dentures: A Laboratory Study. *International Journal of Prosthodontics*, **14**, 231-238.

19

First-Principles Calculations of the Structural, Mechanical and Thermodynamics Properties of Cubic Zirconia

Ibrahim D. Muhammad[1]*, Mokhtar Awang[1], Othman Mamat[1], Zilati Bt Shaari[2]

[1]Mechanical Engineering Department, Universiti Teknologi Petronas, Seri Iskandar, Malaysia
[2]Chemical Engineering Department, Universiti Teknologi Petronas, Seri Iskandar, Malaysia
Email: *ibrahimuhd@gmail.com

Abstract

The structural, mechanical and thermodynamics properties of cubic zirconium oxide (cZrO$_2$) were investigated in this study using *ab initio* or first-principles calculations. Density functional theory was used to optimize the crystal structure of cZrO$_2$ and thereafter, simulations were conducted to predict the lattice parameters and elastic constants. The Zr-O bond distance was calculated as 2.1763 Å with unit cell density of 6.4179 g/cm^3. The data obtained were used to determine Young's modulus, bulk modulus, Poisson's ratio and hardness of cZrO$_2$ as 545.12 GPa, 136.464 GPa, 0.1898 and 12.663 (H$_v$) respectively. The result indicates that cZrO$_2$ is mechanically stable with thermodynamics properties of a refractory material having potential for structural and catalytic applications in various forms as a nanomaterial.

Keywords

Cubic Zirconium Oxide, First-Principles Calculation, CASTEP, Elastic Constants

1. Introduction

Recently, zirconium oxide in cubic polymorph has attracted interest due to its mechanical and thermal properties. These interests have resulted in the use of cZrO$_2$ to various applications such as catalysts, fuel cells, oxygen sensors and others [1]-[4]. Thus cZrO$_2$ has been synthesized at nanoscale in various forms. The sol-gel method

---

*Corresponding author.

was used to produce ZrO_2 nanoparticles with fairly uniform dimension ranging from 50 nm to 90 nm [5]. ZrO_2 nanosheets with thickness in the range of 3.2 - 4.2 nm were produced through bottom-up synthesis by impregnation of graphene oxide in cyclohexane containing Zr-based alkoxides [6]. Also, ZrO_2 nanotubes have been prepared via several methods such as direct anodization, template-assisted depositions and hydrothermal treatments [2]; the nanotubes obtained have different inner and outer diameters, thicknesses and lengths.

For efficient and cost-effective applications of $cZrO_2$, its structural and mechanical properties are required at molecular and atomic scale. But due to complexity and high cost of equipment, only limited experiments have been conducted to characterize required properties of $cZrO_2$ at nanoscale and/or atomic level [2] [4]. This led to computational modeling and simulation of various properties of $cZrO_2$ based on atomistic and continuum approaches [1] [2]. However, most of the simulations already conducted are limited to properties such as electronic structures, phonon dispersion, surface adsorption and diffusion [7]-[13] with limited emphasis on structural and mechanical properties.

Therefore, in this study the first-principles calculations are utilized to numerically predict the structural characteristics, mechanical properties and thermodynamic properties of $cZrO_2$. The values obtained are used to analyze the structural stability of $cZrO_2$ and compared available data for further simulation(s).

2. Computational Methods

The 3D structure of $cZrO_2$ was modeled from available data [2] using the *surface builder* tools of the *Material Studio* software [14]. The crystal structure of$cZrO_2$ was optimized geometrically in order to obtain initial lattice parameters and density based on density function theory (DFT) as implemented in Cambridge Sequential Total Energy Package (CASTEP) code [15]. In this calculation, the Generalized Gradient Approximation (GGA) having Burke-ErnZerhof Potential (PBE) for solids was used [13]. To obtain accurate structures, the calculations were conducted in the irreducible Brillouin zone with $8 \times 8 \times 8$ k point mesh Monk horst-Pack scheme [16]. In order to obtain plane wave expansions, a kinetic energy cut-off value of 380 eV was used. Thereafter the Broyden- Fletcher-Goldfarb-Shannon (BFGS) optimization method was used with fixed basis quality to obtain the symmetric crystal structure of $cZrO_2$. During the optimization process, the total energy was designed to converge to 5×10^{-6} eV and the force per atom diminished to 0.002 eV/Å. The optimized structure of $cZrO_2$ was then simulated to obtain required mechanical and thermodynamics properties.

3. Results and Discussion

3.1. Crystal Structure

The bulk crystal structure of $cZrO_2$ belongs to the *Fm*-3m space [4] [9] and is modeled in various forms as seen in **Figure 1**. The geometric structure has influence on the properties of the material even at nanoscale [2]. After optimization, all the lattice parameters were recorded and compared with values obtained from other simulations and experiments as summarized in **Table 1**. It was observed that the deviation of the lattice parameters is less than 3% from the experimental values, which may be due to the approximation method used during optimization.

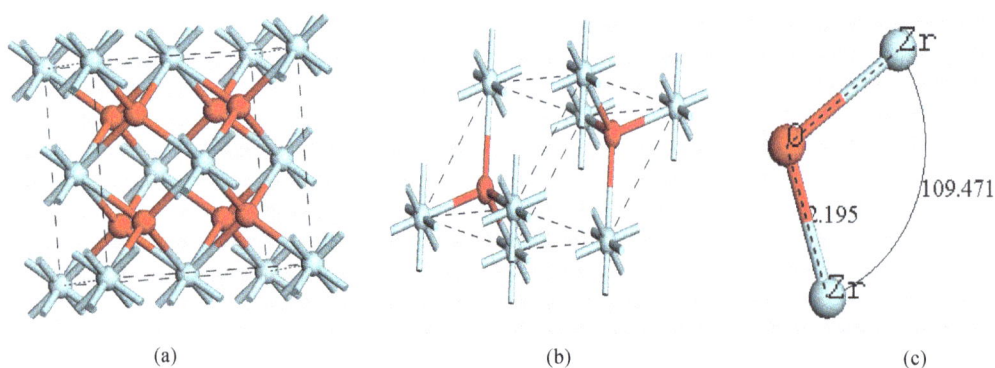

(a)   (b)   (c)

Figure 1. Crystal structure of $cZrO_2$ in various forms: (a) Unit Cell; (b) Primitive Cell; (c) Zr-O Bond (O_2 and Zr atoms are represented by red and light blue balls respectively).

Table 1. Lattice parameters of optimized cZrO$_2$.

Parameters	Simulations		Experimental
	Present Study	Other(s)	
a = b = c (Å)	3.55381	3.58031 [4]	3.6012 [17]
Bond Length (Zr-O), Å	2.1763	2.19537 [2]	2.205 [5]
Volume (Å^3)	31.8808	32.5850 [3]	32.723 [17]
Density (g/cm^3)	6.41791	6.081 [18]	6.151 [18]

3.2. Mechanical Properties

The mechanical properties of cZrO$_2$ are based on the elastic constants of the crystal which indicates response to external forces. There are six components of stress and a corresponding six components of strain for the general 3-D case. Thus, Hooke's law may be expressed as:

$$\sigma_i = C_{ij}\varepsilon_j \tag{1}$$

$$\varepsilon_i = S_{ij}\sigma_j \tag{2}$$

where C = stiffness or elastic constant, S = compliance, σ = stress and ε = strain. In matrix format, the stress-strain relation showing the 36 (6 × 6) independent components of stiffness can be represented as [3]:

$$\begin{pmatrix} \sigma_1 \\ \sigma_2 \\ \sigma_3 \\ \sigma_4 \\ \sigma_5 \\ \sigma_6 \end{pmatrix} = \begin{pmatrix} C_{11} & C_{12} & C_{13} & C_{14} & C_{15} & C_{16} \\ C_{21} & C_{22} & C_{23} & C_{24} & C_{25} & C_{26} \\ C_{31} & C_{32} & C_{33} & C_{34} & C_{35} & C_{36} \\ C_{41} & C_{42} & C_{43} & C_{44} & C_{45} & C_{46} \\ C_{51} & C_{52} & C_{53} & C_{54} & C_{55} & C_{56} \\ C_{61} & C_{62} & C_{63} & C_{64} & C_{65} & C_{66} \end{pmatrix} \begin{pmatrix} \varepsilon_1 \\ \varepsilon_2 \\ \varepsilon_3 \\ \varepsilon_4 \\ \varepsilon_5 \\ \varepsilon_6 \end{pmatrix} \tag{3}$$

The number of independent elastic constants in cZrO$_2$ is three (3), *i.e.* C_{11}, C_{12} and C_{44} which are computed as 596.33, 137.04 and 74.34 GPa respectively. Also the bulk modulus (B) was calculated as 290.134 GPa using the Voigt model, which is comparable to experimental value [17].

With reference to Born-Huang's lattice dynamic theory [19], the mechanical stability of a cubic system is based on the conditions expressed in Equations (4) and (5).

$$C_{11} > 0, C_{44} > 0, C_{11} - C_{12} > 0, C_{11} + 2C_{12} > 0 \tag{4}$$

$$C_{12} < B < C_{11} \tag{5}$$

Based on conditions outlined in Equations (4) and (5), the structure is mechanically stable due to lattice parameters and orientation of the crystal. The stability will make polycrystalline cZrO$_2$ less vulnerable to generating micro cracks [19].

For cZrO$_2$, computed Young's modulus (E) is 545.12 GPa, shear modulus (G) based on Voigh model is 136.464 GPa and the Lame modulus based on Reus model is 222.195 GPa. The G/B ratio indicates the ductility or brittleness of a material [19] and was found out to be 0.469, which indicates that cZrO$_2$ ionic and brittle. This is because for covalent and ionic materials, the typical relations between bulk modulus and shear modulus are as $G \approx 1.1B$ and $G \approx 0.6B$, respectively [20]. The Poisson's ratio (υ) of cZrO$_2$ vary from 0.14 to 0.189, which indicates less metallic and ionic character in the Zr-O bond.

In crystals, hardness quantifies the resistance to deformation and may be predicted using microscopic models [17]. The hardness (Hv) of cZrO$_2$ is determined using a semi-empirical equation defined as follows [21] [22]:

$$H_v = 0.92K^{1.137}G^{0.708}, K = G/B \tag{6}$$

The obtained hardness (Vickers number) for cZrO$_2$ is 12.663, which is consistent with the chemical bonding and elastic modulus as analyzed above and in similar other studies [19]-[22].

3.3. Thermodynamic Properties

The thermodynamic properties of a solid have direct relevance to its phonon characteristics as it indicates the quantum of elastic strain energy [19]. Hence, the phonon dispersion curve of a material elucidates the thermal properties based on the concept of lattice vibrations and interpretation of lattice dynamics [15]. Using the method of linear response with 0.05 1/Å q-vector grid spacing, the phonon dispersion curves (PDC) for $cZrO_2$ was generated as shown in **Figure 2**. The lower section of G-X direction indicates higher longitudinal energy while most of the branches in the X-K direction converge as non-degenerate, thus leading to the expected nine branches for a typical cubic structure [19].

With the aid of data from the phonon dispersion curves (PDC), quasi-harmonic approximation was used to evaluate some temperature dependent properties of $cZrO_2$ such as entropy, enthalpy, free energy and Debye temperature as illustrated in **Figure 3** and **Figure 4**. From **Figure 3**, the zero-point energy was computed to be 0.20385 eV and the heat capacity per unit cell had a maximum value of 16.8957 cal/K (0.07069 kJ/K) at 992.25 K. Debye temperature gives an approximation for the low temperature heat capacity of insulating crystalline solids. The maximum Debye temperature for $cZrO_2$ was found to be 946 K which is similar to 963 K for Y_2O_3 stabilized $cZrO_2$ determined experimentally using neutron powder diffraction [23]. All the computed thermodynamics properties of $cZrO_2$ indicates it has less thermal conductivity despite having high ionic conductivity thus confirming its major characteristic as a refractory material.

4. Conclusion

Using the first principles calculations, the properties of $cZrO_2$ were investigated. The calculated lattice parameters showed conformity with available experimental data. The computed properties based on elastic constants

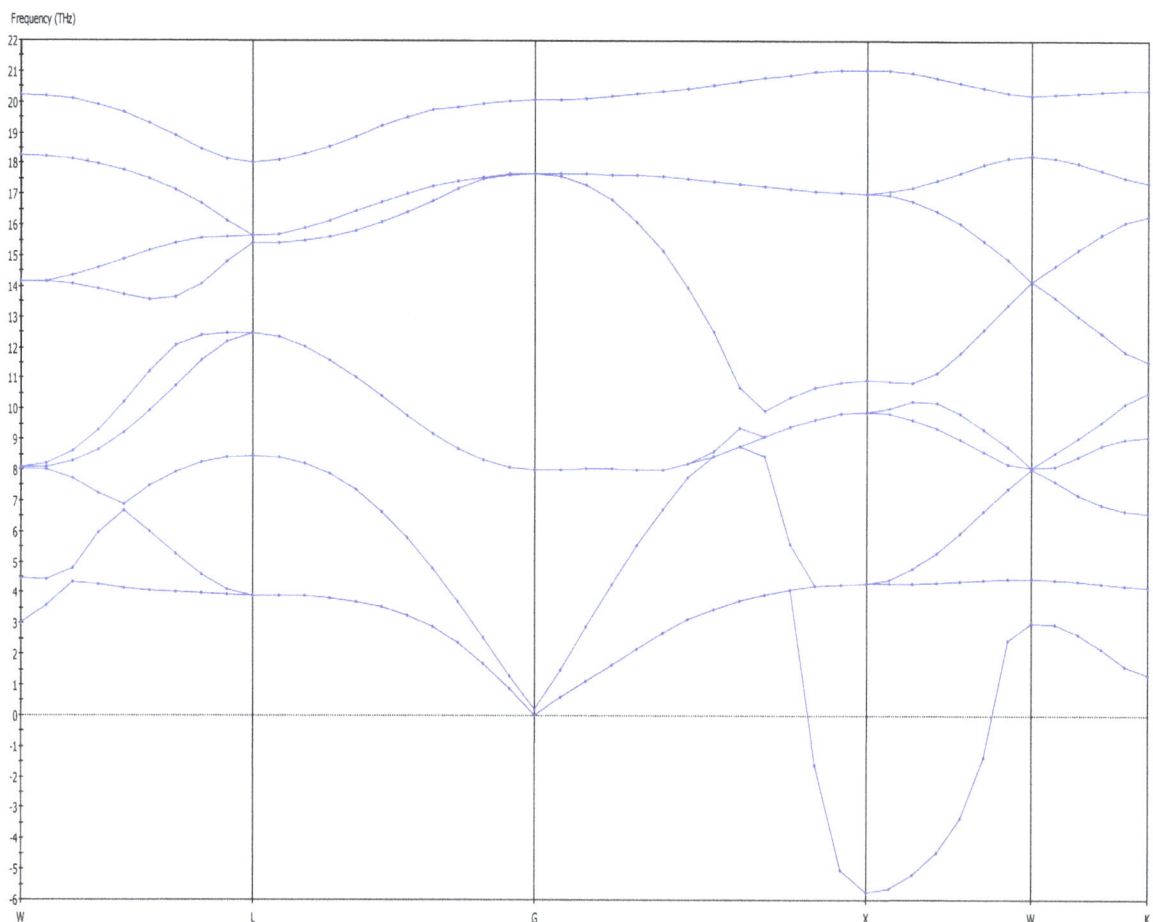

Figure 2. Phonon dispersion curves of $cZrO_2$.

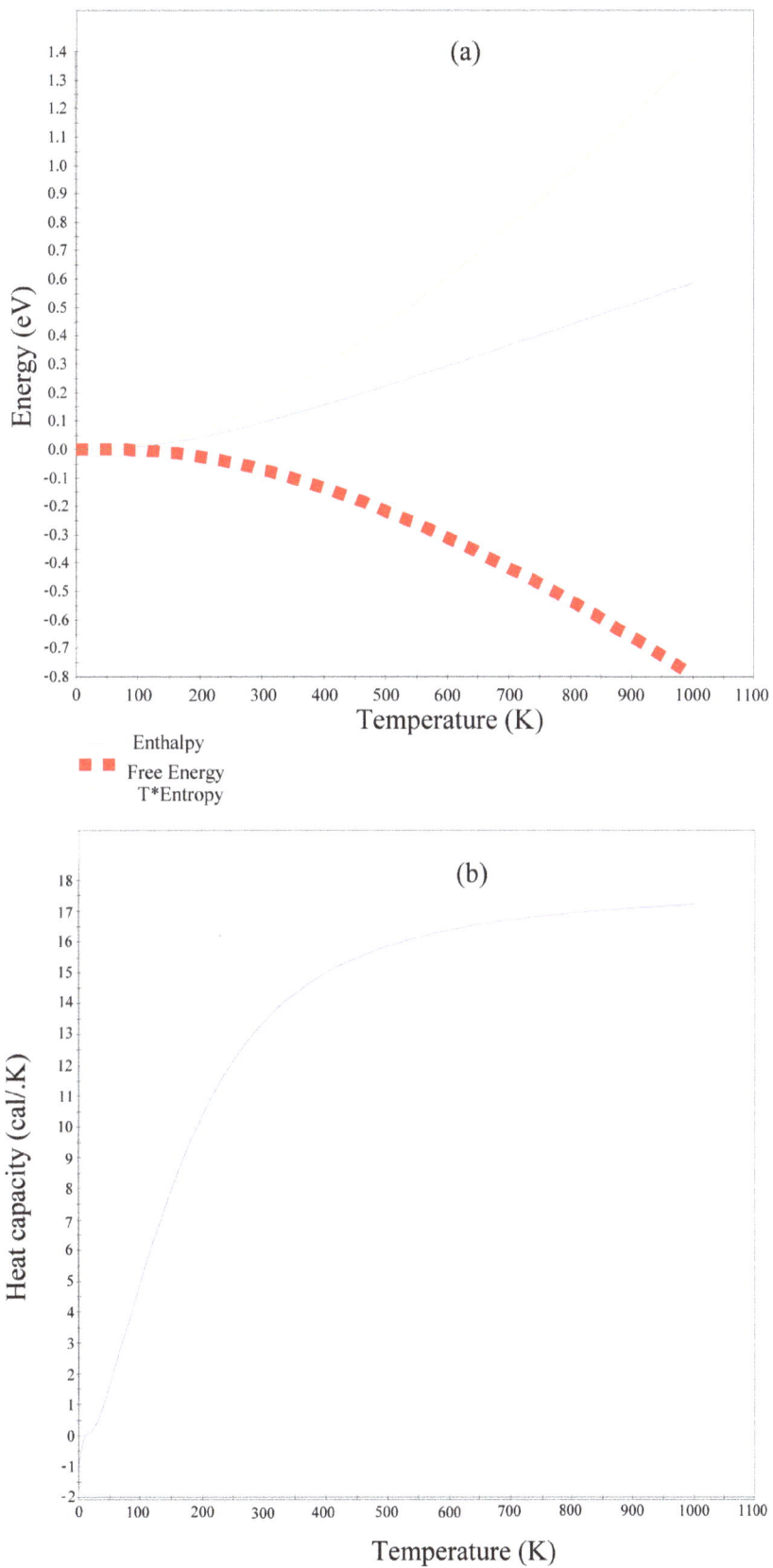

Figure 3. Enthalpy, free energy and entropy (a) and heat capacity of $cZrO_2$.

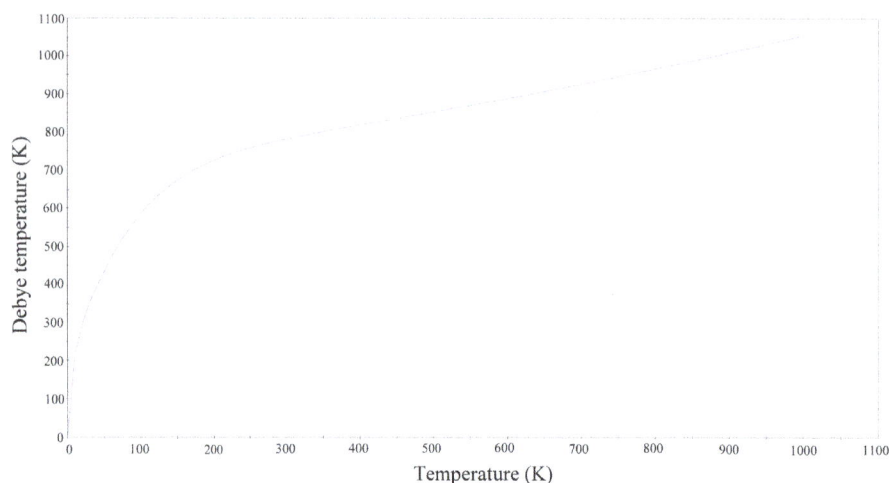

Figure 4. Debye temperature graph of $cZrO_2$.

indicate that $cZrO_2$ has satisfied the mechanical stability requirements; however, the young's modulus and hardness are high when compared to similar material such as cerium oxide. As expected, the thermodynamic properties obtained confirmed the ceramic characteristics of $cZrO_2$. The data obtained may further be used to predict properties in relation to defects and/or dopants and setting parameters for processing nanocomposites related to $cZrO_2$.

Acknowledgements

The authors are grateful for the supports provided by Universiti Teknologi Petronas and Malaysian Ministry of Higher Education (MOHE) through the Long Term Research Grant Scheme (LRGS) for One Baja Research Programme (Project 6).

References

[1]   Xia, X., Oldman, R. and Catlow, R. (2009) Computational Modeling Study of Bulk and Surface of Yttria-Stabilized Cubic Zirconia. *Chemistry of Materials*, **21**, 3576-3585. http://dx.doi.org/10.1021/cm900417g

[2]   Bandura, A.V. and Evarestov, R.A. (2012) *Ab Initio* Structure Modeling of ZrO_2 Nanosheets and Single-Wall Nanotubes. *Computational Materials Science*, **65**, 395-405. http://dx.doi.org/10.1016/j.commatsci.2012.08.001

[3]   Wang, C. (2009) Multiscale Modeling and Simulation of Nanocrystalline Zirconium Oxide. Ph.D. Thesis, University of Nebraska at Lincoln.

[4]   Muhammad, I.D. and Awang, M. (2013) Modelling the Interatomic Potential of Cubic Zirconia. *Applied Mechanics and Materials*, **446-447**, 151-157. http://dx.doi.org/10.4028/www.scientific.net/AMM.446-447.151

[5]   Suciu, C., Gagea, L., Hoffmann, A.C. and Mocean, M. (2006) Sol-Gel Production of Zirconia Nanoparticles with a New Organic Precursor. *Chemical Engineering Science*, **61**, 7831-7835. http://dx.doi.org/10.1016/j.ces.2006.09.006

[6]   Takenaka, S., Uwai, S., Ida, S., Matsune, H. and Kishida, M. (2013) Bottom-Up Synthesis of Titania and Zirconia Nanosheets and Their Composites with Graphene. *Chemistry Letters*, **42**, 1188-1190. http://dx.doi.org/10.1246/cl.130587

[7]   Kulkova, S., Bakulin, A., Hocker, S. and Schmauder, S. (2012) *Ab-Initio* Study of Metal-Zirconia Interfaces. *Materials Science and Engineering*, **38**, 012004. http://dx.doi.org/10.1088/1757-899X/38/1/012004

[8]   Lamperti, A., Cianci, E., Ciprian, R., Sangalli, D. and Debernardi, A. (2013) Stabilization of Tetragonal/Cubic Phase in Fe Doped Zirconia Grown by Atomic Layer Deposition. *Thin Solid Films*, **533**, 83-87. http://dx.doi.org/10.1016/j.tsf.2012.11.127

[9]   Welberry, T.R., Withers, R.L., Thompson, J.G. and Butler, B.D. (1992) Diffuse Scattering in Yttria-Stabilized Cubic Zirconia. *Journal of Solid State Chemistry*, **100**, 71-89. http://dx.doi.org/10.1016/0022-4596(92)90157-Q

[10] Hou, Z.F. (2008) *Ab Initio* Calculations of Elastic Modulus and Electronic Structures of Cubic $CaZrO_3$. *Physical B: Condensed Matter*, **403**, 2624-2628. http://dx.doi.org/10.1016/j.physb.2008.01.025

[11] Zhang, P., Lu, Y., He, C. and Zhang, P. (2011) First-Principles Study of the Incorporation and Diffusion of Helium in Cubic Zirconia. *Journal of Nuclear Materials*, **418**, 143-151. http://dx.doi.org/10.1016/j.jnucmat.2011.06.025

[12] Zhao, X., Shang, S., Liu, Z. and Shen, J. (2011) Elastic Properties of Cubic, Tetragonal and Monoclinic ZrO_2 from First-Principle's Calculations. *Journal of Nuclear Materials*, **415**, 13-17. http://dx.doi.org/10.1016/j.jnucmat.2011.05.016

[13] Miller, S.P., Dunlap, B.I. and Fleischer, A.S. (2012) Cation Coordination and Interstitial Oxygen Occupancy in Co-Doped Zirconia from First Principles. *Solid State Ionics*, **227**, 66-72. http://dx.doi.org/10.1016/j.ssi.2012.07.017

[14] Accelrys Software Inc., San Diego (2012) Materials Studio. http://accelrys.com/products/materials-studio/index.html

[15] Clark, S.J., Segall, M.D., Pickard, C.J., Hasnip, P.J., Probert, M.J., Refson, K. and Payne, M.C. (2005) First Principles Methods Using CASTEP. *Zeitschrift Fuer Kristallographie*, **220**, 567-570.

[16] Perdew, J.P., Burke, K. and Ernzerhof, M. (1996) Generalized Gradient Approximation Made Simple. *Physical Review Letters*, **77**, 3865-3868. http://dx.doi.org/10.1103/PhysRevLett.77.3865

[17] Soo, Y.L., Chen, P.J., Huang, S.H., Shiu, T.J., Tsai, T.Y., Chow, Y.H., *et al.* (2008) Local Structures Surrounding Zr in Nanostructurally Stabilized Cubic Zirconia: Structural Origin of Phase Stability. Faculty Publications—Chemistry Department, Paper 18. http://digitalcommons.unl.edu/chemfacpub/18

[18] Chang, Y., Wang, H., Zhu, Q., Luo, P. and Dong, S. (2013) Theoretical Calculation and Analysis of ZrO_2 Spherical Nanometer Powders. *Journal of Advanced Ceramics*, **2**, 21-25. http://dx.doi.org/10.1007/s40145-013-0036-2

[19] Goldsby, J.C. (2013) Basic Elastic Properties Predictions of Cubic Cerium Oxide Using First-Principles Methods. *Journal of Ceramics*, **2013**, Article ID: 323018. http://dx.doi.org/10.1155/2013/323018

[20] Yang, Z.-J., Guo, Y.-D., Linghu, R.-F. and Yang, X.-D. (2012) First-Principles Calculation of the Lattice, Compressibility, Elastic Anisotropy and Thermodynamic Stability of V_2GeC, *China Physics B*, **21**, 036301 http://dx.doi.org/10.1088/1674-1056/21/3/036301

[21] Tian, Y., Xu, B. and Zhao, Z. (2012) Microscopic Theory of Hardness and Design of Novel Superhard Crystals. *International Journal of Refractory Metals and Hard Materials*, **33**, 93-106. http://dx.doi.org/10.1016/j.ijrmhm.2012.02.021

[22] Chong, X., Jiang, Y., Zhou, R. and Feng, J. (2014) First Principles Study the Stability, Mechanical and Electronic Properties of Manganese Carbides. *Computational Materials Science*, **87**, 19-25. http://dx.doi.org/10.1016/j.commatsci.2014.01.054

[23] Kisi, E. and Yuxiang, M. (2003) Debye Temperature, Anharmonic Thermal Motion and Oxygen Non-Stoichiometry in Yttria Stabilized Cubic Zirconia. *Journal of Physics*: *Condensed Matter*, **10**, 3823-3832. http://dx.doi.org/10.1088/0953-8984/10/17/013

20

Static Crack Propagation of Carbon Nanotube through Non-Bonded Interface of Nanocomposites

Khondaker Sakil Ahmed[*], Ang Kok Keng

Department of Civil & Environmental Engineering, National University of Singapore, Singapore City, Singapore
Email: [*]sakil0104@gmail.com, ceeksa@nus.edu.sg

Abstract

This study presents an analytical shear-lag model to illustrate the interface crack propagation of carbon nanotube (CNT) reinforced polymer-matrix composites (PMCs) using representative volume element (RVE). In the model, a 3D cylindrical RVE is picked to present the nanocomposite in which CNT/polymer chemically non-bonded interface is taken into consideration. In the non-bonded interface, the stress transfer of CNT is generally considered to be controlled by the combined contribution of mechanical interlocking, thermal residual stress, Poisson's contraction and van der Waals (vdW) interaction. Since CNT/matrix interface becomes debonded due to crack propagation, vdW interaction which is a function of relative radial displacement of the CNT/matrix interface makes the modeling of the interface tricky and challenging. In order to solve this complexity, an iterative approach is proposed to calculate the vdW interaction for debonded CNT/matrix interface accurately. The analytical results aim to obtain the characteristics load displacement relationship in static crack propagation for CNT reinforced PMCs.

Keywords

Polymer-Matrix Composites (PMCs), Interface, Computational Modelling, Crack Propagation

1. Introduction

Carbon nanotubes have exceptional mechanical properties such as extremely high strength and stiffness and they have already been considered as superior candidate of reinforcement for mechanically high strength, lightweight and smart nanocomposite [1]-[6]. However, huge strength difference between CNT and most other potential

---

[*]Corresponding author.

polymers that significantly influence composite behaviour makes the CNT/matrix interface more critical [7]-[13]. The key controlling factors at the CNT/polymer non-bonded interface are identified to be mechanical interlocking (friction), thermal residual stress and non-covalent bonding like van der Waals (vdW) interactions [14]-[16].

Evaluating the static crack propagation in nanocomposite between nanotube and matrix at nanoscale is one of many difficult tasks. Very few works have been carried out on investigating the damage behavior, interfacial sliding or crack propagation of CNT in polymer matrix. This can be attributed to the fact that experimental investigation on nanotube crack propagation is quite impossible due to the difficulties arising in gripping, manipulation and stress, strain measurement at the nanoscale. Analytical studies are used to be a shed of light on nanoscale behavior and to obtain information that may not be easily obtained from experiments.

Classical shear-lag model is widely used to obtain interface characteristics for fibre reinforced composite since 1950s. Recently, some researchers have extended the application of the shear-lag model for nanotube as well as nanorope (several CNT as bundle) reinforced composite using representative volume element (RVE) concept [14] [17]-[21]. Though there are some studies based on the interface fracture toughness to investigate interface cracking for fiber reinforced composite using shear-lag model, most of them consider the case of interface cracking of perfectly bonded interface to be debonded interface. However, perfect bonding at the interface is not always common and can be achieved only by the creation of chemical bonding at the interface. The creation of chemical bonding is sometimes not only costly but also difficult to achieve uniformly over the interface. Ang and Ahmed [22] proposed an improved shear-lag model that can be largely used to obtain stress transfer mechanism for chemically non-bonded interface. As far with the author's knowledge, there is no study in the literature that investigates the crack propagation for the chemically non-bonded interface considering vdW interaction.

This study aims to extend their previous study [22] to investigate the interface crack propagation of CNT in polymer Nanocomposites. The key target of this study is to obtain the stress displacement relationship as static crack propagates though the chemically non-bonded CNT/matrix interface. This study can be useful to illustrate the static crack propagation of CNT that may be used as preliminary step before experimental investigation.

2. Analytical Model for Static Crack Propagation

An analytical shear-lag model is proposed to investigate the static crack propagation for non-bonded CNT/matrix interface, as shown in **Figure 1**. The figure includes a 3D representative volume element that comprises a CNT of length $2L$ and outer diameter of $2a$. Generally, the maximum interfacial shear stress of CNT is observed

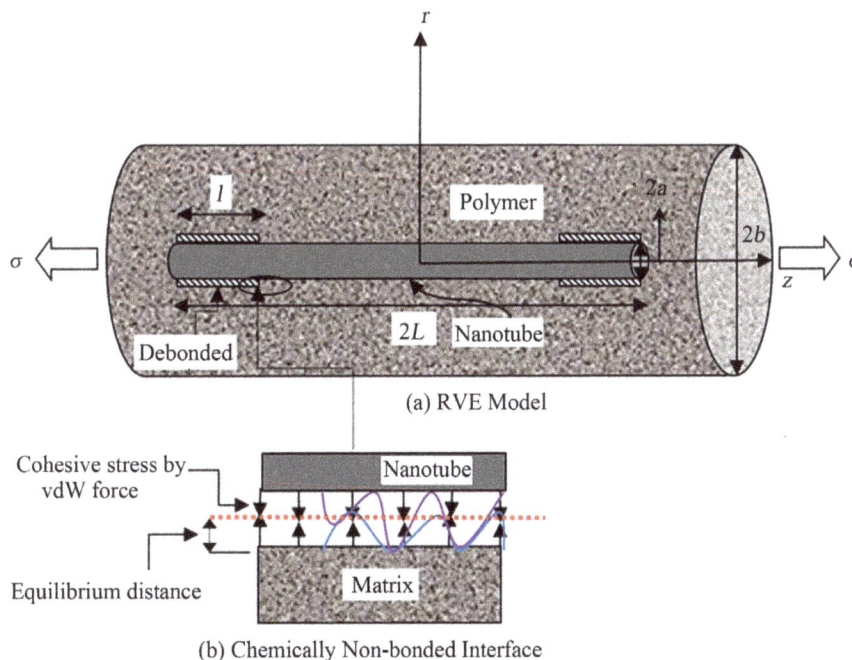

(a) RVE Model

(b) Chemically Non-bonded Interface

Figure 1. Model for static crack propagation of nanotube reinforced composite.

to be near at the end of the nanotube and hence it is considered that debonding starts from the end rather than the center of the RVE.

Since the model is symmetric with respect to its center, it is assumed that the crack will propagate in similar pattern with equal debonded length for both sides of the nanotube. The length of the debonded interface is denoted by l. σ is the applied stress at the remote end of the RVE. The other parameters of the nanocomposite that has been used in the shear-lag model are presented in **Figure 1**.

Similar to our previous study [20], this study also considers that the stress is transferred through matrix to CNT by the combined contribution of thermal residual stress, Poisson's contraction and vdW interactions. The interface debonding crack propagation criterion used in this study is based on fracture mechanics where the strain energy release rate against the debonded length is equated to the interface fracture toughness, G_{it}

$$G_{it} = \frac{1}{2\pi a}\frac{dU_{te}}{dl} \tag{1}$$

U_{te} is the sum of the total strain energy stored in the frictionally bonded region; $0 < z < (L - l)$ and debonded region, $(L - l) < z < L$. Therefore, total strain energy may be obtained from the algebraic sum of the strain energy at the frictionally bonded interface U_{fb} and frictionally debonded interface U_{fd},

$$U_{te} = U_{fb} + U_{fd} \tag{2}$$

The length of the debonded region increases as the crack propagation proceeds. The strain energy due to frictionally bonded interface (U_{fb}) and debonded interface (U_{fb}) can be obtained by integrating over their corresponding stress components over the volume of the respective regions as given in Equation (3a) and (3b).

$$U_{fb} = \int_0^{L-l}\int_0^b \left| \frac{\sigma_{zz}^{f2}}{E_f} + \frac{\sigma_{zz}^{m2}}{E_m} + 2(1+v_m)\frac{\tau_{rz}^{m2}}{E_m} \right| \pi r\,dr\,dz \tag{3a}$$

$$U_{fd} = \int_{L-l}^L\int_0^b \left| \frac{\sigma_{zz}^{f2}}{E_f} + \frac{\sigma_{zz}^{m2}}{E_m} + 2(1+v_m)\frac{\tau_{rz}^{m2}}{E_m} \right| \pi r\,dr\,dz \tag{3b}$$

2.1. Solutions for Frictionally Bonded Interface $0 < z < (l - L)$

Since this study considers chemically non-bonded interface, the stress transfer of CNT through such type of interface is determined by the combined contribution of mechanical interlocking (*i.e.* frictionally bonded), thermal residual stress and vdW interaction. The solutions for axial stress of CNT $\left(\bar{\sigma}_{zz}^f\right)$, matrix $\left(\bar{\sigma}_{zz}^m\right)$ and shear stress $\left(\tau_{rz}^m\right)$ at any radial location of the matrix for the such type of frictionally bonded (chemically non-bonded) interface are recalled from the improved shear-lag model proposed by Ang and Ahmed (2013) [22].

$$\bar{\sigma}_{zz}^f = \frac{(\sigma_o + C)\{\exp(R_1 z + R_2 L) + \exp(R_2 z + R_1 L)\}}{Q} + \frac{C\{\exp(R_2 z) - \exp(R_1 z)\}}{Q} - C \tag{4}$$

$$\bar{\sigma}_{zz}^m = (1+\gamma)(1+v_m)\sigma - \frac{(\sigma_o + C)\{\exp(R_1 z + R_2 L) + \exp(R_2 z + R_1 L)\}}{Q_1} + \frac{C\{\exp(R_2 z) - \exp(R_1 z)\}}{Q_1}$$

$$- k_1\left[q_0 + 2\pi n_p n_c \mathcal{E}\delta^2\left\{\frac{1}{\left(0.4^{\frac{1}{6}} + \frac{O_i}{\delta}\right)^4} - \frac{0.4}{\left(0.4^{\frac{1}{6}} + \frac{O_i}{\delta}\right)^{10}}\right\}\right] \tag{5}$$

$$\tau_{rz}^m = \frac{\gamma(b^2 - r^2)}{2r}\left[\frac{(\sigma_o + C)\{R_1\exp(R_1 z + R_2 L) + R_2\exp(R_2 z + R_1 L)\}}{Q_1} + \frac{C\{R_1\exp(R_2 z) - R_2\exp(R_1 z)\}}{Q_1}\right] \tag{6}$$

where

$$\sigma_0 = \frac{C\left(R_1 - R_2\right)\left\{\exp\left(R_1 L\right) - 1\right\} + \mu\left(q_0 + 2\pi n_p n_c \mathcal{C}\delta^2\left\{\dfrac{1}{\left(0.4^{\frac{1}{6}} + \dfrac{O_i}{\delta}\right)^4} - \dfrac{0.4}{\left(0.4^{\frac{1}{6}} + \dfrac{O_i}{\delta}\right)^{10}}\right\}\right)C_2 Q_1}{\left\{R_2 \exp\left(R_1 L\right) - R_1 \exp\left(R_2 L\right)\right\}} \tag{7a}$$

$$C = v_m\left(1 + \gamma\right)\sigma - k_1\left[q_0 + 2\pi n_p n_c \mathcal{C}\delta^2\left\{\dfrac{1}{\left(0.4^{\frac{1}{6}} + \dfrac{O_i}{\delta}\right)^4} - \dfrac{0.4}{\left(0.4^{\frac{1}{6}} + \dfrac{O_i}{\delta}\right)^{10}}\right\}\right] \tag{7b}$$

$$Q_1 = \exp\left(R_2 L\right) - \exp\left(R_1 L\right) \tag{7c}$$

$$R_1 = \frac{-C_1 + \left(C_1^2 + 4C_2\right)^{\frac{1}{2}}}{2} \tag{7d}$$

$$R_2 = \frac{-C_1 - \left(C_1^2 + 4C_2\right)^{\frac{1}{2}}}{2} \tag{7e}$$

$$C_1 = \frac{ak_1}{\gamma \mu v_m\left(1 + v_m\right)\left(b^2\left(1 + \gamma\right)\ln\dfrac{b}{a} - \dfrac{\left(3b^2 - a^2\right)}{4}\right)} \tag{7f}$$

$$C_2 = \frac{\gamma v_m + \alpha v_f}{\dfrac{\gamma v_m\left(1 + v_m\right)}{2}\left(b^2\left(1 + \gamma\right)\ln\dfrac{b}{a} - \dfrac{\left(3b^2 - a^2\right)}{4}\right)} \tag{7g}$$

$$k_1 = \alpha\left(1 - v_{f?}\right) + 1 + v_m + 2\gamma \tag{7h}$$

$$\alpha = \frac{E_m}{E_f} \tag{7i}$$

$$\gamma = \frac{a^2}{b^2 - a^2} \tag{7j}$$

By the Substitution of Equations (4)-(6) in the Equation (3a) and then double integration over the corresponding region strain energy at the frictionally bonded interface can be obtained.

2.2. Solutions for Frictionally Debonded Interface $(L - l) < z < L$

The template Previously, Ahmed and Ang [23] proposed a shear-lag model for debonded interface to investigate the load transfer mechanism of CNT in polymer composite. The model is capable to provide analytical solutions for the axial stress of CNT $\left(\bar{\sigma}_{zz}^f\right)$, matrix $\left(\bar{\sigma}_{zz}^m\right)$ and shear stress $\left(\tau_{rz}^m\right)$ of the matrix for the debonded interface as presented in Ahmed and Ang [23]. The solutions are recalled as given in Equations (8)-(10)

$$\sigma_{zz}^f = A\left[1 - \exp\left(\frac{2\mu k}{a}\left(L - z\right)\right)\right] \tag{8}$$

$$\sigma_{zz}^{m} = \left(1+\gamma\right)\sigma_{(L-l)}^{f} - A\gamma\left[1-\exp\left(\frac{2\mu k}{a}\left(L-z\right)\right)\right] \qquad (9)$$

$$\tau_{rz}^{m} = A\gamma\mu k\left[\frac{b^{2}-r^{2}}{2r}\right]\exp\left(\frac{2\mu k}{a}\left(L-z\right)\right) \qquad (10)$$

in which

$$k = \frac{\alpha v_{f} + \gamma v_{m}}{\alpha\left(1-v_{f}\right)+1+v_{m}+2\gamma} \qquad (11)$$

Again by the Substitution of Equations (8)-(10) in the Equation (3b) and then double integration over the corresponding region strain energy at the frictionally bonded interface can be obtained. After replacing the total strain energy in Equation (1) and differentiating with respect to debonded length (l) and then after rearranging, the required stress (σ) to cause interface cracking may be derived as follows

$$\sigma = \frac{\sqrt{\left(AP\right)^{2} - Q\left(RA^{2} - G_{it}\right)} - AP}{Q} \qquad (12)$$

in which

$$A = f\left(a,b,\alpha,v_{f},\mu,v_{m},\Delta T,n_{p},n_{c},\mathcal{C},\delta^{2},O_{i}\right) \qquad (13a)$$

$$P = f\left(a,b,E_{m}\right) \qquad (13b)$$

$$R = f\left(a,b,l,L,k,E_{m},E_{f}\right) \qquad (13c)$$

$$Q = f\left(a,b,l,C,E_{m},E_{f},\mu,v_{m},v_{f}\right) \qquad (13d)$$

3. Results & Discussions

When the stress is applied to the nanocomposite and reaches beyond the allowable limit of interfacial shear stress, the interface starts to become debonded as well as the static crack propagation enhances. Based on the analytical solution stated in Equation (12), analytical result for static crack propagation in nanocomposite can be obtained. However, the solution is not straight forward because the debonded interface does not follow strain compatibility which results in relative radial displacement between the CNT and polymer matrix.

The vdW interaction which is a function of initial interface displacement and relative radial displacement due to the application of load varies along the length of the debonded region of the CNT/matrix interface. In this investigation of interface crack propagation along the frictionally bonded interface, the variation of the van der Wall interaction due to the relative radial displacement is accounted in estimating the stress displacement relationship and hence an iterative approach has been used to calculate the vdW interaction for debonded CNT/matrix interface. The available experimental data that has been used in this study are presented in **Table 1**. Since the analytical model for static crack propagation is axisymmetric, only right hand side of the crack propagation is presented.

The characteristics curve for the required applied stress at the remote end of RVE corresponding to the debond length of the embedded nanotube is presented in **Figure 2**. It can be seen from the figure that the maximum and minimum stress required to propagate debonding are found to be approximately 19 GPa and 3.75 GPa, respectively when debonding length is nearly zero and complete debonding, respectively. The figure shows that with the increase of debond length, initially the required applied stress for interface crack propagation sharply decreases before reaching a nearly constant region. For example if the debond length is 25%, the required applied stress decreases by 70% of that required to debond a completely bonded interface. The required stress to debond the last 60% of the embedded length is only 20% of initial requirement.

It is interesting to note that after complete debonding has occurred, the CNT is found to be capable of carrying further stress. This happens due to the fact that after debonding, shear stress due to thermal residual stress and van der Waal interaction will be still active at the debonded interface. In another comparison, it is observed

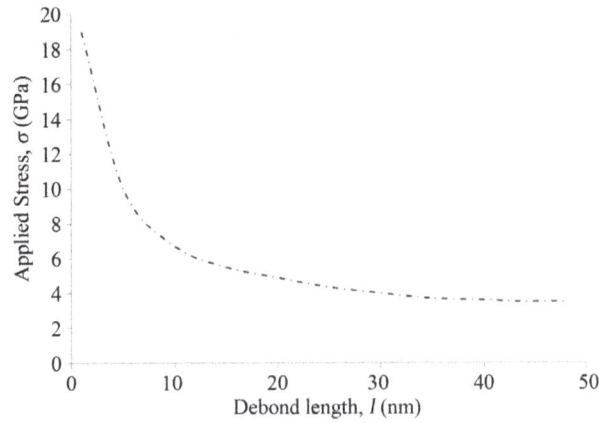

Figure 2. Characteristics curve for interface cracking stress corresponding to debond length.

Table 1. The definition and value of the parameters.

Symbol	Parameter Definition	Value
a	Radius of the CNT	1 nm [(15,15) CNT]
b	Radius of the RVE	4 nm
$2L$	Length of embedded nanotube	100 nm
E_m	Young's modulus of Matrix	10 Gpa $\left(1\,GPa = 10^9\ N/m^2\right)$
E_t	Young's modulus of CNT	1000 GPa
t	Thickness of the nanotube	0.34 nm
ϵ	Bond energy due to van der Waals interaction at the equilibrium distance	0.004656 ev $\left(1\,ev = 1.602 \times 10^{-19}\ j\right)$
v_f	Poisson's Ratio of the Nanotube	0.28
v_m	Poisson's Ratio of the Matrix	0.35
δ	Equilibrium distance between two interface	0.3825 nm
α_f	Coefficient of thermal contraction of CNT	2×10^{-6} nm/nm/°C
α_m	Coefficient of thermal contraction of Matrix	27.1×10^{-6} nm/nm/°C
μ	Coefficient of Friction	0.25
ΔT	Change of temperature after thermal cooling	200°C
n_p	No of polymer molecule per unit volume of matrix	$3.1 \times 10^{28}/m^3$
n_c	No of CNT atom per unit volume of matrix	$3.82 \times 10^{19}/m^2$
O_t	Interface displacement beyond equilibrium distance	0.25 nm
G_{it}	Interface Fracture Toughness	37×10^9 nJ/nm^2

that the stress carrying ability of completely debonded interface is nearly 20% of the stress required to debond the bonded interface.

4. Conclusions

A shear-lag model has also been proposed for investigating the interfacial static crack propagation of CNT reinforced composite. Using the proposed model, closed form analytical solution for required cracking stress corresponding to debonded length is derived. Subsequently, analytical result is presented for static crack propagation with respect to the application of uniform stress. The stress required in causing interface cracking is found to decrease as the debonding length increases. The result also revealed that CNT fiber can take stress even after complete debonding which is mainly due to thermal residual stress and van der Waals interactions. The characteristics curve also shows that after completing debonding has occurred, the CNT is found to be capable of carrying further stress.

One of the key achievements of this study is that the proposed shear-lag model is capable of incorporating the cohesive stress caused by vdW interaction together with the other components. It should be noted that the proposed model is a useful alternative to other more complicated methods such as molecular mechanics and molecular dynamics simulations, which are not only time consuming but also costly.

References

[1]   Ajayan, P.M., Schadler, L.S., Giannaris, C. and Rubio, A. (2000) Single-Walled Carbon Nanotube-Polymer Composites: Strength and Weakness. *Advanced Materials*, **12**, 750-753. http://dx.doi.org/10.1002/(SICI)1521-4095(200005)12:10<750::AID-ADMA750>3.0.CO;2-6

[2]   Ashrafi, B. and Hubert, P. (2006) Modeling the Elastic Properties of Carbon Nanotube Array/Polymer Composites. *Composites Science and Technology*, **66**, 387-396. http://dx.doi.org/10.1016/j.compscitech.2005.07.020

[3]   Bakshi, S.R., Lahiri, D. and Agarwal, A. (2010) Carbon Nanotube Reinforced Metal Matrix Composites—A Review. *International Materials Reviews*, **55**, 41-64. http://dx.doi.org/10.1179/095066009X12572530170543

[4]   Chen, X. (2004) Square Representative Volume Elements for Evaluating the Effective Material Properties of Carbon Nanotube-Based Composites. *Computational Materials Science*, **29**, 1-11. http://dx.doi.org/10.1016/S0927-0256(03)00090-9

[5]   Chen, X.H., Chen, C.S., Xiao, H.N., Liu, H.B., Zhou, L.P. and Li, S.L. (2006) Dry Friction and Wear Characteristics of Nickel/Carbon Nanotube Electroless Composite Deposits. *Tribology International*, **39**, 22-28. http://dx.doi.org/10.1016/j.triboint.2004.11.008

[6]   Manoharan, M.P., Sharma, A., Desai, A.V., Haque, M.A., Bakis, C.E. and Wang, K.W. (2009) The Interfacial Strength of Carbon Nanofiber Epoxy Composite Using Single Fiber Pullout Experiments. *Nanotechnology*, **20**, Article ID: 295701. http://dx.doi.org/10.1088/0957-4484/20/29/295701

[7]   Wang, W., Ciselli, P., Kuznetsov, E., Peijs, T. and Barber, A.H. (2008) Effective Reinforcement in Carbon Nanotube-Polymer Composites. *Philosophical Transactions of the Royal Society A: Mathematical, Physical and Engineering Sciences*, **366**, 1613-1626. http://dx.doi.org/10.1098/rsta.2007.2175

[8]   Salehikhojin, A. and Jalili, N. (2008) A Comprehensive Model for Load Transfer in Nanotube Reinforced Piezoelectric Polymeric Composites Subjected to Electro-Thermo-Mechanical Loadings. *Composites Part B: Engineering*, **39**, 986-998. http://dx.doi.org/10.1016/j.compositesb.2007.12.001

[9]   Qian, D., Dickey, E.C., Andrews, R. and Rantell, T. (2000) Load Transfer and Deformation Mechanisms in Carbon Nanotube-Polystyrene Composites. *Applied Physics Letters*, **76**, 2868-2870. http://dx.doi.org/10.1063/1.126500

[10] Qian, D. (2003) Load Transfer Mechanism in Carbon Nanotube Ropes. *Composites Science and Technology*, **63**, 1561-1569. http://dx.doi.org/10.1016/S0266-3538(03)00064-2

[11] Liao, K. and Li, S. (2001) Interfacial Characteristics of a Carbon Nanotube-Polystyrene Composite System. *Applied Physics Letters*, **79**, 4225. http://dx.doi.org/10.1063/1.1428116

[12] Manoharan, M.P., Sharma, A., Desai, A.V., Haque, M.A., Bakis, C.E. and Wang, K.W. (2009) The Interfacial Strength of Carbon Nanofiber Epoxy Composite Using Single Fiber Pullout Experiments. *Nanotechnology*, **20**, 5701-5705.

[13] Kin, L. and Sean, L. (2001) Interfacial Characteristics of a Carbon Nanotube-Polystyrene Composite System. *Applied Physics Letters*, **79**, 4225-4227.

[14] Haque, A. and Ramasetty, A. (2005) Theoretical Study of Stress Transfer in Carbon Nanotube Reinforced Polymer Matrix Composites. *Composite Structures*, **71**, 68-77. http://dx.doi.org/10.1016/j.compstruct.2004.09.029

[15] Jiang, L., Huang, Y., Jiang, H., Ravichandran, G., Gao, H. and Hwang, K. (2006) A Cohesive Law for Carbon Nanotube/Polymer Interfaces Based on the van der Waals Force. *Journal of the Mechanics and Physics of Solids*, **54**, 2436-2452. http://dx.doi.org/10.1016/j.jmps.2006.04.009

[16] Jiang, Y., Zhou, W., Kim, T., Huang, Y. and Zuo, J. (2008) Measurement of Radial Deformation of Single-Wall Carbon Nanotubes Induced by Intertube van der Waals Forces. *Physical Review B*, **77**, Article ID: 153405. http://dx.doi.org/10.1103/PhysRevB.77.153405

[17] Gao, X. and Li, K. (2005) A Shear-Lag Model for Carbon Nanotube-Reinforced Polymer Composites. *International Journal of Solids and Structures*, **42**, 1649-1667. http://dx.doi.org/10.1016/j.ijsolstr.2004.08.020

[18] Liu, Y.J. and Chen, X.L. (2003) Continuum Models of Carbon Nanotube-Based Composites Using the Boundary Element Method. *Electronic Journal of Boundary Element*, **1**, 20.

[19] Ahmed, K.S. and Keng, A.K. (2012) A Pull-Out Model for Perfectly Bonded Carbon Nanotube in Polymer Composite *Journal of Mechanics of Materials and Structures*, **7**, 753-764.

[20] Ahmed, K.S. and Keng, A.K. (2013) Interface Characteristics of Nanorope Reinforced Polymer Composites. *Computational Mechanics*, **52**, 571-585. http://dx.doi.org/10.1007/s00466-013-0833-z

[21] Ahmed, K.S. and Keng, A.K. (2014) Interface Characteristics of Carbon Nanotube Reinforced Polymer Composites Using an Advanced Pull-Out Model. *Computational Mechanics*, **53**, 297-308. http://dx.doi.org/10.1007/s00466-013-0908-x

[22] Ang, K.K. and Ahmed, K.S. (2013) An Improved Shear-Lag Model for Carbon Nanotube Reinforced Polymer Composites. *Composites Part B: Engineering*, **50**, 7-14. http://dx.doi.org/10.1016/j.compositesb.2013.01.016

[23] Ahmed, K.S. and Ang, K.K. (2009) Load Transfer Mechanism of Nanotube Reinforced Composite Considering Coulomb Friction and van der Waals Interactions. *Proceedings of the 22nd KKCNN Symposium on Civil Engineering*, Chiang Mai, 31 October-2 November 2009, 331-336.

21

High Efficiency SiC Terahertz Source in Mixed Tunnelling Avalanche Transit Time Mode

Pranati Panda, Satya Narayana Padhi, Gana Nath Dash

Electron Devices Group, School of Physics, Sambalpur University, Burla, India
Email: gndash@ieee.org

Academic Editor: Yarub Al-Douri, University Malaysia Perlis, Malaysia

Abstract

High frequency properties of 4H-SiC double drift region (DDR) Mixed Tunnelling Transit Time (MITATT) diodes are studied through computer simulation method. It is interesting to observe that the efficiency of SiC (flat) DDR MITATT diode (16%) is more than 4 times that of Si (flat) DDR MITATT diode (3.59%). In addition, a power output of more than 15 times from the SiC MITATT diode compared to the Si MITATT diode is commendable. A reduced noise measure of 17.71 dB from a low-high-low (lo-hi-lo) structure compared to that of 21.5 dB from a flat structure of SiC is indicative of the favourable effect of tunnelling current on the MITATT diode performance.

Keywords

MITATT, SiC, Terahertz, Tunnelling

1. Introduction

Silicon-based power devices are used in a wide variety of power electronics applications. But, there is a continuous demand for higher current, higher voltage blocking capacity, higher operating temperature and improvement in terms of efficiency, size and weight. This has made us to realise the necessity of better material to replace the existing silicon technology. In this paper the authors have suggested silicon carbide, which exceeds all the limitations of the conventional silicon devices, to be used in new generation of power devices. Silicon carbide is known to be an excellent semiconductor for high-temperature and high-speed electronics [1]-[5]. This material has very high microwave conductivity and is less susceptible to radiation effect [1]. Among all solid state devices, Mixed Tunnelling Avalanche Transit Time diode has recognised itself as a powerful microwave device for operation at high frequency. The basic principle of operation involved in the device offers a great deal of flexibility in the choice of the base material. Yuan *et al.* [6] [7] and Zhao *et al.* [8] have reported some experimental as well as theoretical results for the first time with a 4H-SiC IMPATT (IMPact Avalanche Transit Time) oscillator operating at X and Ka band of frequencies. They have considered the DC and high-power generation aspects of the IMPATT diode. Their results show that this diode exhibits high efficiency and high power as expected, compared to Si and GaAs IMPATT diodes. Pattanaik *et al.* [9] have reported the results of 6H-SiC-based DDR IMPATT diode performance with respect to DC and microwave power as well as noise characteristics and have compared the results obtained with those of Si- and GaAs-based IMPATT diodes under similar operating conditions at the D-band. But, there is no report on performance of SiC Mixed Tunnelling Avalanche Transit Time diode in the THz range except that in [5]. The report at [5] considers only noise behaviour and uses an approximate analytical technique. To improve upon the same we have used here an accurate simulation method for the study of the SiC MITATT diodes. We have obtained excellent results from the simulation, in terms of efficiency and power output as compared to those of a Si-based MITATT diode.

2. Material Parameters of SiC

When a semiconductor is to be used in a power source device the important material parameters which should be considered include the band gap energy, maximum electric field, electron mobility, hole mobility, saturation velocity and thermal conductivity. The values of these parameters for Si and SiC are listed in **Table 1**. The various properties of silicon carbide such as wider band gap, larger critical electric field and higher thermal conductivity let the SiC devices operate at higher temperature and higher voltages offering higher power density and higher current density than the pure silicon devices. Out of numerous poly types of SiC only 4H-SiC and 6H-SiC substrates are commercially available. 4H-SiC is the most widely explored material [10]-[13] for high power devices because its carrier mobility is higher as compared to 6H-SiC. More isotropic nature of electrical properties makes 4H-SiC still more attractive for power device applications [5].

3. Simulation Method

To study the high frequency behaviour of a silicon carbide DDR MITATT diode, a one dimensional model with a doping distribution of the form n^+npp^+ is taken into consideration. The schematic of the MITATT diode considered is shown in **Figure 1**. The diodes are first designed following a MITATT mode DC simulation scheme de-

Table 1. Some material parameters of silicon and silicon carbide.

Properties	Material	
	Silicon	4H-Silicon Carbide
Band gap energy (eV)	1.1	3.260
Max. electric field (MV/m)	41.0	300.000
Electron mobility (m^2/V sec)	0.145	0.090
Hole mobility (m^2/V sec)	0.0480	0.012
Saturation velocity (10^5 m/sec)	0.86	2.000
Thermal conductivity (W/m K)	150.00	500.000

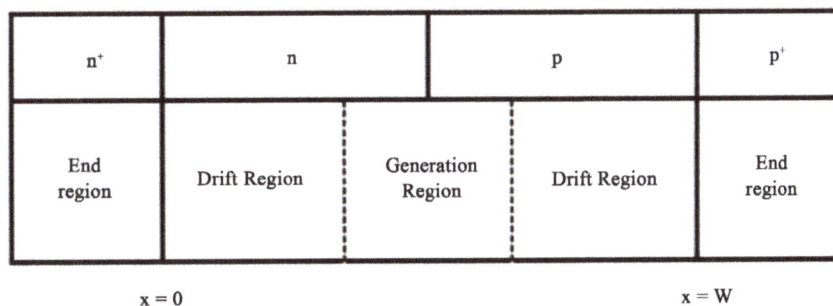

n^+	n		p	p^+
End region	Drift Region	Generation Region	Drift Region	End region

x = 0   x = W

Figure 1. Schematic diagram of MITATT diode.

veloped by Dash *et al.* [14]. The DC analysis also fixes the edges of the diode active layer and determines the DC electric field and carrier current profiles. The DC field profile is used to determine the percentage of tunnelling current by integrating the electron tunnelling generation rate of the form

$$g_{T_n} = A_T E^2 \left(x \right) \exp \left[-\frac{B_T}{E \left(x \right)} \right] \tag{1}$$

(for a definition of the symbols refer to **Appendix**).

The tunnelling generation rate for holes is computed using a simulated energy band diagram [14]. The values of constants A_T and B_T are taken from [15]. The high frequency analysis of the diode is carried out using a small signal simulation method described in [14]. The noise behaviour of MITATT device is computed using MITATT mode noise simulation scheme developed by Dash *et al.* [16]. For this, the diode generation region is treated as consisting of a discrete number of noise generating sources $\left(x' \right)$. The process of impact ionization occurring at different regions contributes to the noise owing to the random nature of the impact ionisation process. A noise source located at x' generates a noise electric field $e(x, x')$ at every point in the depletion layer of the diode. The terminal voltage produced by the noise source located at x' is given by

$$V_t \left(x' \right) = \int_0^w e \left(x, x' \right) \mathrm{d}x \tag{2}$$

From which the transfer impedance can be determined as

$$Z_t \left(x' \right) = \frac{V_t \left(x' \right)}{I_n \left(x' \right)} \tag{3}$$

where the current generated in a space step $\mathrm{d}x'$ around x' is

$$\mathrm{d}I_n \left(x' \right) = q \gamma \left(x' \right) A \mathrm{d}x' \tag{4}$$

Finally, the mean square noise voltage and noise measure (NM) are calculated using the relations [16]

$$\left\langle v^2 \right\rangle = 2q^2 \mathrm{d}f A \int \left| Z_t \left(x' \right) \right|^2 \gamma \left(x' \right) \mathrm{d}x' \tag{5}$$

and

$$\mathrm{NM} = \frac{\left\langle v^2 \right\rangle / \mathrm{d}f}{4 k_B T \left(-Z_R \right)} \tag{6}$$

The noise electric field $e(x, x')$ for a given location of the noise source $\gamma(x')$ is computed by solving the following differential equations [16],

$$\left[D^2 - k^2 + \left(\alpha_n - \alpha_p + 2 r_n k \right) D + 2 \bar{\alpha} k - H - \frac{q r_p}{\bar{v} \in} \left(g'_{T_n} + g'_{T_p} \right) \right] e \left(x, x' \right) = \frac{1}{\bar{v} \in} \left[2 q r_p \gamma \left(x' \right) \right] \tag{7}$$

where

$$H = \left(\alpha'_p - \alpha'_n \right) D E_0 + \frac{2 \bar{\alpha}' J_0}{\bar{v} \in}, \quad k = \frac{1}{\bar{v}} \frac{\partial}{\partial t}, \quad \bar{v} = \left(v_n v_p \right)^{\frac{1}{2}}$$

$$D = \frac{\partial}{\partial x}, \quad \bar{\alpha} = \frac{\alpha_p v_p + \alpha_n v_n}{2 \bar{v}}, \quad r_n = \frac{v_n - v_p}{2 \bar{v}}, \quad r_p = \frac{v_n + v_p}{2 \bar{v}}$$

The primes on α and g denote their field derivatives.

The computation starts by putting the noise source at the beginning of the generation region. The noise electric field $e(x, x')$ corresponding to the location of the noise source is computed by solving Equation (7), from which the terminal voltage and transfer impedances are determined by using Equations (2) and (3). The noise source γ is then shifted to the next space step and the process is repeated until γ covers the whole generation region. Then the mean square noise voltage and noise measure are determined using Equations (5) and (6).

4. Results and Discussion

We have applied the simulation method of analysis to different types of 4H-SiC MITATT DDR diode structures (flat and low-high-low) for operation at 0.5 THz. For a comparative study of the results obtained we have also considered a flat profile DDR diode based on Silicon. The total width in each of the two SiC diodes has been taken to be 296 nm whereas for the Silicon diode it has been taken as 104 nm (as width of diode taken depends upon the value of saturation drift velocity of charge carriers of the material). The doping concentration has been adjusted for an optimum punch through factor in each side as given in **Table 2**. The bias current density in each of the diode structures is taken to be 2.7×10^9 A/m^2 and the diode area (A) is taken to be 10^{-10} m^2 for all structures.

The dc and small signal characteristics of both the SiC DDR diodes and those of Silicon DDR diode at a frequency of 0.5 THz. are presented in **Table 3**. We observed that the maximum electric field for SiC MITATT (flat) is much greater as compared to Silicon MITATT (flat) diode. The breakdown voltage (V_B) for SiC DDR diode is more than 10 times compared to that of Silicon DDR diode. This result is in accordance with the experimental results of Yuan et al. [7]. We found that SiC MITATT has much higher breakdown voltage as compared to silicon. This is due to larger band gap and greater saturation drift velocity of SiC as compared to that of Si. As a result the power output of SiC flat profile DDR MITATT is found to be about 15 times higher than that of Silicon flat profile DDR MITATT. Again the efficiency of SiC flat profile DDR MITATT is about 4 times that of Si flat profile DDR MITATT. The efficiency of SiC MITATT has been further increased by taking lo-hi-lo type doping profile. The change in the nature of the electric field profile with change in doping pattern from flat to lo-hi-lo type may be noticed from **Figure 2**. The stiff rise in electric field localises the generation region and enhances the efficiency of the latter type diode. The peak negative conductance of SiC (flat) diode is found to be less than that of the Si (flat) diode (**Figure 3**). But, the breakdown voltage of SiC being much higher compared to that of Silicon the power output of SiC (flat) MITATT diode becomes 15 times more compared to the Si (flat) MITATT diode. Comparing the simulation result of SiC (flat) DDR MITATT diode with that of GaN (flat) DDR diode at 0.5 THz (results not shown here) we found that the efficiency of GaN (flat) diode (20.6%) is better than that of Si (flat) diode (3.59%) and SiC (flat) diode (16%). But the power output of GaN (flat) diode (1.36 W) is nearly 2.5 times than that of Si diode (0.529 W) and nearly one sixth that of SiC (flat)

Table 2. Design Parameters of the Si and 4H-SiC MITATT diodes considered in this paper.

Structures	Widths (nm)						Doping Concentrations ($\times 10^{24}$ m^{-3})					
	n-side			p-side			n-side			p-side		
	low	high	low	low	high	low	low	high	low	low	high	low
Si (flat)	-	-	52	52	-	-	-	-	1.4	1.4	-	-
SiC (flat)	-	-	148	148	-	-	-	-	1.94	1.94	-	-
SiC (lo-hi-lo)	104	22	22	22	22	104	1.6	5.0	1.6	1.6	5.0	1.6

Table 3. Microwave properties of Si and 4H-SiC MITATT diodes at $J_0 = 2.7 \times 10^9$ A/m^2 and design frequency of 0.5 THz.

Properties	Structures		
	Si (flat)	SiC (flat)	SiC (lo-hi-lo)
E_0 ($\times 10^8$ V/m)	1.225	5.7875	6.01
V_B (V)	8.42	90.8	90.0
η (%)	3.59	16.0	17.5
$-G_p$ ($\times 10^7$ S/m^2)	29.9	7.73	7.94
$-Z_R$ ($\times 10^{-11}$ m^2)	2.49	4.94	3.87
P_{RF} (W)	0.529	7.96	8.03
J_T/J_0 (%)	26.98	24	31.67

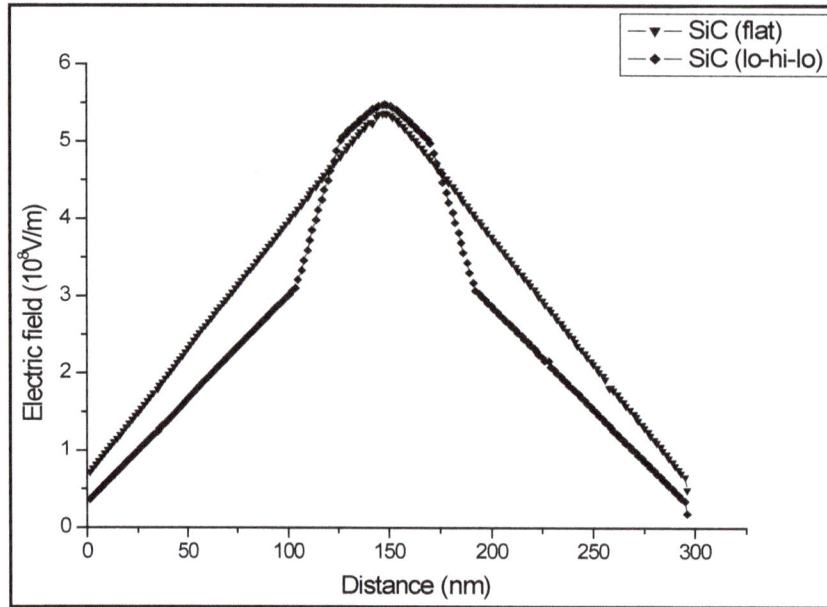

Figure 2. Electric field profile of SiC (flat) and SiC (lo-hi-lo) DDR MITATT diodes.

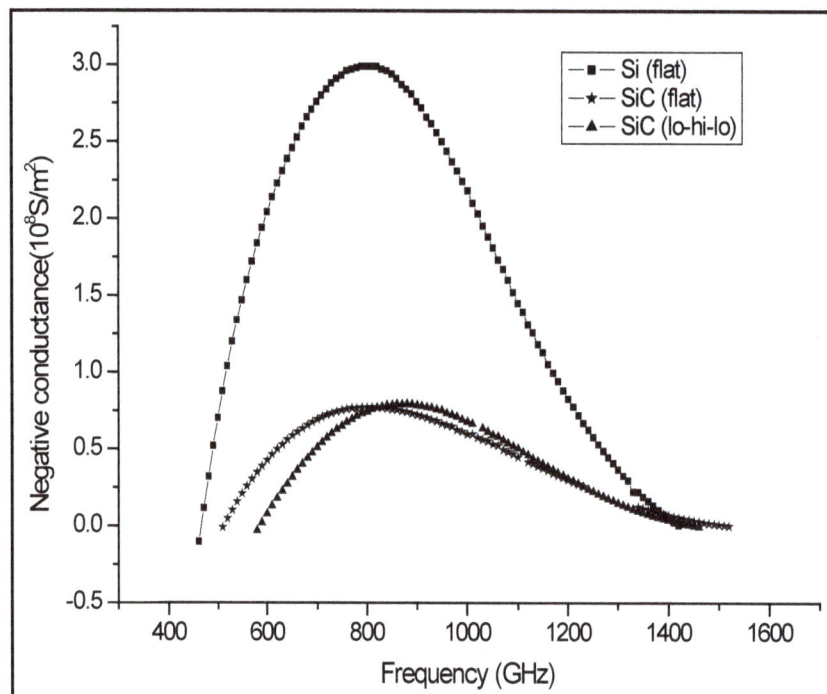

Figure 3. Variation of device negative conductance as a function of frequency for Si (flat), SiC (flat) and SiC (lo-hi-lo) MITATT diodes.

diode (7.96 W). Thus, from the above properties, it can be observed that SiC is a much superior material compared to Si and GaN as a terahertz source operating in MITATT mode.

In MITATT diodes avalanching is a noisy process while tunnelling is a quiet process. The percentage of tunnelling current in a MITATT diode can be increased either by increasing the junction field with a suitable structural modification or by decreasing the operating bias current density. We have varied the structure of diode to study the noise in SiC MITATT at 0.5 THz operating frequency. We have computed the mean square noise voltage in each case. We found that as the percentage of tunnelling current increases the height of the noise spike decreases. **Figure 4** shows this effect very clearly. The value of noise measure gives the value of noise to power ratio. The values of minimum NM is found to be 12.71 dB (at 0.98 THz), 21.50 dB (at 1.25 THz) and 17.71 dB (at 1.3 THz) for the Si (flat), SiC (flat) and SiC (lo-hi-lo) MITATT diode respectively. **Figure 5** shows the plots of NM with frequency. The lowest value of NM for SiC is found to decrease with increase in tunnelling current. This is due to decrease in percentage of avalanche component of current. The minimum NM value is found to be 27.2 dB (at 0.71 THz) for GaN (flat) diode. The noise measure value of GaN (flat) diode is much higher as compared to Si (flat) and SiC (flat) diodes. This is due to very small value of percentage of tunnelling current (3.85%) in case of GaN (flat) diode which renders it operate in nearly IMPATT mode where noise is generally higher.

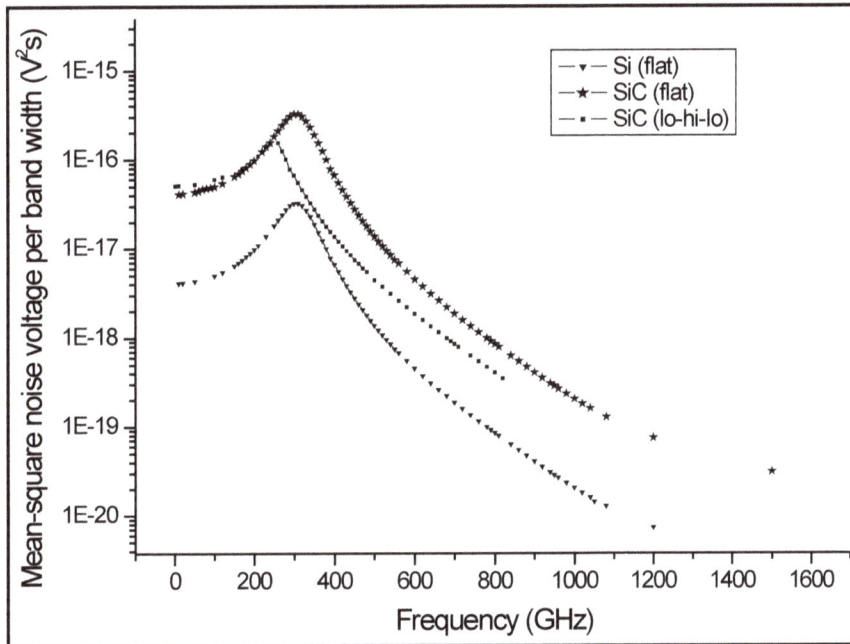

Figure 4. Mean-square noise voltage per band width versus frequency of Si (flat), SiC (flat) and SiC (lo-hi-lo) MITATT diodes.

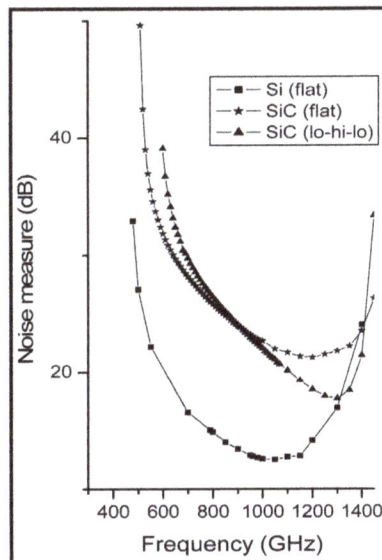

Figure 5. Noise measure versus frequency for Si (flat), SiC (flat), SiC (lo-hi-lo) MITATT diodes.

5. Conclusion

Mixed mode analysis of 4H-SiC MITATT diodes has been carried out. The paper, while establishing the superior performance of 4H-SiC over Si, unveils some interesting properties of mixed mode operation. DC to microwave conversion efficiency of 17.5% from the SiC lo-hi-lo MITATT diode is noteworthy. The advantage of mixed mode operation is clearly demonstrated in the tunnelling-assisted noise reduction from 21.5 dB (SiC flat) to 17.7 dB (SiC lo-hi-lo). Thus, we conclude that 4H-SiC has a great potential for application as a MITATT diode even at the terahertz frequency.

References

[1]   Brezeanu, G. (2007) High Performance Power Diodes on Silicon Carbide and Diamond: *The Publishing House of the Romania Academy*, **8**, 1-14.

[2]   Singh, K., Cooper, J.A., Meloch, M.R., Chow, T.P. and Palmour, J.W. (2004) Silicon Carbide Power Schoottky and Pin Diodes. *IEEE Transactions on Electron Devices*, **49**, 665-672. http://dx.doi.org/10.1109/16.992877

[3]   Traplee, M.C., Madangagly, V.P., Zhang, Q. and Surdarsan, T.S. (2001) Design Rules for Field Pate Edge Termination in SiC Schottky Diodes. *IEEE Transactions on Electron Devices*, **48**, 2659-2664. http://dx.doi.org/10.1109/16.974686

[4]   Sheridan, D.C., Niu, G., Merrett, J.N., Cresller, J.D., Ellis, C. and Tin, C.C. (2000) Design and Fabrication of Planar Guard Ring Termination for High Voltage Silicon Carbide Diodes. *Solid-State Electronics*, **44**, 1367-1372. http://dx.doi.org/10.1016/S0038-1101(00)00081-2

[5]   Karan, D.K., Panda, P. and Dash, G.N. (2013) Effect of Tunneling Current on Noise Characteristics of a 4H-SiC Read Avalanche Diode. *Journal of Semiconductors*, **34**, Article ID: 014001. http://dx.doi.org/10.1088/1674-4926/34/1/014001

[6]   Luo, Y., Melloch, M.R., Cooper, J.A. and Webb, K.J. (2000) Silicon Carbide IMPATT Oscillator for High-Power Microwave and Millimeter-Wave Generation. *IEEE/Cornell Conference on High Performance Devices*, Ithaca, 7-9 August 2000, 158-167. http://dx.doi.org/10.1109/CORNEL.2000.902533

[7]   Yuan, L., Melloch, M.R., Cooper, J.A. and Webb, K.J. (2001) Experimental Demonstration of a Silicon Carbide IMPATT Oscillator. *IEEE Electron Device Letters*, **22**, 266-268. http://dx.doi.org/10.1109/55.924837

[8]   Zhao, J.H., *et al.* (2000) Monte Carlo Simulation of 4H-SiC IMPATT Diodes. *Semiconductor Science and Technology*, **15**, 1093-1100. http://dx.doi.org/10.1088/0268-1242/15/11/314

[9]   Pattanaik, S.R., Dash, G.N. and Mishra, J.K. (2005) Prospects of 6H-SiC for Operation as an IMPATT Diode at 140 GHz. *Semiconductor Science and Technology*, **20**, 299-304. http://dx.doi.org/10.1088/0268-1242/20/3/008

[10] Zhang, C.X., *et al.* (2011) Effects of Bias on the Irradiation and Annealing Responses of 4H-SiC MOS Devices. *IEEE Transactions on Nuclear Science*, **58**, 2925-2929. http://dx.doi.org/10.1109/TNS.2011.2168424

[11] Imhoff, E.A., *et al.* (2011) High Performance Smoothly Tapered Junction Termination Extensions for High Voltage 4H-SiC Devices. *IEEE Transactions on Electron Devices*, **58**, 3395-3400. http://dx.doi.org/10.1109/TED.2011.2160948

[12] Zhang, H., Tolbert, L.M. and Ozpinec, B. (2011) Impact of SiC Devices on Hybrid Electric and Plug-In Hybrid Electric Vehicles. *IEEE Transactions on Industry Applications*, **47**, 912-921. http://dx.doi.org/10.1109/TIA.2010.2102734

[13] Panda, A.K. and Rao, V.M. (2009) Modeling and Comparative Study on the High Frequency and Noise Characteristics of Different Polytypes of SiC Based IMPATTs. *IEEE Asia Pacific Microwave Conference*, Singapore, 7-10 December 2009, 1569-1572. http://dx.doi.org/10.1109/APMC.2009.5384396

[14] Dash, G.N. and Pati, S.P. (1992) A Generalized Simulation Method for MITATT Mode Operation and Studies on the Influence of Tunnel Current on IMPATT Properties. *Semiconductor Science and Technology*, **7**, 222-230. http://dx.doi.org/10.1088/0268-1242/7/2/008

[15] Sze, S.M. (1987) Physics of Semiconductor Devices. 2nd Edition, John Wiley & Sons, New York.

[16] Dash, G.N., Mishra, J.K. and Panda, A.K. (1996) Noise in Mixed Tunneling Avalanche Transit Time Diode. *Solid-State Electronics*, **39**, 1473-1479. http://dx.doi.org/10.1016/0038-1101(96)00054-8

Appendix: Definitions of Symbols

A : area of the diode

$E(x)$: electric field at x

E_0 : maximum electric field

$e(x, x')$: noise electric field at x due to noise source at x'

$\mathrm{d}f$: frequency interval

G_p : peak value of diode conductance

g_{T_n} : tunnelling generation rate for electrons

$\mathrm{d}I_n(x')$: current generated in a space step $\mathrm{d}x'$

J_T : tunnelling current density

J_0 : total current density

k_B : Boltzmann constant

NM: noise measure

P_{RF} : RF power output

q : electronic charge

T : absolute temperature

V_B : breakdown voltage

v_n : drift velocity for electrons

v_p : drift velocity for holes

$\langle v^2 \rangle$: mean-square noise voltage

V_t : terminal voltage caused by noise source $\gamma(x')$

W : width of the depletion layer

x : general symbol to define distance in active layer

x' : position of the noise element

Z_R : real part of device negative resistivity

Z_t : transfer impedance

α_n : ionisation rate for electrons

α_p : ionisation rate for holes

$\in$: permittivity of the semiconductor

γ : noise generation rate

Permissions

List of Contributors

Anima Johari, Vikas Rana and M. C. Bhatnagar
CARE, Physics Department, IIT Delhi, New Delhi, India

Anoopshi Johari
THDC Institute of Hydropower Engineering and Technology, Tehri, India

Yendrapati Taraka Prabhu, Kalagadda Venkateswara Rao and Vemula Sesha Sai Kumar
Centre for Nano Science and Technology, IST, Jawaharlal Nehru Technological University, Hyderabad, India

Bandla Siva Kumari
Botany Department, Andhra Loyola College, Vijayawada, India

Ken-ichi Saitoh
Department of Mechanical Engineering, Kansai University, Suita, Japan

Youhei Sameshima
Graduate School of Science and Engineering, Kansai University, Suita, Japan

Syuhei Daira
Graduate School of Science and Engineering, Kansai University, Suita, Japan
Denso Techno, Co. Ltd., Obu, Japan

Vivek Kant Jogi
School of Studies in Electronics, Pt. Ravi Shankar Shukla University, Raipur, India

Mari Ishihara, Ryuji Hirase and Hideki Yoshioka
Hyogo Prefectural Institute of Technology, Kobe, Japan

Mohammad M. Uonis, Bassam M. Mustafa and Anwar M. Ezzat
Department of Physics, College of Science, Mosul University, Mosul, Iraq

Branislav Radjenović and Marija Radmilović-Radjenović
Institute of Physics, University of Belgrade, Zemun, Serbia

Bilel Hafsi
IEMN Laboratory, University of Lille1, Avenue Poincaré, 59652 Villeneuve d'Ascq Cedex, France
Microelectronics and Instrumentation Laboratory, Faculty of Sciences of Monastir, University of Monastir, Monastir, Tunisia

Rabii Elmissaoui
Research Unit on Study of Industrial Systems and Renewable Energies, National Engineering School of Monastir, Monastir, Tunisia

Adel Kalboussi
IEMN Laboratory, University of Lille1, Avenue Poincaré, 59652 Villeneuve d'Ascq Cedex, France

Ting Tian
School of Perfume and Aroma Technology, Shanghai Institute of Technology, Shanghai, China

Jing Hu and Zuobing Xiao
School of Perfume and Aroma Technology, Shanghai Institute of Technology, Shanghai, China
Shanghai Research Institute of Fragrance & Flavor Industry, Shanghai, China

Aly H. Atta
Department of Chemistry, Faculty of Education, University of Dammam, Dammam, KSA
Department of Chemistry, Faculty of Science, Suez University, Suez, Egypt

Ahmed I. El-Shenawy
Department of Chemistry, Faculty of Education, University of Dammam, Dammam, KSA
Department of Chemistry, Faculty of Science, Benha University, Benha, Egypt

Fathy A. Koura
Department of Chemistry, Faculty of Education, University of Dammam, Dammam, KSA
Department of Chemistry, Faculty of Science, Al-Azhar University, Cairo, Egypt

Moamen S. Refat
Department of Chemistry, Faculty of Science, Taif University, Al-Hawiah, KSA
Department of Chemistry, Faculty of Science, Port Said University, Port Said, Egypt

Hiroki Fukatsu
Graduate School of Science and Engineering, Shizuoka University, Hamamatsu, Japan
Technical Solution Center, Polyplastics Co., Ltd., Fuji, Japan

Masaaki Kuno and Yasuhiro Matsuda
Graduate School of Engineering, Shizuoka University, Hamamatsu, Japan

Shigeru Tasaka
Graduate School of Science and Engineering, Shizuoka University, Hamamatsu, Japan
Graduate School of Engineering, Shizuoka University, Hamamatsu, Japan

Eman A. N. Al-Lehaibi
Mathematics Department, College of Science and Arts —
Sharoura, Najran University, Najran, KSA

Hamdy M. Youssef
Mechanics Department, Faculty of Engineering, Umm Al-
Qura University, Makkah, KSA

Yuri Hendrix, Alberto Lazaro, Qingliang Yu and Jos Brouwers
Department of the Built Environment, Eindhoven
University of Technology, Eindhoven, The Netherlands

Ramasamy Thirunakaran
CSIR-Central Electrochemical Research Institute,
Karaikudi, India

Gil Hwan Lew and Won-Sub Yoon
Department of Energy Science, Sungkyunkwan University,
Suwon, Republic of Korea

Abdelaali Fargi, Neila Hizem, Adel Kalboussi and Abdelkader Souifi
Department of Physics, Faculty of Sciences of Monastir,
Monastir, Tunisia

Xuyao Xu, Xiaosong Zhou, Lin Ma, Miaoyan Mo, Cuifen Ren and Rongkai Pan
School of Chemistry Science & Technology, Institute of
Physical Chemistry, and Development Center for New
Materials Engineering & Technology in Universities of
Guangdong, Zhanjiang Normal University, Zhanjiang,
China

Mohamed Ashour Ahmed
Department of Prosthodontics, Faculty of Dental
Medicine, Taif University, Taif, KSA

Mohamed I. Ebrahim
Department of Restorative Dentistry, Taif University,
Taif, KSA
Department of Dental Biomaterial, Al-Azhar University,
Cairo, Egypt

Ibrahim D. Muhammad, Mokhtar Awang and Othman Mamat
Mechanical Engineering Department, Universiti Teknologi
Petronas, Seri Iskandar, Malaysia

Zilati Bt Shaari
Chemical Engineering Department, Universiti Teknologi
Petronas, Seri Iskandar, Malaysia

Khondaker Sakil Ahmed and Ang Kok Keng
Department of Civil & Environmental Engineering,
National University of Singapore, Singapore City,
Singapore

Pranati Panda, Satya Narayana Padhi and Gana Nath Dash
Electron Devices Group, School of Physics, Sambalpur
University, Burla, India

www.ingramcontent.com/pod-product-compliance
Lightning Source LLC
Chambersburg PA
CBHW080258230326
41458CB00097B/5117